瞿昙寺组群剖面图 1-1
Section 1-1 of Qutan Temple

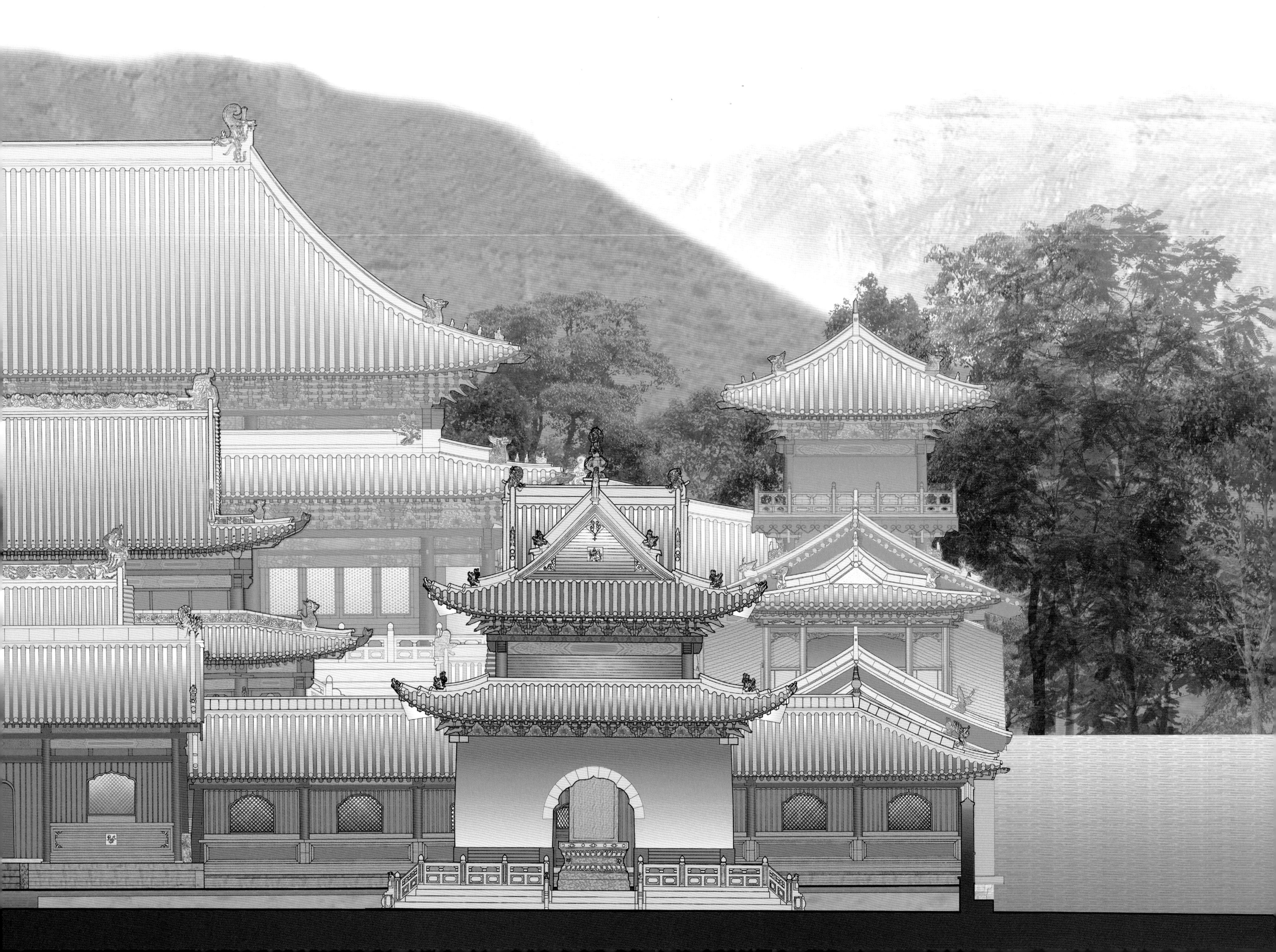

瞿昙寺组群剖面图 2-2
Section 2-2 of Qutan Temple

瞿昙寺组群剖面图 3-3
Section 3-3 of Qutan Temple

瞿昙寺组群剖面图 4-4
Section4-4 of Qutan Temple

瞿昙寺组群剖面图 5–5
Section5-5 of Qutan Temple

瞿昙寺组群剖面图 6-6
Section 6-6 of Qutan Temple

瞿昙寺组群剖面图 7—7
Section 7-7 of Qutan Temple

瞿昙寺组群剖面图 8-8
Section 8-8 of Qutan Temple

瞿昙寺组群东立面图
East Elevation of Qutan Temple

瞿昙寺中轴线东立面图
East Elevation of Central Axis of Qutan Temple

国家出版基金项目
NATIONAL PUBLICATION FOUNDATION

『十二五』国家重点图书出版规划项目

中国古建筑测绘大系·宗教建筑

瞿昙寺

天津大学建筑学院　青海省文物考古研究所　合作编写
王其亨　吴　葱　主编

中国建筑工业出版社

Traditional Chinese Architecture Surveying and Mapping Series:
Religious Architecture

QUTAN TEMPLE

Compiled by School of Architecture, Tianjin University
Qinghai Provincial Bureau of Cultural Relics
Edited by WANG Qiheng , WU Cong

China Architecture & Building Press

Editorial Board of the Traditional Chinese Architecture Surveying and Mapping Series

『中国古建筑测绘大系』编委会

Contents

目录

地　　址：青海省海东市乐都县瞿昙镇新联村
始建年代：1391 年（明洪武二十四年）
占地面积：3.4 公顷
主管单位：瞿昙寺文物管理所
测绘单位：天津大学建筑学院
测绘时间：1993—1997 年

Address: Xinlian Village, Qutan Town, Ledu County, Haidong City, Qinghai Province

Age of construction：1391（the 24th year of Hongwu in Ming Dynasty）

Site area: 3.4 hectors

Competent organization: The Office of Cultural Relics of Qutan Temple

Survey organization: School of architecture, Tianjin University

Time of survey: 1993-1997

Introduction

Qutan Temple is located at the Xinlian Village, Qutan Town(Fig.1), 2400 meters above sea level, about 21.5 kilometers south to Ledu County east to Xining City, the provincial capital of Qinghai Province. The whole building complex is built upon the topography of mountains and rivers nearby, with an axis facing the southeast (Fig.2). The complex is divided into three courtyards, the front, the middle and the back, with the monk dormitories lying on both sides of the temple. The total land area is about 3.4 hectors (Fig.3, Fig.4).

The temple was originally built in 1391, with its formal name Qutan granted by emperor ZHU Yuanzhang two years later. Since 1408, plenty of money and builders were sent to the project, supervised by eunuchs and officials as imperial envoys of the followed three emperors ZHU Di, ZHU Gaozhi, ZHU Zhanji. With a continuous construction through the next nineteen years, the whole temple complex was completed in 1427. With the integration of both local and royal building craftsmanship and techniques, the Qutan Temple became the top Tibetan Buddhist temple complex, which is not only splendid but also sumptuous, featured with the large-scale Han corridor-plus-courtyard layout in the western region of China during the Ming Dynasty.

导言

瞿昙寺位于青海省会西宁市迤东乐都县南四十三里（约21.5公里）的瞿昙镇新联村（图一），海拔2400多米，全寺背山临流，朝向东南（图二），前后三进院落循地势往北渐次升高，两侧另有囊谦和僧舍，供僧侣起居。占地共3.4公顷（图三、图四）。

寺院初创于明洪武二十四年（1391年），洪武二十六年由明太祖朱元璋敕赐寺名，永乐六年（1408年）后，明成祖朱棣、明仁宗朱高炽、明宣宗朱瞻基相继钦派太监、朝官携帑率匠董工，踵事增华十九个春秋，至宣德二年（1427年）告竣，融汇了当时的地方建筑和北京皇家建筑技艺，恢宏壮丽，成为彪赫明代西陲的汉式藏传佛教名刹。

图一　瞿昙寺区位

图二　瞿昙寺的环境格局

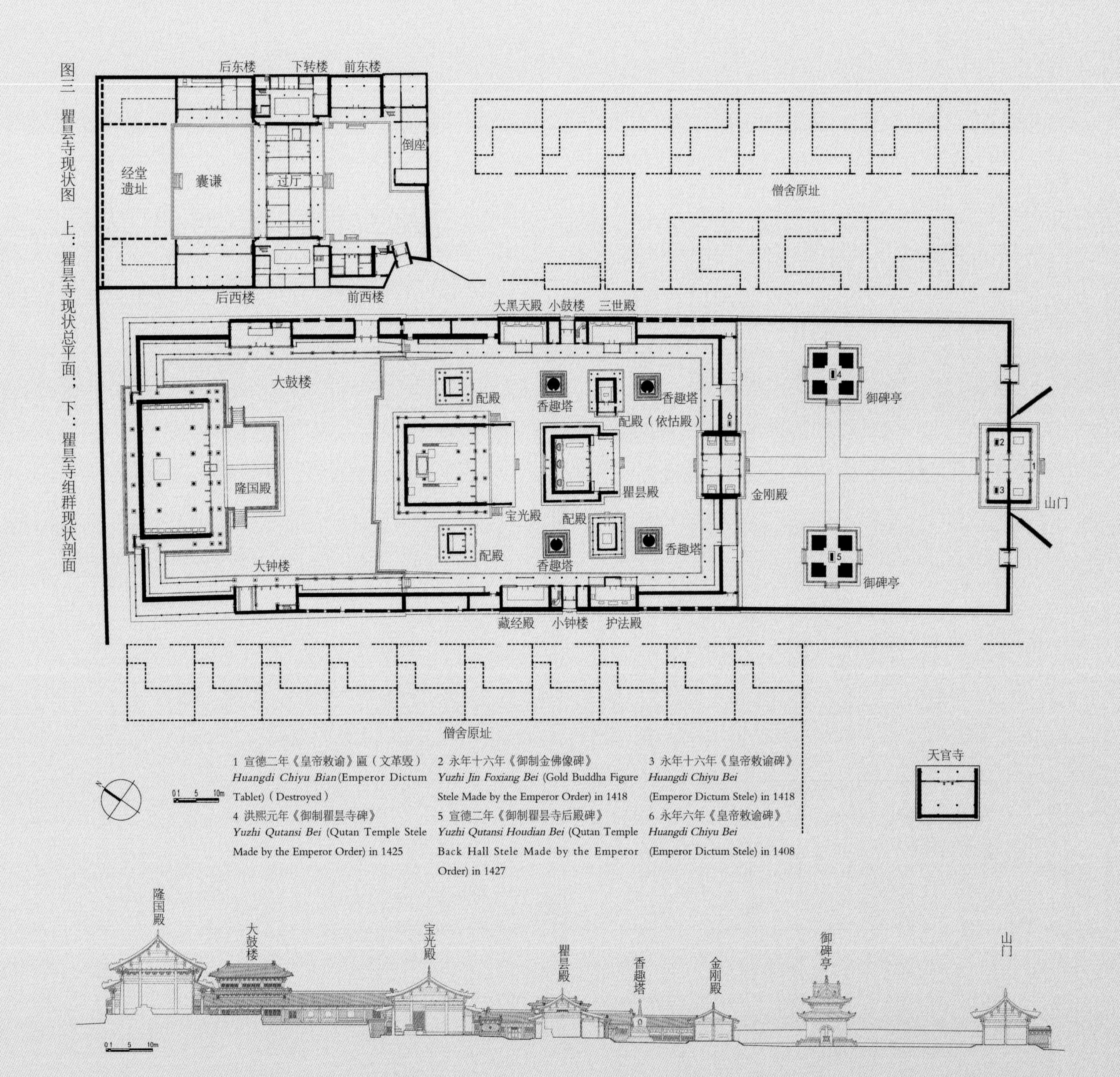

图三　瞿昙寺现状图　上：瞿昙寺现状总平面；下：瞿昙寺组群现状剖面

Fig.1 Location of the Qutan Temple
Fig.2 Environment layout of the Qutan Temple
Fig.3 Current Situation Map of Qutan Temple upper: General plan of the current Qutan Temple;lower: Section of the current Qutan Temple complex

图四 1980年代瞿昙寺鸟瞰

Fig.4 Bird-eye-view photo of the Qutan Temple taken in 1980s

Through the ages of six centuries, the main buildings in the Qutan Temple remain to be completely reserved, including its architecture, stucture color painting, decoration, furnishings, frescoes and sculptures. Within this important corridor-plus-courtyard layout architecture heritage, the building complex of Longguo Hall was built in a scale-reduced blueprint of the Fengtian Hall renamed as Taihe Hall in early Qing Dynasty in 1420, with double eaves and hip roofs (Fig.5), together with the Wen Building (Tiren Tower in Qing Dynasty) and the Wu Building (Hongyi Tower in Qing Dynasty) on both sides, both featured in two-floored hip roofs and connected by ascending side corridors, which has become a precious historical image of palace architecture in today's Beijing. It is therefore that the Qutan Temple has been eulogized by local people of different nationalities as "the Forbidden City on the plateau". Due to its distinguished cultural relics value, Qutan Temple was listed in the Second Batch of National Key Cultural Relics Protection Units announced by the State Council in 1982.

I.Geographical and Historical Backgrounds of Qutan Temple

Ledu County, where Qutan Temple is located, belongs to Qinghai Province. It is an area where a main tributary of the Yellow River, the Huang River, merges into the former. It used to be the eastern gateway of the Qinghai-Tibetan Plateau, which belonged to the Ando Tibetan Region in history. Due to the frequent exchange and communication with central China, it had been deeply influenced by central China's culture. During the Ming Dynasty, it was regarded as *an important military base, which could safeguard the Hexi Corridor Area*, controlling all armed escorts beyond the Great Wall, *preventing intrusions from Mongolia and foreign tribes from the southern region*.

Tibetan Buddhism had been developed very fast in China because it was highly respected by the royal court since the Yuan Dynasty. One may find that there are many Tibetan Buddhist temples in today's Tibet, Qinghai, Gansu, Ningxia, Inner Mongolia, Xinjiang, Yunnan, Sichuan, Shanxi, Hebei, Beijing，Liaoning within China and as well as Mongolia. Their architectural layouts and forms are very rich and colorful, which can be divided into the following types: Indian type, Tibetan type, combined Tibetan-Nepal type, combined Han-Tibetan type, paralleled Han-Tibetan type and Han type(Fig.6). Being built by the first four emperors of the early Ming Dynasty, Qutan Temple is a typical representative of the Han type styled Tibetan Buddhist temples.

历经六百多年沧桑，瞿昙寺的主体建筑，包括结构、彩画、装修、陈设、壁画、雕塑等，大多都完整保存；其中的隆国殿组群，曾直接取仿明初永乐十八年（1420年）落成的北京皇宫奉天殿（图五）；以抄手斜廊连缀着东西分列的文楼、武楼，今天已成为北京宫殿建筑的珍贵历史映像。瞿昙寺也被当地各族民众传颂为『高原小故宫』。基于突出的文物价值，1982年被国务院公布为第二批全国重点文物保护单位。

一、瞿昙寺的地理环境和历史背景

瞿昙寺所在的青海乐都县，处于黄河及其重要支流湟水交汇区域，夙为青藏高原的东方门户，原属历史上的安多藏区，与内陆交往频繁，深受中原文化影响。在明代，这里东卫关陇，西控『塞外诸卫』，『北拒蒙古，南捍诸番』，遂成『用武之重地，河西之捍卫』。

元代以降，中国的藏传佛教因备受朝廷尊崇而迅速发展，在今我国西藏、青海、甘肃、宁夏、内蒙古、新疆、云南、四川、山西、河北、北京、辽宁和蒙古国等地，都有藏传佛教寺院分布，幅凑中原。其建筑形式丰富多彩，大体可分为印度式、藏式、藏尼结合式、汉藏结合式、汉藏并列式和汉式（图六）。由明初四朝皇帝敕建的瞿昙寺，则是汉式藏传佛教寺院的典型代表。

明代为稳定西陲，对西藏地区，朝廷承袭并发展元代羁縻方略，建立了『政教合一』的僧官体系；在安多地区，行政建置则由朝廷任命扶植，尤其礼敬、优渥高僧，广赐名号，授以地方教务管理职权，

In order to stabilize the west border region, especially the Tibetan region, the Ming court followed the Yuan court's strategy in comforting the minority regions along the border, including setting up the unification of the state and the religion system, in which the officials and monks had different functional divisions. In the Ando Region, in particular, the local administration establishment was directly appointed and fostered by the imperial court, which gave special respects and generosities to eminent monks. The latter were granted with alias and local religious administration duties. And a great amount of money were invested to build temples on the awarded lands. Qutan Temple is the most remarkable fruit of such a policy.

According to the notes carved on the beam of the main hall in the Qutan Temple, the temple was started to build in 1391 at the latest. The founder of the temple is Sanluo Lama (Fig.7), who is praised as the mahasiddha of the Ando Region in the Tibetan Buddhism history. According to the third living Buddha Chos-kyi-nyi-ma during the Qianlong's reign, the school that Sanluo Lama worshiped has been the Sakya (Stripe hat) Sect or the Kaggu (White) Sect. The original name of the temple was called Zhuocang Buddhist Hall, or the Gautama Yard of the Zhuocang Shrine, which is pronounced as Qutan Temple of south mountain in mandarin Chinese. In 1393, Sanluo Lama, the most vonerable person in Ando Region went to Nanjing to pay the tribute to the imperial court. ZHU Yuanzhang, the first emperor of the Ming Dynasty granted the family name of Sakyamuni, Qutan (Gautama), as the formal name of the temple. Accordingly, the nearby area was also renamed as Qutan (today's Qutan Town). Meanwhile, Department of Monks was set up at Xining Prefecture and Sanluo Lama was appointed by the Emperor to become the department head, in charge of local religious affairs. Continuous imperial edictums and orders were made by the successive six emperors to appoint great monks as successors of Sanluo Lama from Qutan Temple to become the department head, with even more respected names such as the (Great) Adviser of the State and the West Heaven Buddha (Fig.8). Hence, the temple has become the one that received the highest position and fame, and the most awards in the Ando Region.

Within the Ledu County, most Tibetan Buddhist temples are located on the riverbanks of Huang River, which is called by locals as South Mountain (Zhuocang in Tibetan language) and North Mountain. The Qutan Temple is located at the Xinlian Village, Qutan Town, which belongs to the upper part of the Shenjia Channel on the South Mountain side. North to the elegant Luohan Mountain, with the Qutan River, the branch of the Huang River, running in the front, and south towards the Phoenix Mountain, with snow mountains lying

更不惜重金大建寺院，封赏土地。瞿昙寺就正是这一政策的显著结果。

按瞿昙殿梁架题记和清同治四年（1865年）智观巴·贡却乎丹巴饶吉《安多政教史》等藏文史料，该寺始建于明洪武二十四年（1391年），开山祖师三罗喇嘛（图七）在藏传佛教界被尊为大成就者；他信奉的教派，据清乾隆朝三世土观活佛曲吉尼玛称，是萨迦派（花教）或噶举派（白教），强调了他广结善缘的宽容。寺院初称『卓仓佛殿』，或『卓仓神庙乔达摩院』，即汉语的『南山瞿昙寺』。洪武二十六年，德望饮誉安多地区的三罗喇嘛赴南京朝贡，明太祖以佛祖姓氏『瞿昙』赐为寺名，地名也改称『瞿昙』。同时，朱元璋建置西宁僧纲司，钦命三罗喇嘛为『都纲』，管理当地教务。嗣后明朝六位皇帝，又频频敕谕、诰命，赐授三罗喇嘛及传承其衣钵的瞿昙寺大德为『都纲』，诰封『国师』『大国师』『西天佛子』等尊号（图八），瞿昙寺成为安多地区受封职位名号最高、受赏最多的寺院。

在乐都境内，藏传佛教寺院多分布在湟水两岸的南山和北山。瞿昙寺就坐落在南山沈家峡上游的瞿昙镇新联村，北倚形势端庄的罗汉山，前临湟水支流瞿昙河，南朝凤凰山，可遥望乐都十二景之一的雪山，实属环境雄浑而又清幽的形胜之地。

瞿昙寺曾周匝夯土城墙，寺院坐落在西部的内城，俗称『旧城』；东、南两面地势较低，包绕外城，俗称『新城』，内为寺主族人和佃户居住的里坊，又称『新城街』。城墙设上城门、下城门、新城门，

图五-1 万历五年（1577年）之前，皇极殿及两旁抄手斜廊（徐显卿《宦迹图·皇极侍班图》）

图五-2 康熙三十四年（1695年）重建太和殿，为防延烧，两翼抄手斜廊改建为跌落式防火墙（江藻《太和殿纪事·太和殿图》）。

图五 自永乐十八年（1420年）奉天殿落成至康熙十八年（1679年）末太和殿被焚前，曾四次重建，两旁连缀抄手斜廊夙为传统。

Fig.5-1 Image of Huangji Hall and its ascending side corridors before 1577 (XU Xianqing, *Huanji Tu - Huangji Shiban Tu*)

Fig.5-2 Taihe Hall after the reconstruction completed in 1696, changing the roof shape of the ascending side corridors into a stepped down fire-proof wall to prevent from fire spreading of the fire (Spreading JIANG Zao, *Taihe Dian Jishi - Taihe Dian Tu*)

Fig.5 Between the completion of the Fengtian Hall in 1420 and the burnt down of the Taihe Hall in 1679, the hall has been rebuilt for four times, always joined by the ascending side corridors on each as a tradition.

图六 藏传佛教寺院的建筑平面形式

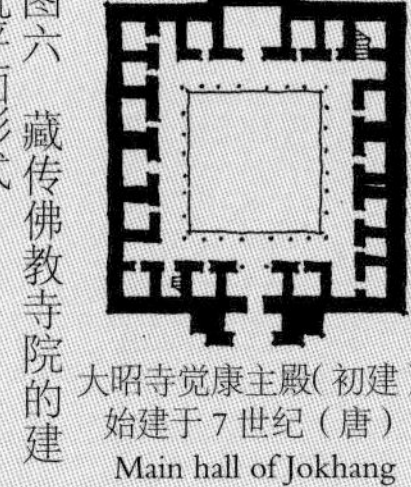

大昭寺觉康主殿（初建）
始建于 7 世纪（唐）
Main hall of Jokhang Temple, built in the 7th century (Tang Dynasty)

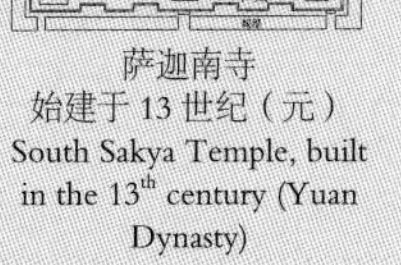

萨迦南寺
始建于 13 世纪（元）
South Sakya Temple, built in the 13th century (Yuan Dynasty)

印度式
India style

桑耶寺
始建于 8 世纪（唐）
Samye Temple, built in the 8th century (Tang Dynasty)

藏式
Tibetan style

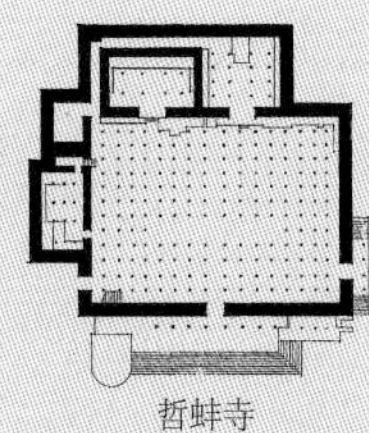

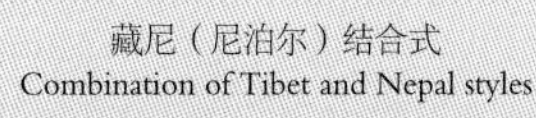

哲蚌寺
始建于 15 世纪（明）
Zhebang Temple, built in the 15th century (Ming Dynasty)

藏尼（尼泊尔）结合式
Combination of Tibet and Nepal styles

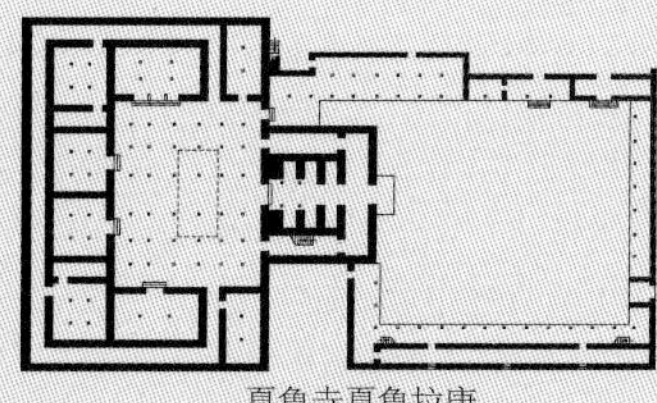

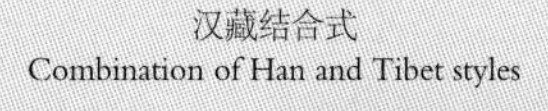

夏鲁寺夏鲁拉康
始建于 11 世纪（宋）
Xialu Lakang in Xialu Temple, built in the 11th century (Song Dynasty)

汉藏结合式
Combination of Han and Tibet styles

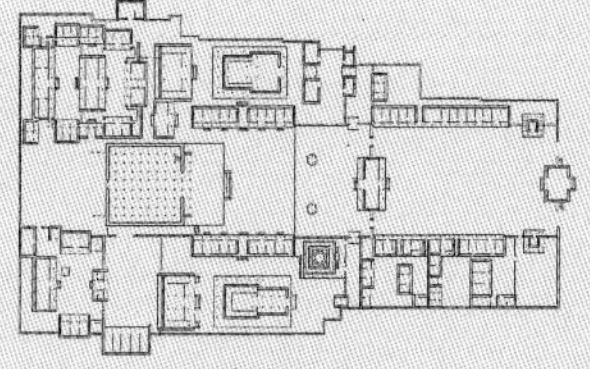

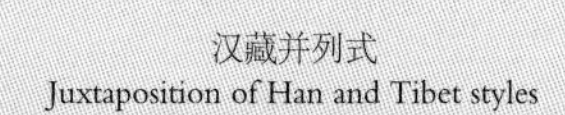

席力图召
始建于 16 世纪（明）
Xilitu Temple, built in the 16th century (Ming Dynasty)

汉藏并列式
Juxtaposition of Han and Tibet styles

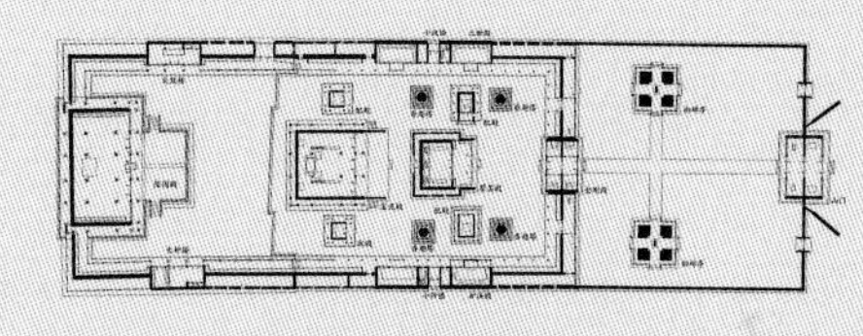

瞿昙寺
始建于 14 世纪（明）
Qutan Temple, built in the 14th century (Ming Dynasty)

汉式
Han style

图八 明朝皇帝颁赐给瞿昙寺诸大德的印信
1.明永乐十年四月颁给班丹藏卜『灌顶净觉弘济大国师印』篆文银镀金印（礼部造智字一百十号）；2.明永乐十年十一月颁给锁南坚参『慈光普照』象牙图章；3.『宣德二年裕如日』赐喇嘛绰失吉领占『真修无碍』象牙印章；4.明成化二十二年十二月赐『广慧悟法净觉妙善翊国衍教灌顶戒定西天佛子大国师印』；5.明弘治二年五月颁给尼嘛藏布『都纲之印』铜印

1

2

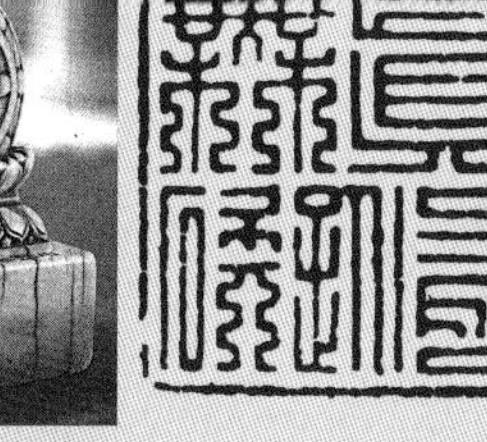

3

4

5

图七 瞿昙寺宝光殿内开山祖师三罗喇嘛塑像（刘彤彤、向永德摄）

Fig.6 Different architectural plans of Tibetan Buddhist temples

Fig.7 Sculpture of Sanluo Lama, the founder of the Qutan Temple standing in the Baoguang Hall (photo taken by LIU Tongtong and XIANG Yongde)

Fig.8 Seals granted by the Ming emperors to masters at the Qutan Temple

1 The gilded seal with the inscription of "guanding-jingjue-hongji-daguoshi-yin (the seal of empowered great state adviser with wisdom, awareness, and willingness to help all others)", No. 110, the Ministry of Rites, granted to Bandan Zangbu in April, 1412

2 The ivory stamp with the inscription of "ci-guang-pu-zhao (illuminating all things with kindly lights)", granted to Suonan Jiancan in November, 1412

3 The ivory seal granted to Lama Chos-rgyal-rin-chen on the Yuru day in 1427, with and inscription of "true practice is unhindered"

4 The seal granted in December, 1486, with an inscription of "*guanghui-wufa-jingjue-miaoshan-yiguo-yanjiao-guanding-jieding-xitianfozi-daguoshi-yin* (the seal of the empowered great state adviser, who is with great wisdom and deep understanding of Buddhism, with exquisite beauty and kindness, who is assisting the state and spreading the religion, disciplined and mediated, and the son of the western heaven)"

5 The copper seal with the inscription of "*dugang-zhi-yin* (the seal for the head of the Department of Monks)", granted to Nimazangbu in May, 1489

far away, also as one of the twelve scenes of Ledu, the location of the temple is really a good *fengshui* site, a vigorous, quiet and beautiful place with obvious geographical advantages.

Qutan Temple used to be built within an inner-city, also called the old city, which was located in the west part of a rammed earth castle. The eastern and southern parts were lower, surrounding the inner city as the outer city, which was called the new city. Neighborhoods were built within the new city fore the clan families and tenant farmers to live in, which is called the new street. Along the city wall there were upper city gate, lower city gate and new city gate. In order to get into the Qutan Temple, people have to go through a barbican entrance.

In the socio-political life of the west China area during the Ming Dynasty, Qutan Temple had played an important role. In fact, as early as in 1368, the first year of the Ming Dynasty, the founder of the temple, Sanluo Lama succeeded in calling the Mongolia tribes to surrender by writing a letter, which greatly helped to stabilize the Ando region. Such an action was highly praised and honored by Emperor ZHU Yuanzhang and his successors, who decided to make Qutan Temple the model for all Tibetan Buddhist temples which are loyal to the Ming imperial court, through which impressive and dignified manner as well as grand special favor of the imperial court could be spread. As the special tie in connecting the central government and the Tibetan Buddhist monks and believers, such a decision was indeed good for the long term peace and stability of the whole country.

For example, Emperor ZHU Di followed the suggestion of Sanluo Lama, who sent envoys to Tibet several times during the first year of his reign, inviting Tsong-kha-pa, the founder of the Gelug (Yellow) Sect, to attend the imperial court. In December (referring to the Chinese lunar month and similarly hereinafter) 1414, Sha-skya-ye-shes, a brilliant disciple of Tsong-kha-pa, led several hundreds of lamas to Nanjing to present themselves to the emperor. Two years later, Sha-skya-ye-shes returned to Tibet with the appointed title of Western Heaven Buddha and Great Adviser of the State. In 1419, Sha-skya-ye-shes founded the Sera Temple with the money and goods awarded by the emperor, making it one of the three most famous temples in Lhasa. In 1434, this great adviser of the state went to Beijing, the new capital again to present himself to Emperor ZHU Zhanji, who granted him a new title of the Great Mercy Buddha. In the next October, Sha-skya-ye-shes passed away on his way back to Tibet. Following

进入瞿昙寺，还须穿过一道瓮城。

在明代西陲，瞿昙寺曾发挥重要作用。早在洪武初年，三罗喇嘛曾以其威望『为书招降罕东（蒙古族）诸部』，促成安多地区臻向安定，受到朱元璋及后嗣皇帝褒奖，刻意将瞿昙寺打造成效忠明王朝的藏传佛教寺院楷模，彰扬皇家的威仪和恩典，作为联系朝廷与藏区僧众的纽带，以利国家长治久安。

例如，永乐初年，朱棣曾采纳三罗喇嘛建议，屡派使臣赴拉萨，礼请藏传佛教格鲁派即黄教创始人宗喀巴入朝。永乐十二年（1414年）十二月，宗喀巴的高徒释迦也失率领数百名喇嘛抵南京觐见。两年后，受封为西天佛子大国师的释迦也失返藏，并于永乐十七年，用皇帝赏赐帑物创建了拉萨三大寺之一的色拉寺。宣德九年（1434年），这位大国师再赴北京觐见明宣宗，翌年十月，被朱瞻基赐封为大慈法王的释迦也失在返藏途中圆寂，朝廷下令就地建造塔院，收藏其舍利，赐名『弘化寺』。成化九年（1473年）七月，塔院岁久损坏，明宪宗敕命镇守等官修葺，并加筑城堡如瞿昙寺制。

往后，瞿昙寺继续发挥了这一重要纽带作用。略如清顺治八年（1651年）瞿昙寺《节奉敕诰代辈相传亲供底册》（下称《底册》）记载：『成化十七年十一月，瞿昙寺灌顶国师班卓尔藏布奉谕密赍乌思藏阐教王，礼部编发进供敕谕一道，勘合二十道，赏给沿途盘费……仍取藏王印信番字奏本回部讫。』

瞿昙寺作为明初各朝接续经营的样板，受到安多地区藏传佛教寺庙和广大信众的企踵景慕，典型如甘肃省永登县连城镇鲁土司的家寺妙因寺，据《安多政教史》的传奇性记载，就曾虔敬吸纳瞿昙寺的汉式建筑规制和法物。

the order of the imperial court, a new pagoda-temple was built right at the place where he passed away, with a given name as Honghua Temple to enshrine his relics (sarira). In July 1473, when the pagoda-temple became dilapidated due to long term neglect, Emperor Xianzong announced an imperial order to demand local officials to repair the temple, and to build an extra rammed earth castle around it, in the same layout as that of Qutan Temple.

Later on, the Qutan Temple continued to play the role as an important link between the Ando region and the imperial court. For example, according to *Jiefeng chigao daibei xiangchuan qingong dice* (Archived Copy of Inherited Royal Grants by Generations) of the Qutan Temple completed in 1651 (hereinafter referred to as Archived Copy): In November, 1481, the empowered state adviser Banjo Zang bu based in Qutan Temple was sent by the emperor to contact the King of Teaching in Dhus-Gtsang secretly. Together with the edictum issued by the Ministry of Rites, twenty special passes (*kan he*) and travel expenditures were all provided. ... And a sealed memorial from the King of Teaching should be submitted to the Ministry to complete the whole procedure.

As a model of the early reigns of the Ming Dynasty, the Qutan Temple had been highly respected by Tibetan Buddhist temples and their believers in Ando region. A typical example was the Miaoyin Temple, a family-based temple built by the Chieftain LU in Liancheng Town, Yongdeng County, Gansu Province. According to legendary documentations recorded in *The Political and Religious History of Ando* (1865), this Miaoyin Temple had largely modelled on the architectural layout and building code from the Qutan Temple featured in the Han style, together with the latter's religious instruments and equipment.

The ancestor of Chieftain LU is Tuohuan, a descendant of Genghis Khan, the great grandnephew of Kublai Khan, who surrendered to the Ming Court and was appointed by Emperor Taizu to reside in Liancheng to become the first generation of the Chieftain in 1370. In 1423, Shijia, the third generation of the Chieftain, was bestowed by Emperor Yongle with LU as the family name due to his meritorious military service, which became a very influential chieftain family in Ando region. In 1427, the Chieftain LU built a family-based temple with the imperial edict by Emperor Xuande, starting that the Datong Temple west to Zhuanglang area was given the name of Miaoyin, to be protected by the imperial court. As a proof for such a description in

鲁土司的始祖脱欢，本是元太祖成吉思汗后裔，元世祖忽必烈的侄重孙，明洪武三年（1370年）降明，被明太祖安置在连城，为一世土司。永乐二十一年（1423年），三世土司失伽因战功蒙明成祖钦赐鲁姓，成为安多地区极有影响的土官家族。宣德二年（1427年）鲁土司兴建家寺，明宣宗敕谕：『今以庄浪地面西大通寺，赐寺名曰妙因，颁敕护持。』印证《鲁氏家谱》这一记载，尚有妙因寺主殿万岁殿脊枋题记『大明国宣德二年岁次丁未秋七月六日』，至今完好遗存（图八）。

不仅妙因寺万岁殿建筑缩仿瞿昙殿，殿内壁画等也刻意拟照；另如《安多政教史》提到，其中还特地从瞿昙寺迎来青铜三世佛像、银汁抄写的《甘珠尔》以及朱书的《丹珠尔》大藏经等。而且一如瞿昙寺，妙因寺万岁殿内也供奉皇帝万万岁牌位，每年正月初八，鲁土司率领下属官员，到万岁殿向牌位虔诚上香。

时移世易，瞿昙寺盛极而衰。起因是万历六年（1578年），在青海湖东的仰华寺，格鲁派黄教活佛索南嘉措与蒙古土默特部俺答汗结盟，被尊为达赖喇嘛三世。藏、蒙地区黄教自此大兴，政治经济实力远胜其他教派，明王朝『断羌胡之交』政策瓦解，噶举派名刹瞿昙寺也改宗格鲁派。清代，格鲁派更受

图九 瞿昙殿（左）与万岁殿（右）比较

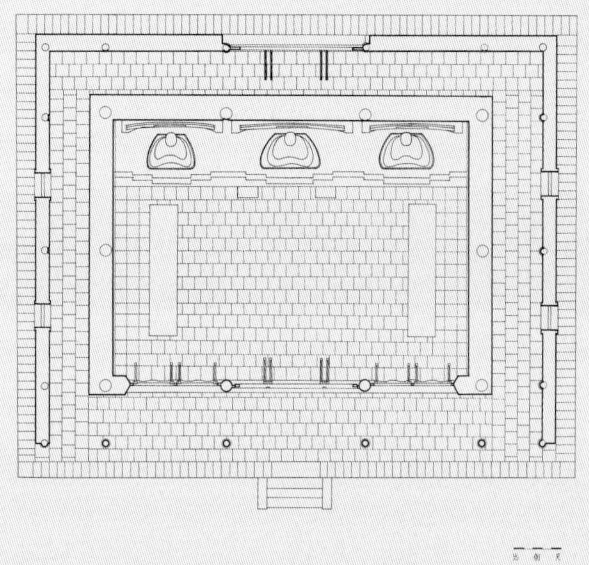

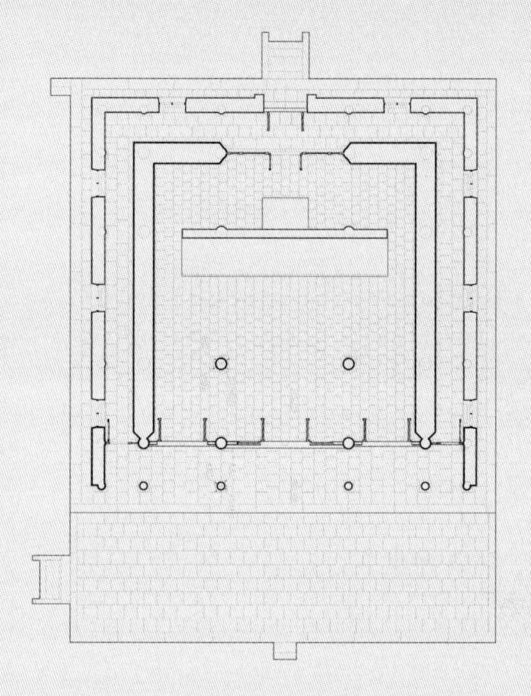

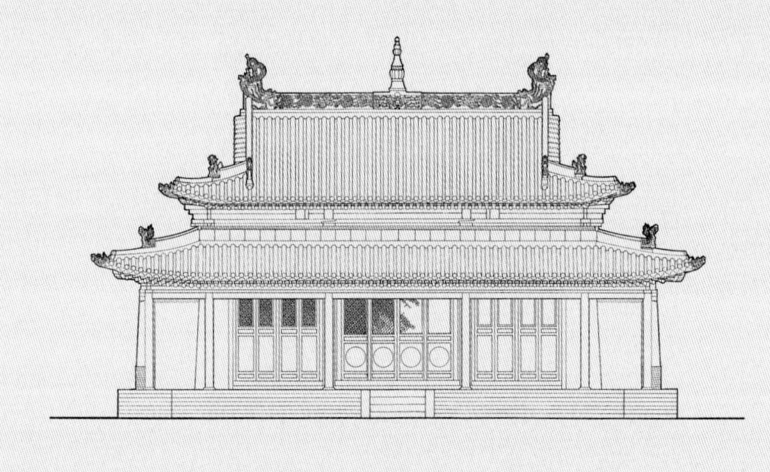

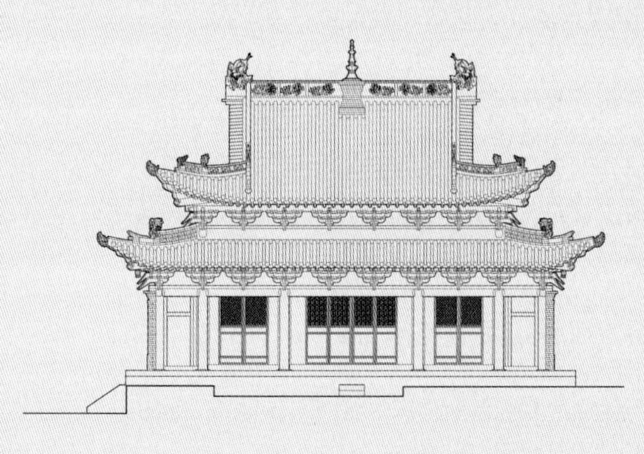

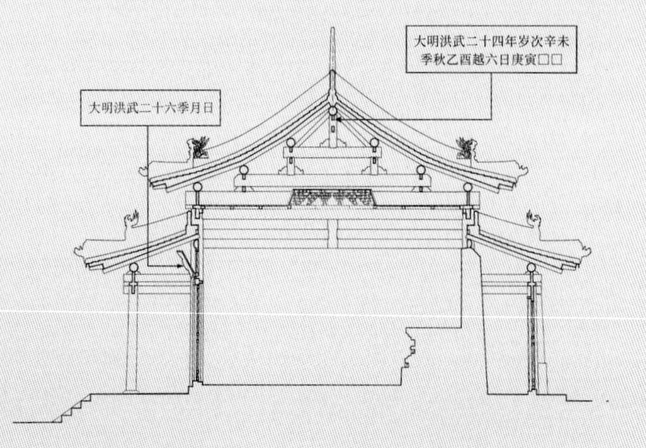

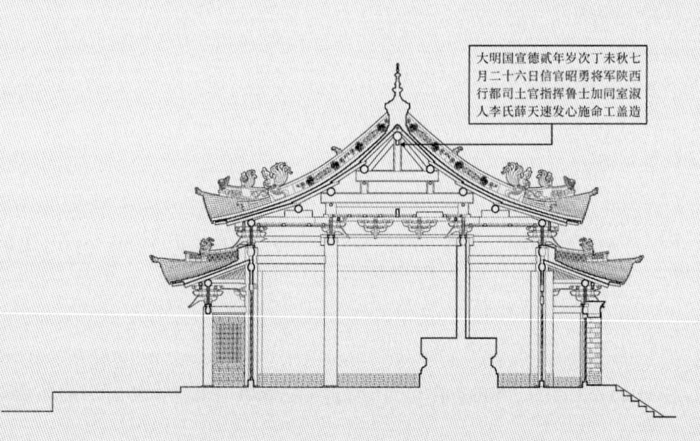

Fig.9 Comparison between the Qutan Hall (left) and the Wansui Hall (right)

Lu Family's Records, the exact date of construction can be found in the inscription on the ridge tiebeam of the Wansui Hall, the main hall of the Miaoyin Temple, which remains intact till today (Fig.9) .

Not only that the architecture of the Wansui Hall in the Miaoyin Temple has modelled on the Qutan Hall, so are the frescos were created in purpose. As stated in *The Ando Political and Religious History*, bronze statues of the three-period Buddha were respectfully invited from the Qutan Temple, along with them were Tibetan Tripitaka including *Kangyur* copied in silver ink and *Tengyur* copied in cinnabar. And same as the Qutan Temple, there was also a memorial tablet of Long Live the Emperor enshrined and worshiped in the Wansui Hall in the Miaoyin Temple. The Chieftain Lu would lead his subordinate officers to light incense and pray for the emperor on January the 8th every year.

As time changed quickly, and so was the society. The special status and fame of the Qutan Temple was no longer available during the late period of Ming Dynasty. In 1578, the Living Buddha Sonam Gyatso of the Gelug Sect formed an alliance with Altan Khan from the Tumer Tribe of Mongolia at Yanghua Temple, east to the Qinghai Lake. The former was respectfully called Dalai Lama, who was known as the Third Dalai. Since then, the Gelug Sect had become widely spread in Tibetan and Mongolia areas, with many more temples built for this sect, with much stronger political and economical strength than all other sects. As the Ming imperial court's policy of "appeasing Tibet to resist Mongolia" was disintegrated by such an alliance, Qutan Temple, the representative of the Kaggu Sect, was turned into the Gelug Sect. The Gelug Sect received even more supports from the imperial court during the Qing Dynasty. Qutan Temple there lost its fame and was much desolated, with continuous internal conflicts. Due to its involvement in the Loeb Zandazin's rebellion event took place in 1723, Zonci Sambu, the master of the temple had been in prison for seven years, which led to the loss of resources in supporting the Qutan Temple ever since. Except for regular repairing work, very little transformation and expansion projects took place in years afterwards.

2. Construction and Evolution of Qutan Temple

(1) Initial Construction of Qutan Temple

As recorded in *Ming Taizu shilu (Memoir of Emperor Taizu of Ming)*, the Tibetan Buddhist

到朝廷大力扶持，瞿昙寺倍受冷遇，内讧不断；嗣因雍正元年（1723年）青海蒙古和硕特部首领罗卜藏丹津叛乱，寺主阿旺宗哲卷入其中，平叛后入狱七年，瞿昙寺从此一蹶不振。其后，除了修葺，仅有少量改建和添建。

二、瞿昙寺的建置沿革

（一）肇建瞿昙

瞿昙寺肇建，如《明太祖实录》记载：『洪武二十六年二月廿七日，西宁番僧三剌贡马；先是，三剌为书招降罕东诸部，又创佛刹于碾白南川，以居其众；至是，始来朝，因请护持及寺额；上赐名曰「瞿昙寺」。』

这位三剌，就是三罗喇嘛，朱元璋尊重其心志，以佛祖姓氏『瞿昙』赐为佛寺嘉名。题款『大明洪武二十六季月日立』的匾额（图十），至今高悬瞿昙殿。寺中现存永乐六年（1408年）明成祖《皇帝敕谕碑》、洪熙元年（1425年）明仁宗《御制瞿昙寺碑》和宣德二年（1427年）明宣宗《御制瞿昙寺后殿碑》，也都郑重铭记此事。

应指出的是，近年曾有一份明洪武三年八月十五日的珍贵印信契书见世，提到『开荒耕种瞿昙寺……戎马水旱地』，按全文，瞿昙寺命名不晚于洪武三年，但仅涉圈划领地，拢聚信众包括『外户人』开荒耕种，积累资财，以筹建佛寺。

瞿昙殿脊枋题记『大明洪武二十四年岁次辛未季秋乙酉朔越六日庚寅□□』（图十一），为上梁典礼所书，明徵此时肇建。清乾隆十三年（1748年）松巴堪布·益希班觉《佛教史》按藏文史料说洪武二十五年建立基业，是寺院和僧众聚落初具规模。清同治十年（1871年）《补修瞿昙寺募化疏引》（下称《疏引》）申言：『瞿昙寺……兼有左右小寺，四面高塔与护法大殿，俱系明洪武二十六年修。』凡此表明，瞿昙寺初创，历时三年，陆续建有瞿昙殿，按藏传佛教尊奉曼荼罗图式配置的东西配殿、三世殿、护法殿和四隅小塔，以及寺旁僧舍、西南侧藏传佛教宁玛派（旧派红教）的天官寺、东面居住族人和佃户的里坊（图十二、图十三）。

monk Sanla came to the imperial court to tribute horses and asked for imperial support and a name for the newly built temple, who had persuaded Mongolia tribes to surrender and built a temple to comfort the locals. The emperor therefore named the temple Qutan.

The above mentioned Sanla is the same person named Sanluo, who is the respectful monk of the Kaggu Sect within the Tibetan Buddhism. The temple that he created was given the name of the Buddha, "Qutan", by Emperor ZHU Yuanzhang. One may find the inscription "erected in 1393" on the horizontal tablet hanging over the Qutan Hall (Fig.10). Existing steles include: *Huangdi Chiyu Bei* (Emperor Dictum Stele) by ZHU Di in 1408, *Yuzhi Qutansi Bei* (Qutan Temple Stele Made by the Emperor Order) by ZHU Gaochi in 1425, and *Yuzhi Qutansi Houdian Bei* (Qutan Temple Back Hall Stele Made by the Emperor Order) by ZHU Zhanji in 1427, all of which had recorded the naming of the temple in a serious way.

It should be pointed out that there had been a precious sealed agreement dated on Aug 15th, 1370, in which it was mentioned that "land near the Qutan Temple was to be reclaimed and cultivated, ... soldiers participated in transforming the dry land into paddy field". According to the content of this agreement, the naming of the Qutan Temple took place no later than year 1370, but mainly about designating the boundary of the temple, gathering the believers including the outsiders, reclaiming and cultivating the land on one hand, and accumulating funding and treasures on the other hand, in order to prepare building the temple.

According to the existing text found on the ridge tiebeam of the Qutan Hall (Fig.11), which should be written before the beam lifting ceremony, the Qutan Hall was started to be built before the date of September 6th, 1391. According to the book *Fojiao shi* (*History of Buddhism*), written by Sunbakanbu Yasibanjo in 1748, the temple was more or less completed and all monks had settled in 1392 based on the related Tibetan literature. As stated in *Buxiu qutansi muhua shuyin* (*Supplementary Explanation for Collecting Alms in Building Qutan Temple*) (hereinafter to be referred as *Explanation*) completed in 1871, "the Qutan Temple ... include two small temples on both sides, with four tall pagodas and the main hall, which were built in 1393." This suggests that the early construction of Qutan Temple lasted for three years, with the completion of Qutan Hall, side halls, Trikalea Buddhas Hall, Guardian Hall and four pagodas, together with monk dormitories on both sides of the complex, Tianguan Temple of the Nyingma (Red) Sect in its southeast direction, as well as residential neighborhoods for clansmen and tenant peasants in its east direction (Fig.12, Fig.13).

(2) Construction of the Baoguang Hall Complex

After ZHU Di was enthroned, he made special notes in the text of *Emperor Dictum Stele* written in 1418, which says, Bandan Zangbu, the empowered *jingjue hongji* great state adviser, built a Buddhist temple in the area near Xining, with the granted name of Baoguang (glory) by the emperor. Bandan Zangbu was in fact the nephew of Sanluo Lama, who was entitled as *jingjue hongji* State adviser in October, 1410 and ascended to empowered *jingjue hongji* great state adviser in April, 1412. After Sanluo Lama passed away in April, 1414, Bandan Zangbu became the new master of Qutan Temple, who started to build more structures for the temple, which are also recorded carefully in the above mentioned two stele texts made by emperor ZHU Gaochi and ZHU Zhanji respectively.

As mentioned in the Explanation, the construction of Baoguang Hall, Bell and Drum Towers and two small side halls were started in 1408, all granted for the great state adviser Bandan Zangbu by the emperor. Based on the *Ming Taizong shilu* (*Memoir of Emperor Taizong of Ming*), Bandan Zangbu was bestowed with abundant of money on October 18th 1407 and July 7th 1408 respectively, which were all put into the reconstruction of the Buddhist halls. According to the Archived Copy, the court has sent officials and eunuchs as the project envoys. It took ten years for the construction of the Baoguang Hall, the Small Bell Tower, the Small Drum Tower and their side halls to be completed on January, 1418 (Fig.14, Fig.15). As mentioned in the *Yuzhi Jin Foxiang Bei* (Gold Buddha Figure Stele Made by the Emperor Order) written by ZHU Di in the same year, and the *Archived Copy*, the copper gilded Buddha figure (Fig.16), the *xumizuo* made of blotch stone, bronze tripod and bottles, were all employed by the emperor, which remain till today to witness such a history.

(3) Continued construction of Longguo Hall Complex

After the completion of Baoguang Hall complex, more constructions started one after another on an even larger scale (Fig.17, Fig.18). As mentioned in the *Archived Copy*, eunuch MENG and commander TIAN Xuan were sent by the emperor as special envoys to build Baoguang Hall and Longguo Hall during the Yongle period, both with stele records. The *Explanation* further points out that the construction of Longguo Hall, side buildings connecting to new Bell and Drum Towers, as well as the front and back gate of the temple, were all completed by 1427, as a gift to be granted to the then great state

（二）重作宝光

朱棣践祚后，如其《皇帝敕谕碑》强调：兹者灌顶净觉弘济大国师班丹藏卜，于西宁迦伴虎蓝满都儿都地面起盖佛寺，特赐名曰『宝光』。班丹藏卜即三罗喇嘛之侄，永乐八年十月赐封净觉弘济国师，永乐十年四月升灌顶净觉弘济大国师，永乐十二年四月三罗喇嘛圆寂后主持瞿昙寺。就此兴作，明仁宗《御制瞿昙寺碑》和明宣宗《御制瞿昙寺后殿碑》也都虔敬纪述。

前引《疏引》指出：宝光殿、钟鼓楼并左右小寺，固系永乐六年修，时敕赐大国师班丹藏卜与焉。据《明太宗实录》载，永乐五年十月十八日赐班丹藏卜钞币，翌年七月初七又赐白金、钞币，旋『重作佛殿』。按前述《底册》，工程尚钦派朝官和内廷太监董理，到永乐十六年正月，宝光殿、小钟楼、小鼓楼并左右配殿竣工（图十四、图十五），历时十年。同年朱棣《御制金佛像碑》和《底册》提到的铜鎏金佛像（图十六），宝光殿内镌刻『大明永乐施』的御用花斑石须弥座及铜鼎、壶等，至今遗存为历史见证。

（三）继作隆国

宝光殿等竣工后，规模空前的建设接踵展开（图十七、图十八）。如《底册》强调：永乐年间节奉钦差孟太监、指挥田选等奉旨修建宝光、隆国二殿，立有碑记。《疏引》更指出：隆国殿、钟鼓楼并两

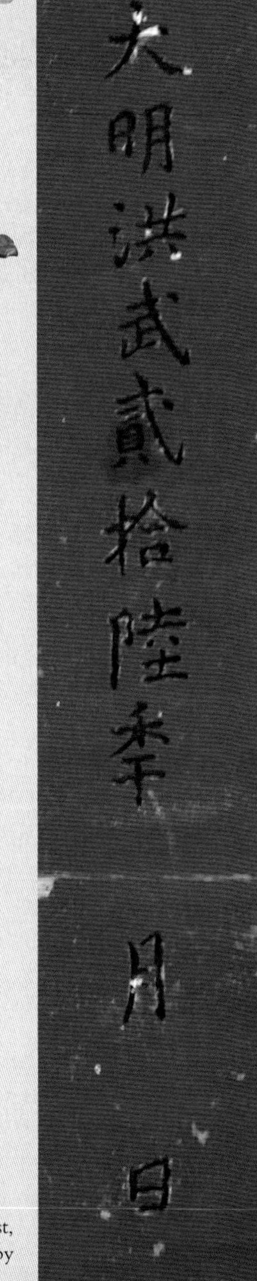

图十　洪武二十六年瞿昙寺匾额

图十一　瞿昙殿脊枋题记，自东往西：大明洪武二十四年岁次辛未季秋乙酉朔越六日庚寅口口（刘彤彤、张峻崚摄）

Fig.10 The horizontal tablet of the Qutan Temple, dated in 1393

Fig.11 Inscription on the ridge tiebeam of the Qutan Hall, with texts from east to west, meaning that (the building is constructed on) the 6th of September, 1391 (photo by LIU Tongtong and ZHANG Junling)

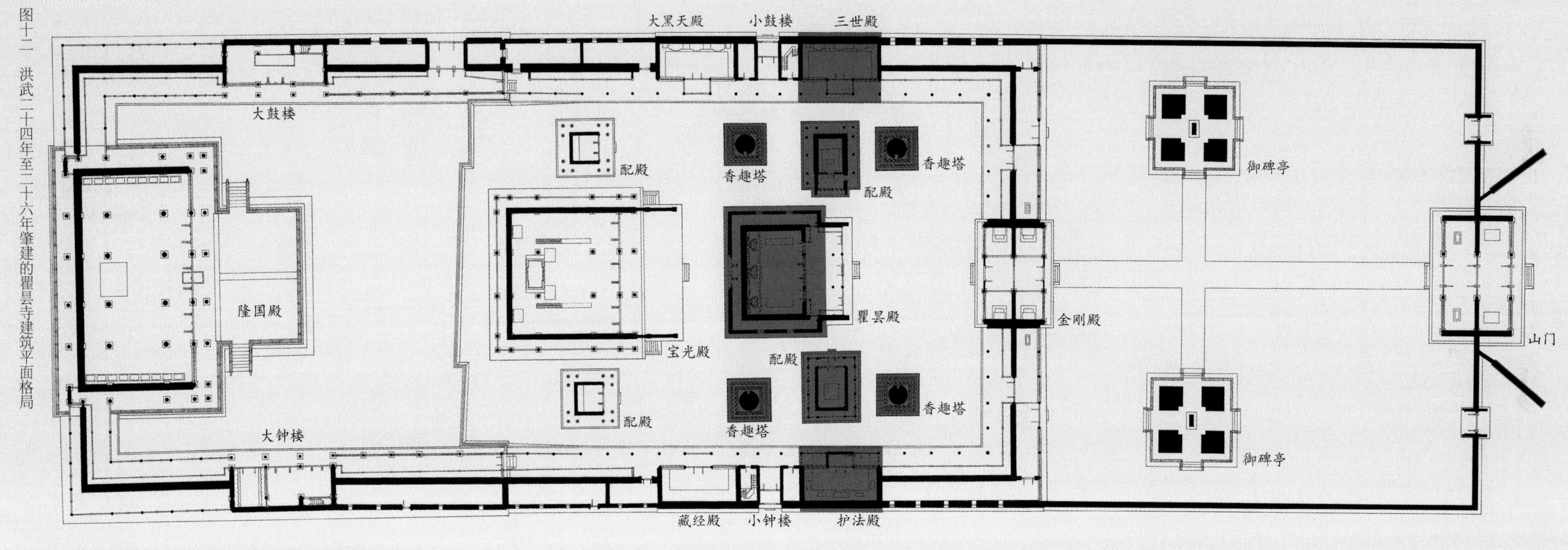

图十二　洪武二十四年至二十六年肇建的瞿昙寺建筑平面格局

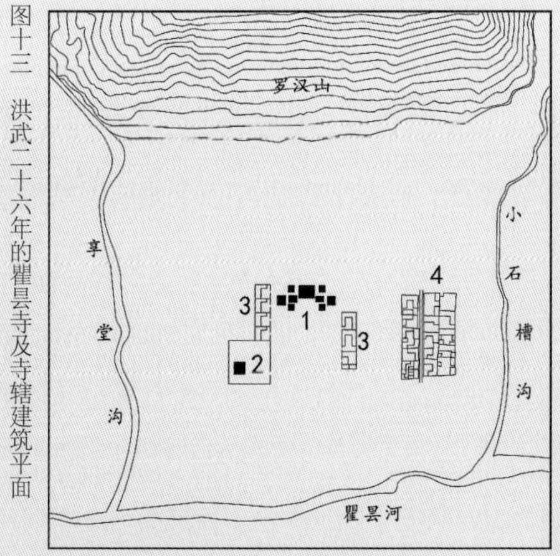

图十三　洪武二十六年的瞿昙寺及寺辖建筑平面

1瞿昙寺　2天官寺　3僧舍　4街坊

Fig.12 Layout Plan of the early Qutan Temple built between 1391–1393

Fig.13 Site plan of the Qutan Temple and buildings under its supervision in 1393
1 Qutan Temple　2 Tianguan Temple　3 monks dormitory　4 neighborhood block

图十四　永乐六年至十六年增建的宝光殿建筑群平面

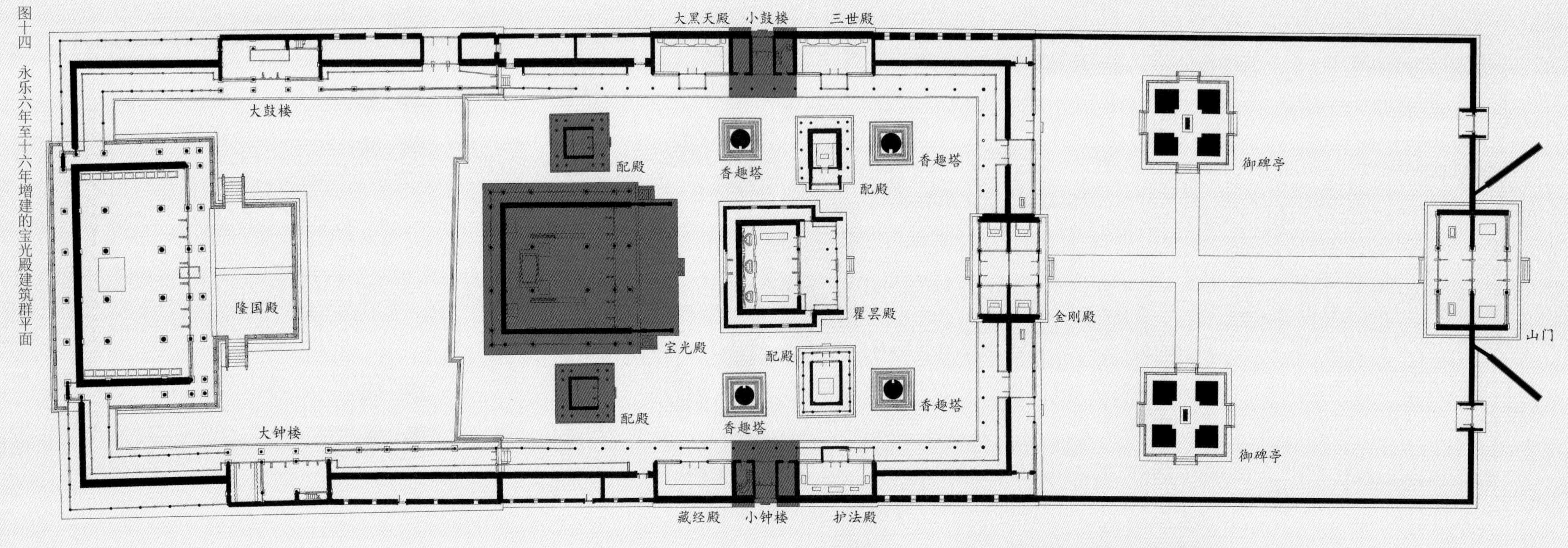

图十六　永乐十六年朱棣布施灌顶净觉弘济大国师班丹藏卜的铜鎏金佛像

图十五　永乐十六年的瞿昙寺平面

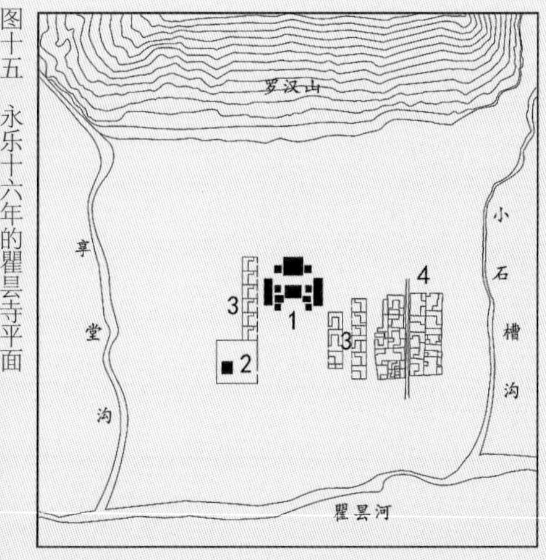

1. 瞿昙寺　2. 天官寺　3. 僧舍　4. 街坊

Fig.14 Baoguang Hall building complex built between 1408 and 1418

Fig.15 Site plan of the Qutan Temple in 1418

1 Qutan Temple　2 Tianguan Temple　3 monks dormitory　4 neighborhood block

Fig.16 The copper gilded Buddha figure granted by Emperor ZHU Di to Bandan Zangbu, the empowered *jingjue hongji* great state adviser Details of the copper gilded Buddha figure completed in 1418

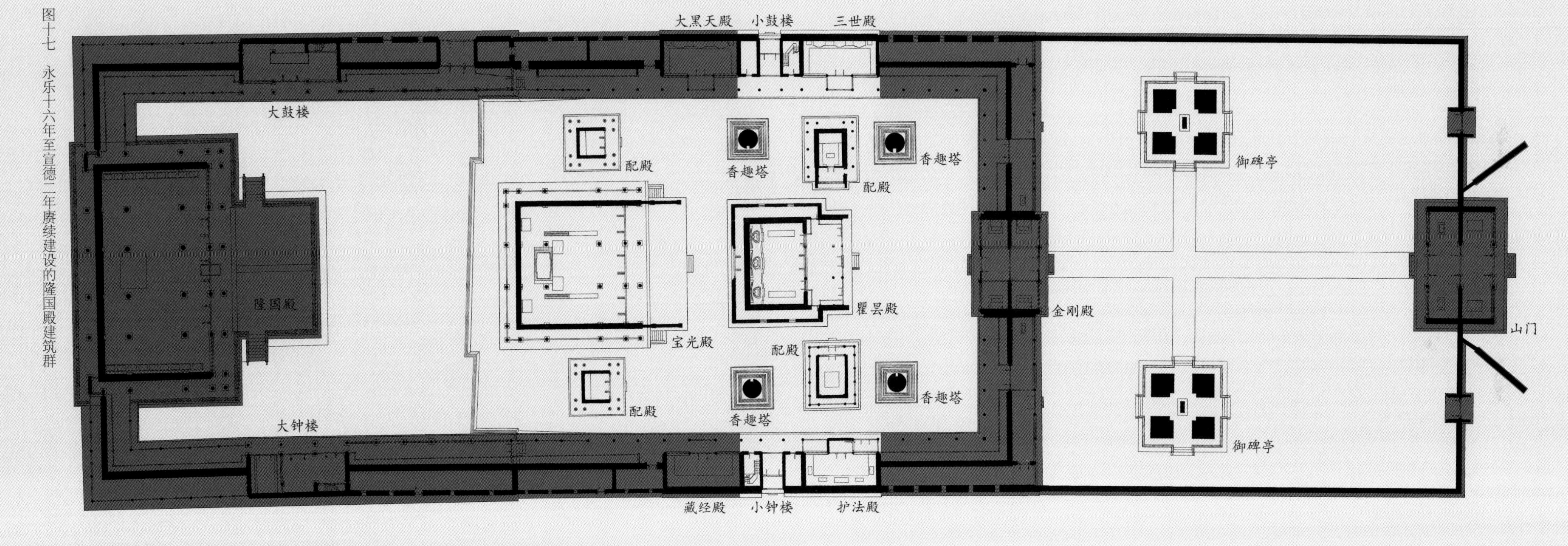

图十七　永乐十六年至宣德二年赓续建设的隆国殿建筑群

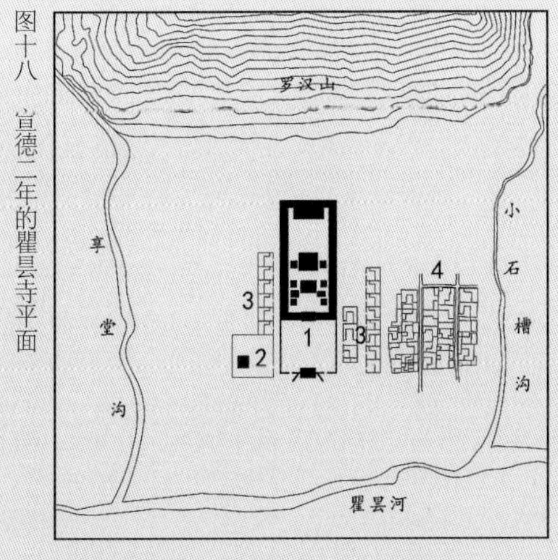

图十八　宣德二年的瞿昙寺平面

1瞿昙寺　2天官寺　3僧舍　4街坊

Fig.17 Plan of the Longguo Hall Complex built between 1418–1427

Fig.18 Site plan of the Qutan Temple in 1427

1 Qutan Temple　2 Tianguan Temple　3 monks dormitory　4 neighborhood block

adviser Sandan Zangbu. According to *Ming Chengzu shilu (Memoir of Emperor Chengzu of Ming)* , he was granted by the emperor to inherit the respected title of his uncle Bandan Zangbu, the empowered *jingjue hongji* great state adviser, on February 21, 1424, at the climax of the project construction.

On the backside of the memorial tablet of the Long Live the Emperor in the Longguo Hall, it writes, setting up by imperial supervision eunuchs MENG Ji, SAHNG Yi, CHEN Heng and YUAN Qi on February 9, 1427. Compared with the Archived Copy, we known that they were ordered by Emperor ZHU Di to supervise the projects, starting from the construction of Baoguang Hall in 1408 until the completion of Longguo Hall in 1427, nineteen years in total. During this period of time, they had accompanied the great state adviser from Qutan Temple to return to the capital. Their achievements were also recorded in *Ming Xuanzong shilu (Memoir of Emperor Xuanzong of Ming)* and *Ming Yingzong shilu (Memoir of Emperor Yingzong of Ming)* . For example, MENG Ji and SHANG Yi were sent by the emperor to meet ZHENG He to be in charge of the final phase of the Nanjing Dabaoen Temple. SHANG Yi was sent by emperors of both Zhengtong and Jingtai Periods to supervise the construction of Dagongde chan Temple, Daxinglong Temple and Dalongfu Temple in Beijing ,working together with the reputed master builder KUAI Xiang (Fig.19). Especially MENG Ji, who had worked on the project for more than nineteen years with deep affection to the place. Although he returned to Beijing after the project completion, he decided to bury himself in the Daxia Village, Gaodian Town, north to the Qutan Temple after he passed away. In front of his tombs, huge stone tablet, stone sheep and stone men were erected in order to honor his contributions, in comply with the ceremonial system for the secondary official post.

The continued construction of Qutan Temple included the Longguo Hall, the big bell tower, the big drum tower, the corridor, the front gate and the back gate. Because two emperors had passed away and two new emperors had ascended to throne, there were too many complicated taboos associated with such rituals and ceremonies. And also because of the transportation difficulties and long freezing winters, the whole construction had been much delayed. As a result, it took nine years for the continued construction to be completed until 1427. February 9th, 1427, as the date inscribed on the *Yuzhi Qutansi Houdian Bei* (Qutan Temple Back Hall Stele Made by the Emperor Order) by ZHU Zhanji, the tablet Longguo Hall, the memorial tablet of the Long Live the Emperor, the inscription text “constructed in Daming Xuande year” on the vessel, pot and bell all made in copper, as well as the title of the interior furnishing drawings referring to the installation

厢廊及前、后山门，咸系宣德二年修，时敕赐大国师三丹藏卜与焉。三丹藏卜如《明成祖实录》载，永乐二十二年二月二十一日，正值工程高潮，其叔父班丹藏卜圆寂，三丹藏卜被钦命嗣领灌顶净觉弘济大国师尊号。

至于太监孟继、尚义等，又见隆国殿内『皇帝万万岁』牌位背刻：『大明宣德二年二月初九日，御用监太监孟继、尚义、陈亨、袁琦建立。』按《底册》，他们由朱棣钦派，迄隆国殿告竣，董工十九年，期间屡扈寺主大国师赴京。后绩还见录明宣宗和英宗《实录》等，如孟继、尚义，宣德三年又奉旨绾会郑和主持南京大报恩寺收尾工程，正统、景泰朝，尚义还被钦派督修北京大功德禅寺、大兴隆寺，董领杰出哲匠蒯祥等鼎建大隆福寺等名刹（图十九）。尤其是孟继，长期操劳瞿昙寺大工而情愫缱绻，死后竟葬在寺北高店镇大峡村；为表彰其贡献，朝廷按二品官制，在坟前树立巨碑、石羊和石人。

瞿昙寺增华时，际遇两位皇帝丧葬和两朝皇帝登基大礼的避忌，再加边地山路崎岖，每季寒冬漫长，施工滞碍，以致耗时九年告成。落款『宣德二年二月初九日』的朱瞻基《御制瞿昙寺后殿碑》、隆国殿匾额和『皇帝万万岁』牌，铭文『大明宣德年施』的铜鼎、壶和大钟，墨书『大明宣德二年二月初九日西宁瞿昙寺安奉大持金刚佛像图样』的陈设图样，无不辉映出当年竣工庆典的隆重（图二十、图二十一）。

隆国殿工程崇宏。略如中、后两院周匝厢廊，尤其是隆国殿和两翼大鼓楼、大钟楼及双面抄手斜廊，就刻意缩仿永乐十五年（1417年）始建的北京大内重檐庑殿顶的奉天殿、单檐庑殿顶的二层文楼、武楼，以抄手斜廊连缀，不过覆盖布瓦而非黄琉璃瓦，与皇宫金碧流溢的豪华毕竟有别。深谙皇考继作瞿昙寺宏图的朱高炽，尽管祚短，未遑目睹大工告竣胜貌，却禁不住在其《御制瞿昙寺碑》热切赞颂道：『重作奉佛之殿，崇高附丽于日星，五章辉灼于霞彩，瀜深闳伟，超出尘外，香云缭布，如现鹫峰。』

嗣后，如《明宣宗实录》载：宣德二年十月……西宁卫净觉弘济大国师三丹藏卜，以修完寺宇，差喇嘛完卜捕黑般等进马谢恩。《底册》又纪：同年三月，皇帝还下令调拨西宁卫五十二名官军长驻瞿昙寺，洒扫巡视寺宇。

（四）后世变迁

明初四位皇帝倾心投入，历时三十六年，瞿昙寺完成了从椎轮到大辂的华丽变身。对这一西陲巨刹，明代后嗣皇帝也曾予增修或缮葺。

如《明英宗实录》载，景泰年间（1450—1456年）曾钦差内官、内使大规模修缮瞿昙寺殿宇。天顺

of the Vajradhara Buddha Sculpture on the same date, all reflecting the grandness of such a completion ceremony then (Fig.20, Fig.21) .

The Longguo Hall complex project was an extensive, grand and magnificent one. Its main layout, i.e. those of the middle and back courtyard complex, especially the Longguo Hall, the Great Drum Tower, the Great Bell Tower and the ascending side corridors, had copied the overall layout, shapes and structures of the Fengtian Hall complex (including the double eaves Fengtian Hall, the single eave two-floored Wen Building and Wu Building, as well as the ascending side corridors) at the core of the Forbidden City in Beijing in 1417. The only difference lies in the building scale and roof tile materials as the Fengtian Hall complex was much larger and using glazed yellow-colored roof tiles while the Longguo Hall complex was smaller and the roof tiles were common ones. This is why emperor ZHU Gaozhi, who knew quite well about the grand plan of his deceased father in the continued construction of the Qutan Temple, though too short on his throne to witness the completion of the Longguo Hall, still praised highly in his *Yuzhi Qutansi Bei* (Qutan Temple Stele Made by the Emperor Order) that, "The reconstruction of the hall to worship the Buddha is as grand as the sun and the star, with its flame more brilliant than the rays of morning and evening sunlight. And its deepness and grandness are beyond this world, surrounded with fragrant clouds as if it were built on top of the high mountain."

Later, as recorded in the *Ming Xuanzong shilu (Memoir of Emperor Xuanzong of Ming)*, "In October 1427, the *jingjue hongji* great state adviser Sandan Zangbu has completed the construction of the temple and sent lama wan-bu-bu-hei-ban to pay the tribute to the royal court with horses. As written in the Archived Copy, in March of the same year, the emperor had ordered to send 52 government soldiers to reside in the Qutan Temple, helping to clean and scout the temple.

(4) Transitions in Later Ages

With highly concerns and devotions of the first four emperors of the Ming Dynasty, the Qutan Temple had completed its flamboyant change from its initial stage to its final shape. The descendant emperors of the Ming Dynasty also contributed to its extension and renovation.

As recorded in the *Ming Yingzong shilu (Memoir of Emperor Yingzong of Ming)* that,

图十九　明景泰三年（1452年）尚义董建的北京隆福寺『天龙八部罗叉女众』水陆画（美国克利夫兰美术馆藏）

Fig.19 part of the painting of "The Demi-Gods and Semi-Devils and the Rosai women", produced by SHANG Yi in 1452 (collection of Cleveland Art Museum, USA)

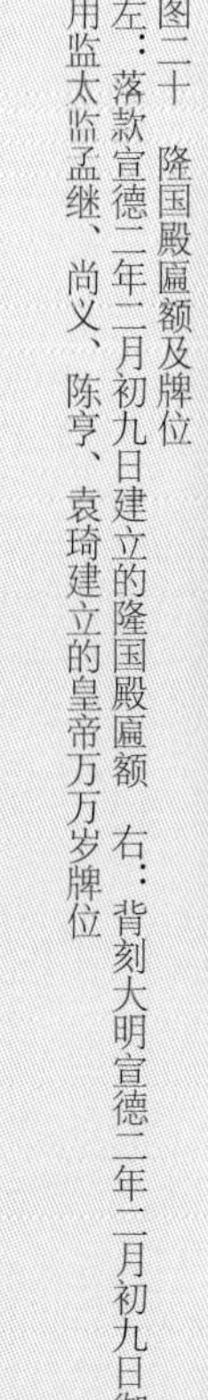

图二十　隆国殿匾额及牌位
左：落款宣德二年二月初九日建立的隆国殿匾额　右：背刻大明宣德二年二月初九日御用监太监孟继、尚义、陈亨、袁琦建立的皇帝万万岁牌位

图二十一　隆国殿及大钟鼓楼文物
左：隆国殿内象背云鼓　中：大鼓楼内云鼓
右：大钟楼内铭文宣德二年的黄铜大钟

Fig.20　Tablets of Longguo Hall
Left: Inscription of the tablet hanging over the Longguo Hall, mentioning the date of February 9th, 1427; Right: The Memorial tablet of Long Live the Emperor set up by the imperial supervision eunuchs MENG Ji, SHANT Yi, CHEN Heng and YUAN Qi with the same date February 9th, 1427, inscribed on its back

Fig.21　Cultural relics in Longguo Hall，Great Drum Tower and Great Bell Tower
left: Cloud drum on the back of the stone elephant in Longguo Hall; middle: Cloud drum in the Great Drum Tower; right: Bronze Bell inscribed with manufacturing year, located in the Great Bell Tower

六年（1462年）三月三十日，钦准灌顶净觉弘济大国师领占藏卜请求，修浚瞿昙寺墙及沟壕，以防鞑靼蒙古侵掠，这就是今称旧城的瞿昙寺城堡内城（图二十二）。

此后，按前述《疏引》载，明宪宗成化二年（1466年），瞿昙寺前院又增建东、西碑亭，荫护宣德二年（1427年）《御制瞿昙寺后殿碑》和洪熙元年（1425年）《御制瞿昙寺碑》（图二十三）。

据《西宁府新志》记载，明嘉靖三十七年（1558年），游牧青海的鞑靼蒙古劫掠了瞿昙寺，所幸建筑未毁。为此，明神宗万历元年（1573年），钦准瞿昙寺城堡东、南两面加筑城墙，以护卫寺辖族人、佃户居住的里坊，俗称『新城』（图二十四）。

按《底册》所记，明熹宗天启末年（1627年），瞿昙寺东北隅又建起包括经堂在内的囊谦，供嗣承灌顶净觉弘济大国师和灌顶广智弘善国师的高僧日常居住、诵经和礼佛（图二十五）。

往后，清乾隆四十七年（1782年）五月，瞿昙殿前添建抱厦（图二十六、图二十七）。

按前引《疏引》，清同治十年（1871年）又修葺已倾圮的小塔和碑亭等。

during the years between 1450 and 1456, the emperor had sent his envoys to supervise the large-scale repairing work towards the building complex of the Qutan Temple. On March 30, 1462, as requested by the empowered *jingjue hongji* great state adviser Ling-zhan-zang-bo, it was proved by the royal court to repair the wall and dredge the ditch around the temple in order to prevent it from Mongolian's intrusion. This wall was the boundary in defining the inner city of the Qutan Temple castle, which is called the old city today (Fig.22) .

According to previously mentioned *Explanation*, it is recorded that two pavilions were added on east and west side of the front courtyard in 1466, protecting the two imperial steles completed in 1427 and 1425 respectively (Fig.23) .

According to the records in *Xining Fu Xin Zhi (the New Archive of the Xining Prefecture)*, the Tarta Mongolians plundered the Qutan Temple when they moved into the Qinghai region in 1558, luckily the buildings remain intact. Therefore, the imperial court approved the request of the Qutan Temple to build walls on its east and south sides in 1573, protecting its people and the neighborhood, which was later called the new city (Fig.24) .

According to the *Archived Copy*, more buildings including a scripture hall were built at the northeast corner of the Qutan Temple in 1627, providing the everyday living space for the succeeding empowered *jingjue-hongji* great state adviser and the empowered *guangzhi-hongshan* state adviser, including reciting passages from scriptures and praying to Buddha (Fig.25) .

A new entrance hall was added to the front facade of the Qutan Hall in May, 1782 (Fig.26, Fig.27) .

According to the *Explanation*, the collapsed small tower and the stele pavilion were restored in 1871.

After the damage caused by the earthquake in 1944, large-scale repairing work had been carried out upon the Qutan Temple for about two years, which was recorded in detail on the horizontal stele of the Baoguan Hall. During this process, some matured local craftsmanship techniques were adopted in renovating the main gate, the Qutan Hall, and the stele pavilion. They were all together called the Hezhou Crafting Method, which had been gradually formed and had become popular in the Gansu-Qinghai region during the Qing Dynasty. Small corner towers of the Qutan Hall were also rebuilt into the shape and structure of the Tibetan style scent tower.

3. Architecture in the Qutan Temple

(1) Layout of the Architectural Complex

The building complex of the Qutan Temple was in comply with the *fengshui* principles, resulted in an organic composition with the environment to form a natural scenery. Its main axis lies in the direction of southeast and northwest, with a total length of 225 meters and a height difference of 7.6 meters, composing the front, the middle and the back courtyards with rich spatial arrangements (Fig.28) .

The front courtyard is surrounded by red walls, with the main gate in the middle towards the south, forming the entrance to the temple. The 八 -shaped screen walls are extended on its side and a pair of flag posts in the front, and along its two wings, there are festooned doors directly attached to the walls. Within the front courtyard, on top of the *xumizuo*-style base surrounded by stone sculptured railings, there are two double-eave and cross-shaped ridge imperial stele pavilions facing each other on the east and west axis. The space within the front courtyard is wide and open with only five buildings including the two corner gates, making the two pavilions become even more prominent.

The entry building to the middle courtyard is the Jinggang Hall, followed by the main buildings such as Qutan Hall and Baoguang Hall, standing one after another. On both sides, there are four smaller side halls and scent towers respectively. The enclosed corridors on both sides are open to the courtyard and extending to the back courtyard. The space facing the side hall of Qutan Hall is the Sanshi Hall on the east side, with the Small Drum Tower built on top of the three-bay corridor to its north. Symmetrically, the Small Bell Tower and the Hufa Hall are built on the west side. Compared with the front courtyard complex, the building complex of the middle courtyard are more concentrated with more intricate space arrangements.

North to the Baoguang Hall, the terrain ascends by 2.6 meters, an upside down U-shaped brick wall is built as a retaining wall, with a beautiful hollowed pattern as its roof top. Following ascending terrain and corridor path north to the Small Drum Tower and the

民国33年（1944年），瞿昙寺遭大地震破坏，又经两年大规模修缮，宝光殿匾额题记曾详细载述。修缮中，山门、瞿昙殿、御碑亭等局部融入了清代成型的当地所谓河州工艺做法，瞿昙殿四隅小塔也被改修成藏式香趣塔的形制。

三、瞿昙寺的建筑

（一）组群布局

瞿昙寺的建筑组群布局，遵循风水形势，同景物天成的自然环境有机结合，沿轴线从东南向西北展开，纵深逾225米，前后高差7.6米，组成空间层次十分丰富的前、中、后三进院落（图二十八）。

前院周匝红墙，南面山门居中，为全寺入口，两侧延出砖雕八字影壁墙，前立幡杆；山门两翼还各辟随墙垂花门。院内东、西相向，围以石雕栏楯的须弥座台基上，峙立着重檐十字脊的御碑亭。整个前院，包括角门，仅五座建筑，空间舒朗，两座御碑亭格外突出。

中院入口为金刚殿，后为中院主体建筑瞿昙殿和宝光殿，左右各有小配殿，瞿昙殿四隅还分建香趣塔。院旁围合廊庑敞向院内，并向后院延伸。东廊对位瞿昙殿东配殿为三世殿，迤北三间上构小鼓楼；西廊对称，建有护法殿和小钟楼。和前院比较，中院建筑更为密集，空间层次丰富。

宝光殿北，地势高起2.6米，横亘门形平面的砖砌挡土墙，以空花宇墙结顶，循中院两侧小鼓楼和

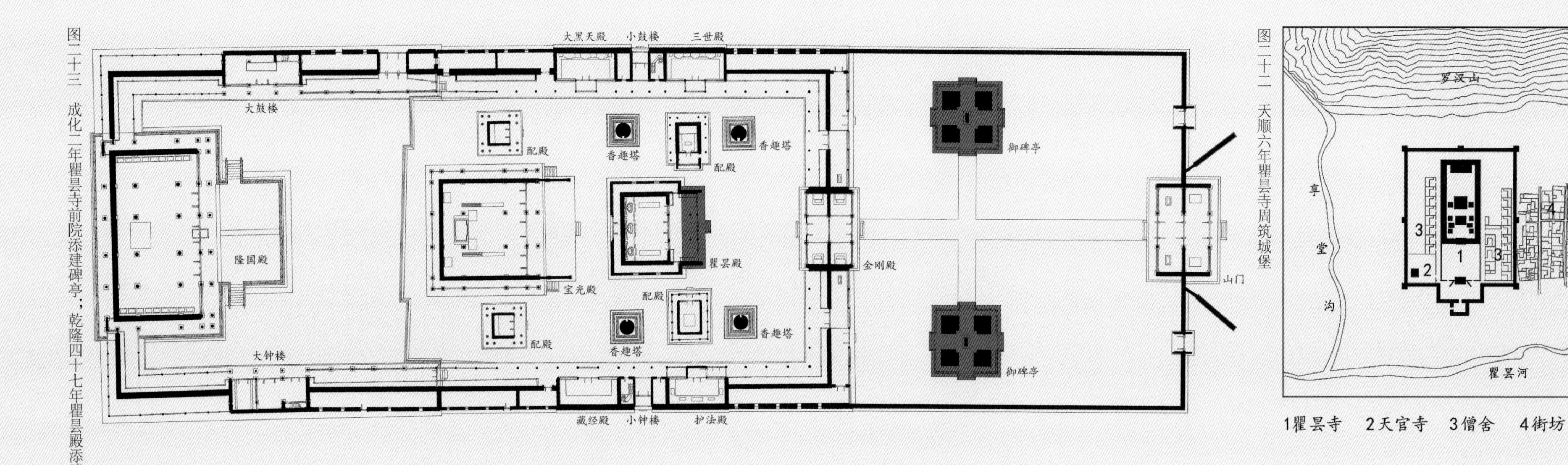

图二十三 成化二年瞿昙寺前院添建碑亭；乾隆四十七年瞿昙殿添建前抱厦

图二十二 天顺六年瞿昙寺周筑城堡

1瞿昙寺　2天官寺　3僧舍　4街坊

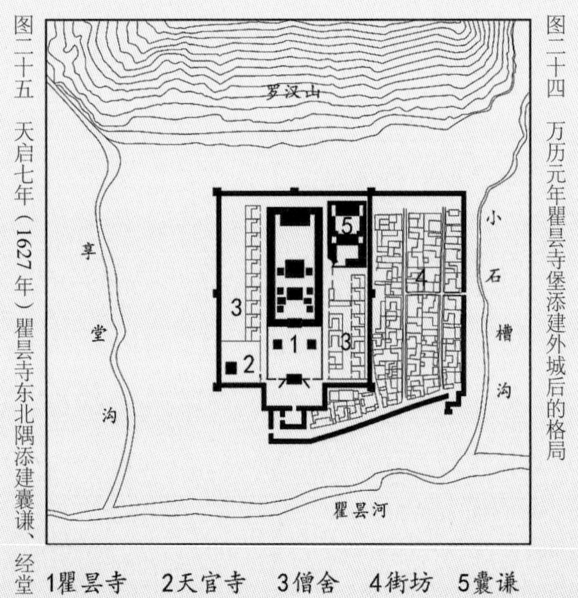

图二十五 天启七年（1627年）瞿昙寺东北隅添建囊谦、经堂

1瞿昙寺　2天官寺　3僧舍　4街坊　5囊谦

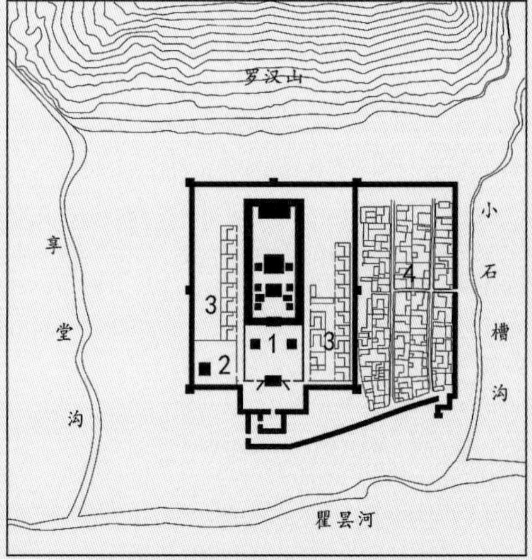

图二十四 万历元年瞿昙寺堡添建外城后的格局

1瞿昙寺　2天官寺　3僧舍　4街坊

Fig.22 Walls built around the Qutan Temple to form the castle in 1462
1 Qutan Temple　2 Tian guan Temple　3 monks dormitory　4 neighborhood block

Fig.23 Stele pavilions added in the front courtyard of the Qutan Temple in 1466; the front hall attached to the Qutan Hall completed in 1782

Fig.24 Site plan of the Qutan Temple after the new city wall were added in 1573
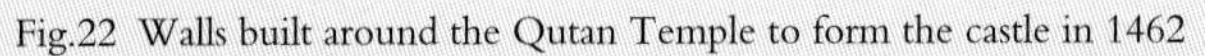
1 Qutan Temple　2 Tianguan Temple　3 monks dormitory　4 neighborhood block

Fig.25 Site plan of the Qutan Temple, with newly built complex outside the northeast corner of the temple complex
1 Qutan Temple　2 Tianguan Temple　3 monks dormitory　4 neighborhood block　5 nangqian

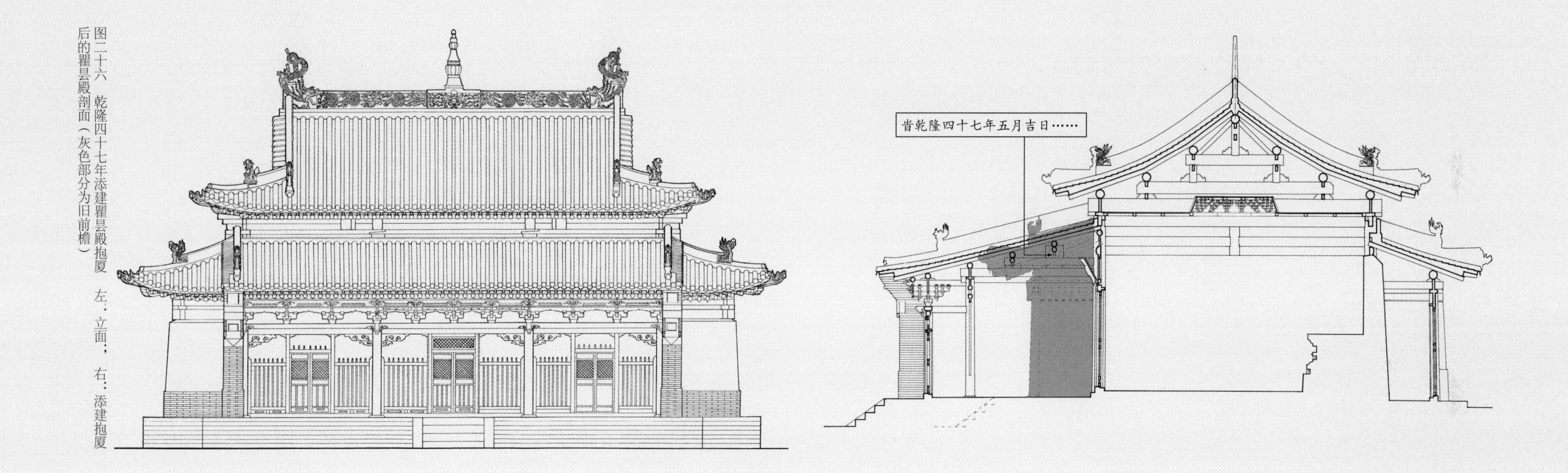

图二十六 乾隆四十七年添建瞿昙殿抱厦 左：立面；右：添建抱厦后的瞿昙殿剖面（灰色部分为旧前檐）

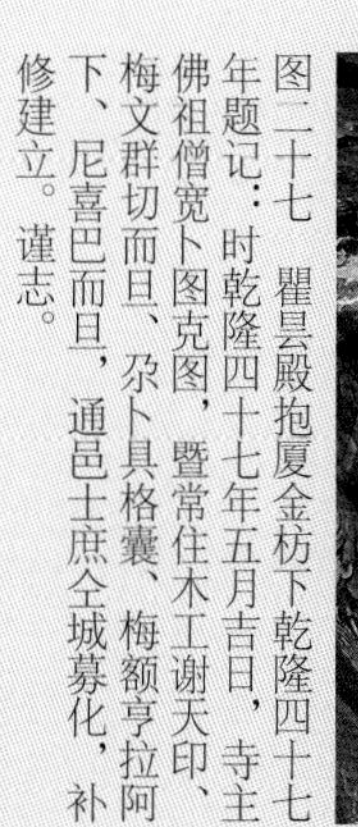

图二十七 瞿昙殿抱厦金枋下乾隆四十七年题记：时乾隆四十七年五月吉日，寺主佛祖僧宽卜图克图，暨常住木工谢大印、梅文群切而旦、尕卜具格囊，梅额亨拉阿下、尼喜巴而旦，通邑士庶仝城募化，补修建立。谨志。

Fig.26 The Qutan Hall built in 1782 left: New Facade of the Qutan Hall; right: Section of the Qutan Hall after the new entrance hall was added (grey part refers to the original front eave area)

Fig.27 The inscription text on the bottom side of the purlin tiebeam in the new entrance hall of the Qutan Hall, which stated the date and main figures who participated in the extension project in 1782

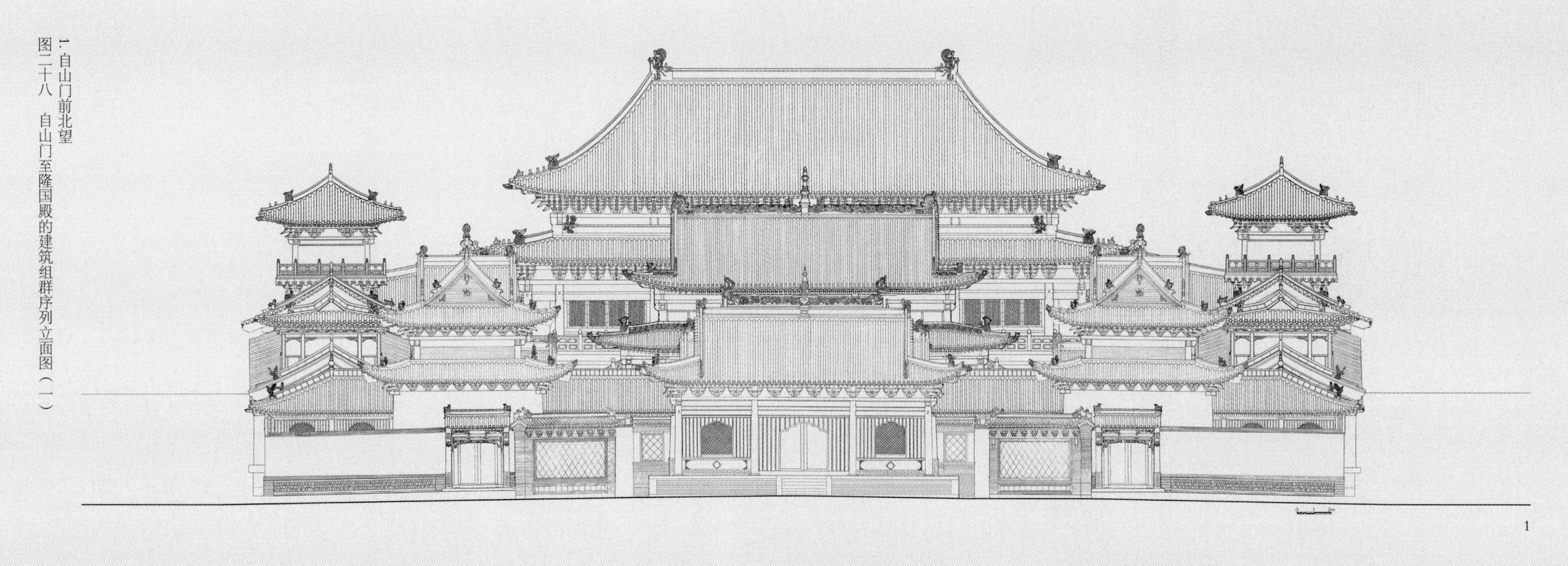

1

2

图二十八　自山门至隆国殿的建筑组群序列立面图（一）

1. 自山门前北望

2. 自御碑亭前北望

Fig.28 The spatial sequence of Qutan Temple building complex from the main gate to Longguo Hall (1)

1 North view from the main gate

2 North view from the imperial pavilions

图二十八　自山门至隆国殿的建筑组群序列立面图（二）

3. 自金刚殿前北望

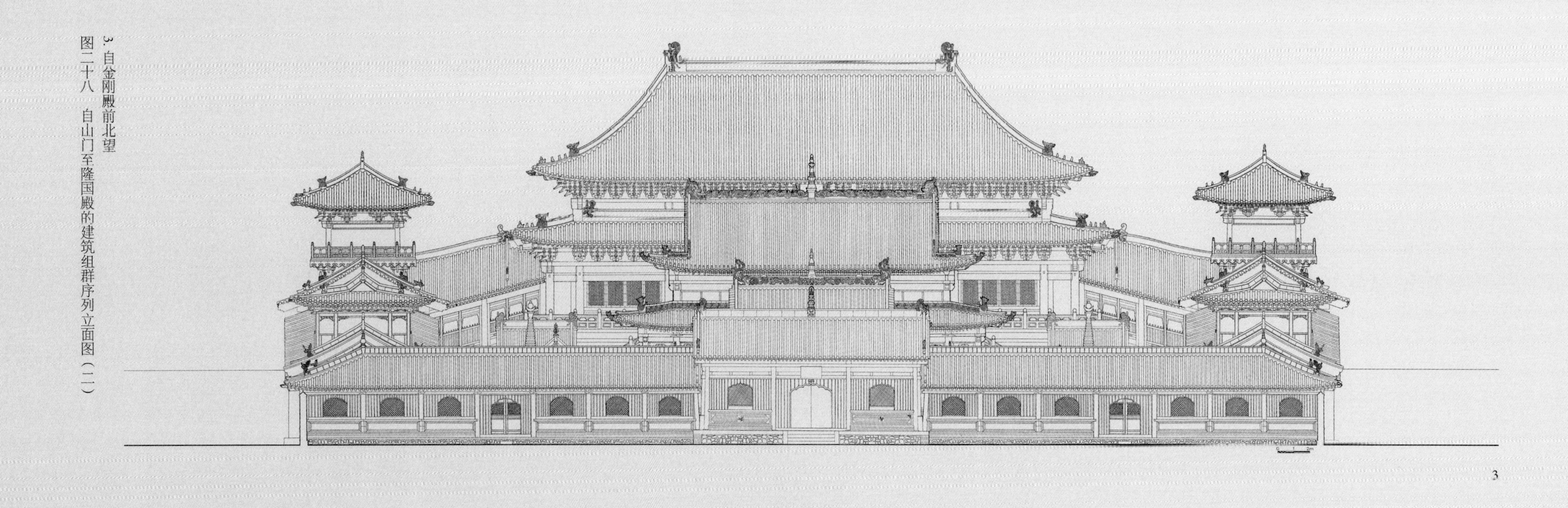

3

4. 自瞿昙殿前北望

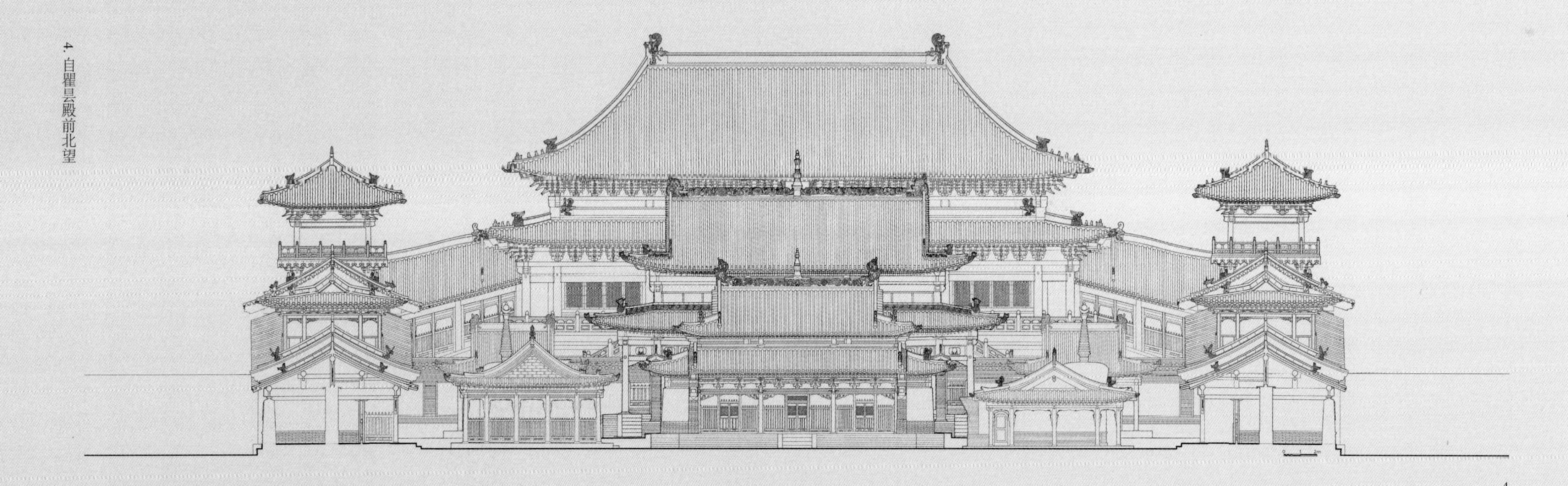

4

Fig.28 The spatial sequence of Qutan Temple building complex from the main gate to Longguo Hall (2)
3 North view from the Jingang Hall
4 North view from the Qutan Hall

图二十八　自山门至隆国殿的建筑组群序列立面图（三）

5. 自宝光殿前北望

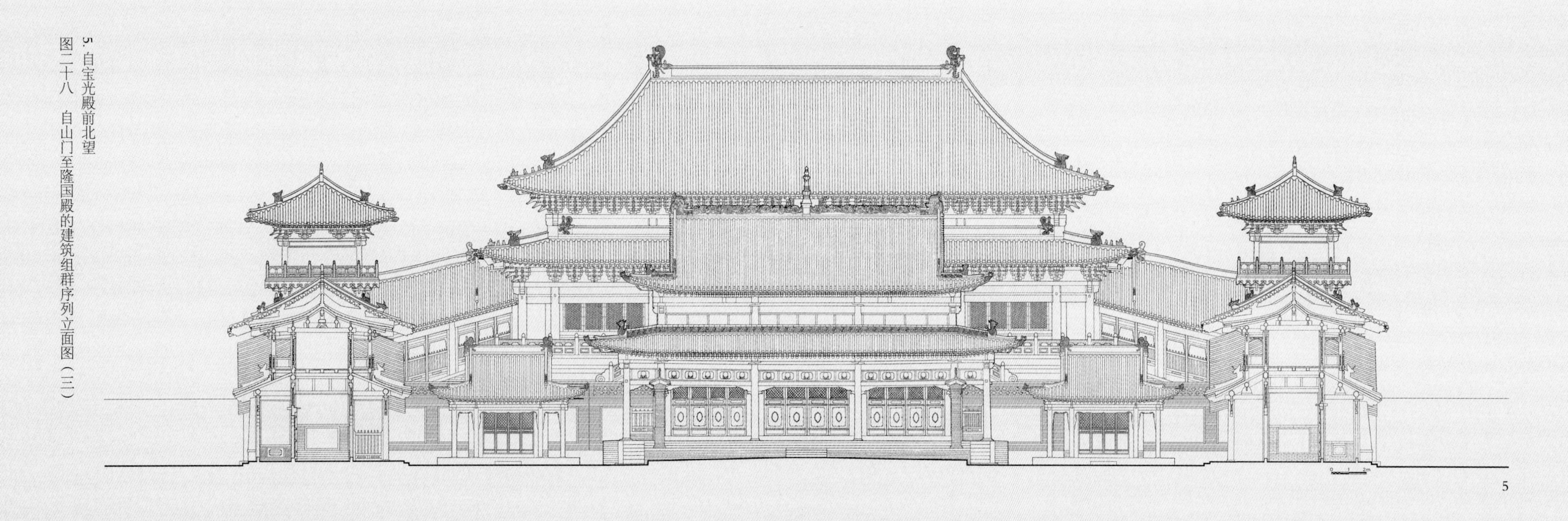

5

6. 自大鼓楼、大钟楼北望

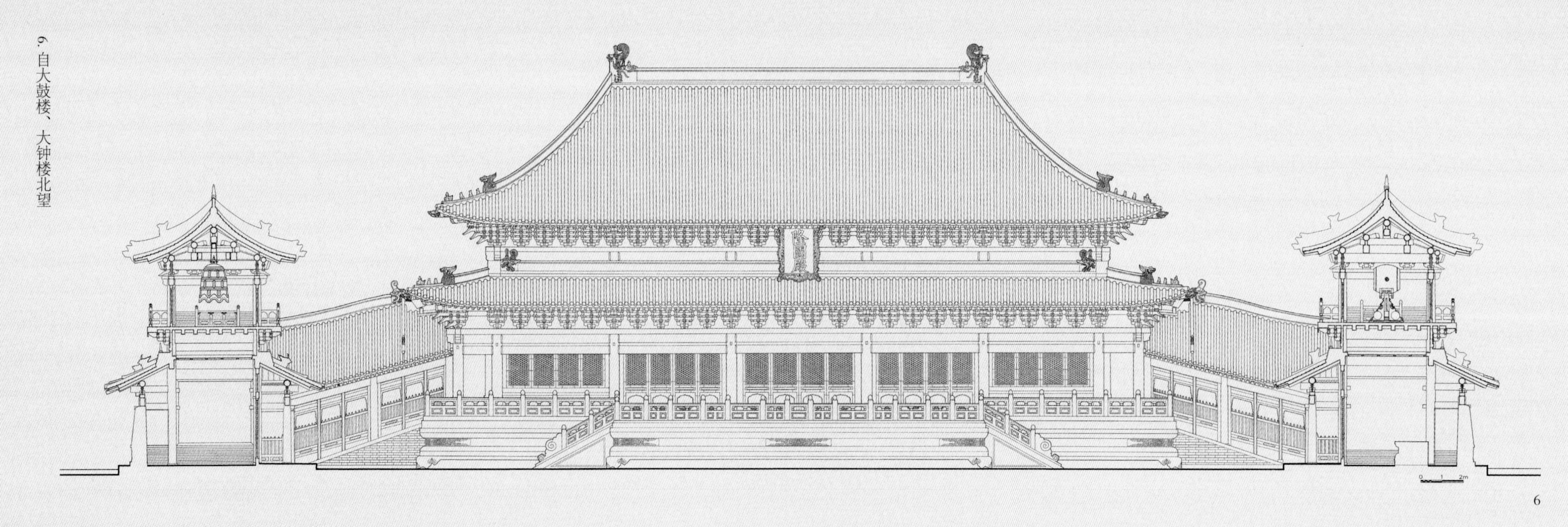

6

Fig.28 The spatial sequence of Qutan Temple building complex from the main gate to Longguo Hall (3)
5 North view from the Baoguang Hall
6 North view from the Great Drum and Bell Tower

图二十八 山门至隆国殿的建筑组群序列立面图（四）
7. 隆国殿及两翼抄手斜廊

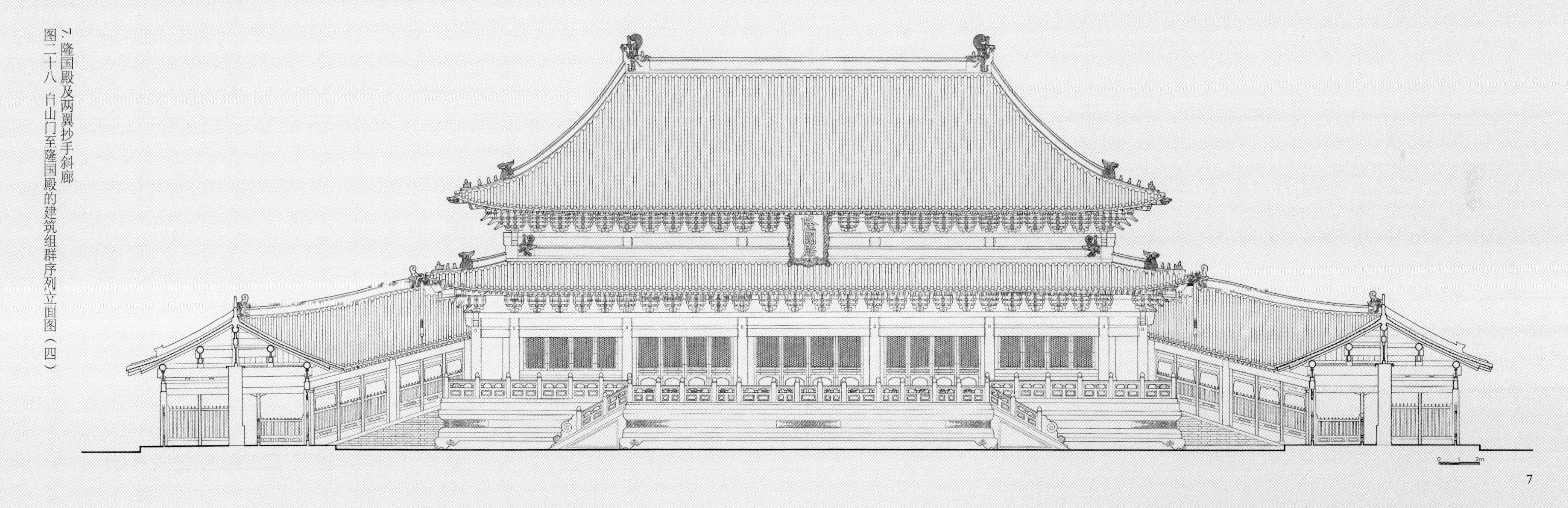

Fig.28 The spatial sequence of Qutan Temple building complex from the main gate to Longguo Hall (4)
7 Longguo Hall and the ascending side corridors on its wings

小钟楼北随地势上升的斜廊前行，可至后院。宽敞的庭院北面，居中高达2.3米的石雕须弥座台基前出大月台，周匝石雕望柱栏板，隆国殿耸峙为全寺空间序列的终结，一派皇家殿堂气象，冠表全寺。东西厢廊上拔起造型端庄的大鼓楼和大钟楼，各以双面抄手斜廊呈向上朝拱之势连缀隆国殿，有力强化了隆国殿作为建筑组群重心的宏伟气势（图二十九）。

寺院东北隅，为活佛居所囊谦，两进院落坐向与寺院平行，出于风水『朝山』即远方对景的考虑，垂花门式样的入口南向偏西。前院中路为单层大过厅，左右建二层厢房，院南倒座为伙房和库房。过厅两端各有小楼院，称『上转楼』和『下转楼』。后院左右各建配楼，尽头重檐歇山顶的大经堂，民国时焚毁，现存围护土墙。整个囊谦，尺度亲切，遍施精致木雕和砖雕，在全寺庄重的梵天氛围中，创造了世俗情趣浓郁的日常起居环境。

按传统，瞿昙寺的组群布局，必经缜密选址规划设计。如明仁宗《御制瞿昙寺碑》申言，该寺灏深闳伟，缘自明太祖以来『命官相土，审位面势』；出自《尚书》《周礼》等坟典的这寥寥八字，本谊就是选址规划设计。

Small Bell Tower on both sides of the middle courtyard, one may arrive at the back courtyard. In the wide northern side of the courtyard, lies the Longguo Hall, which was standing on top of the 2.3-meter-high stone carving *xumizuo* base, with a large platform surrounded by stone carving baluster columns and breast boards. Being the most magnificent building of the entire temple, the Longguo Hall shows the most royal style and manner, marking the end of the building complex and the climax of the spatial arrangements. On its east and west sides, there are the Great Drum Tower and the Great Bell Tower, respectively built in a graceful way. Moreover, by connecting the two Towers with the Longguo Hall through the ascending side corridors, the latter has been illuminated as the core of the entire temple complex with the grandeur momentum (Fig.29).

Northeast to the main building complex of the temple, a double-courtyard building complex is built as the dwelling place of the living Buddha, built parallel to the temple complex in general. But according to the *fengshui* principle of facing mountain, in other terms, the need of an opposing scenery, the location of its main gate is built in the festooned style, at the southwest corner of the complex instead of in the middle of the southern side. In the middle of the front courtyard lies a one floored large hallway and two floored side buildings in a symmetric way. The kitchen and storage rooms are located on its southern side. On the end of each side of the hallway, there is a small tower respectively named as the Upper Tower and the Lower Tower. There are also side buildings in the back courtyard, and the Great Scripture Hall with the gable and hip roof and double eaves was built in the middle of the northern side, which had been burnt down during the Republic era, with only the earthed walls survived. The whole area are built in a quite pleasant scale, with delicate wooden and brick carving everywhere, creating a special environment with rich secular and everyday life environment within the rather serious and sacred Brahma atmosphere.

Following the building tradition, the layout of the Qutan Temple must have gone through a careful process of site selection, planning and design. As inscribed on the imperial stele of emperor ZHU Gaochi, the grandeur of the temple started with the process "*ming guan xiang tu, shen wei mian shi* (asking high-rank officials to survey the land, and studying the orientation and position)". The original meaning of these eight characters, quote from the classics such as *Shang Shu* and *Zhou Li*, is the site selection, planning and design.

According to the imperial construction practice concepts and principles of the Ming Dynasty, the integration of the Qutan Temple and the mountain and river environment around it has to do with the *xingshi* theory focusing on the exterior space design. *Xing* and *shi* respectively refer to the different spatial compositions and visual effects, which are to be produced by the near and the far, the small and the big, the single and the complex, the partial and the whole, the detail and the outlook. In comply with such a theory, the building scale, volume and visual distance are to be controlled by the principle of "*qian chi wei shi, bai chi wei xing*" (the scale of 1000 *chi* helps to compose the overall outlook, and the scale of 100 *chi* helps to compose the detailed image). The overall distance of a building complex normally will not exceed 1000 *chi*, which is often between 230 to 350 meters. For a single building or a courtyard, the distance is normally within 100 *chi*, which is between 23 to 35 meters. By doing so, it helps visitors to obtain a continuous perception group experience that can be easily switched among the near, the middle and the far views, which fit into people's audio-visual experience, rich and complete, full of human kindness.

In fact, if we use the traditional "*ping-ge*" modular network, which is featured with earth surface DEM character, to examine the overall plan of the Qutan Temple, and if we then compare it with those found in the archived *yang-shi-lei* drawings of the imperial tomb of the Qing Dynasty (Fig.30), we may easily find the similarity in following this "*qian chi wei shi, bai chi wei xing*" theory in their planning and design.

Moreover, after SHANG Yi returned to Beijing, he started to prepare constructing the Great Longfu Temple, the first Tibetan Buddhist temple to be built in the Han-style courtyard typology, serving the royal family in the capital. Deeply influenced by the Qutan Temple project, the same methods were adopted by following the *ping-ge* module and other maps (Fig.31). The traditional way of integrating the building elevation into the plan helps to better express the horizontal as well as vertical design intentions and images, from single building to building complex, all following the principle of "*qian chi wei shi, bai chi wei xing*".

In short, the building complex of the Qutan Temple contains the wisdom and elegant inspirations of the traditional planning and design, which is definitely worth to be explored and inherited.

(2) Building Rules and Regulations

Except for the four brick towers, buildings in the Qutan Temple all adopt the traditional Han-style wood structure, which is featured with black brick walls, tiled roof, brick carving ridge decorations, colorful paintings on beams and walls. Some major building

根据明代皇家建筑的实践和理念，瞿昙寺融入山水环境，必然讲究外部空间设计理论性的『形势』说。形与势，即近与远、小与大、个体与群体、局部与总体、细节与轮廓等对立性的空间构成及其感受效果；各建筑尺度、体量和视距按『千尺为势，百尺为形』控制；组群一般不逾千尺，即230～350米，单体乃至院落，多在百尺之内，即23～35米，力使行进其间所获得的近、中、远景和相互转换的知觉群的连续体验，切合人际交流的视听感受，丰富深刻、流畅而完整，充满人性情味。

事实上，以具有地表数字高程模型性质的传统『平格』模数网观照瞿昙寺总平面，比对清代样式雷图档中典型如定东陵平格样（图三十），两者『千尺为势，百尺为形』规划设计意向的高度契合，显而易见。

无独有偶，前述尚义返京后董建大隆福寺，作为皇家香火院、京师首座汉地廊院式藏传佛教巨刹，曾深受瞿昙寺影响，按平格模数审视相关舆图（图三十一），其结合建筑立面的传统画法，从单体到组群，以『千尺为势，百尺为形』统筹平面和竖向设计的意象，尤其彰明较著。

总之，瞿昙寺的建筑，蕴涵着传统建筑规划设计的藻思，当然值得发掘和继承。

图二十九 瞿昙寺两翼带抄手斜廊的隆国殿（前）同北京现存同期建造的故宫神武门（中）、明长陵祾恩殿（后）形制比较

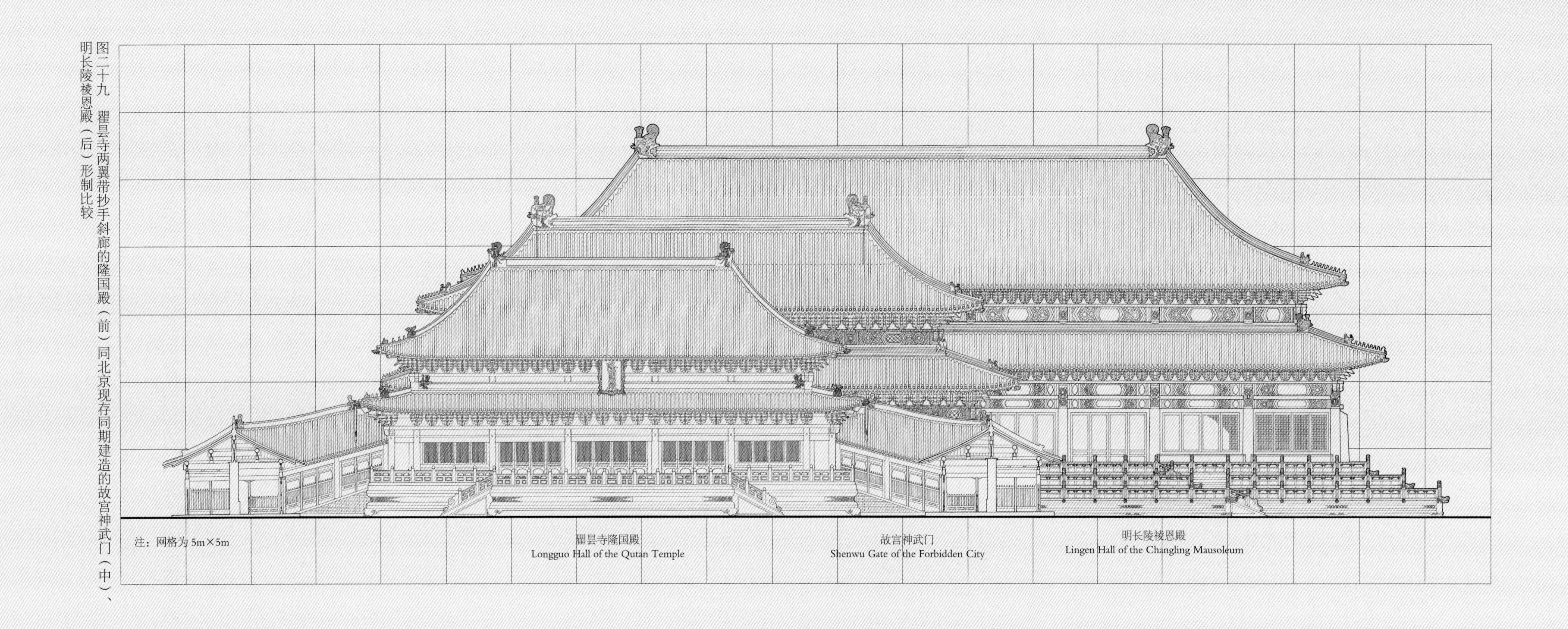

Fig.29 Comparison among the three buildings built in the same period of the Ming Dynasty

the Longhuo Hall joined with ascending side corridors of the Qutan Temple (front), the Shenwu Gate of the Forbidden City (middle), and the Lingen Hall of the Changling Mausoleum (back)

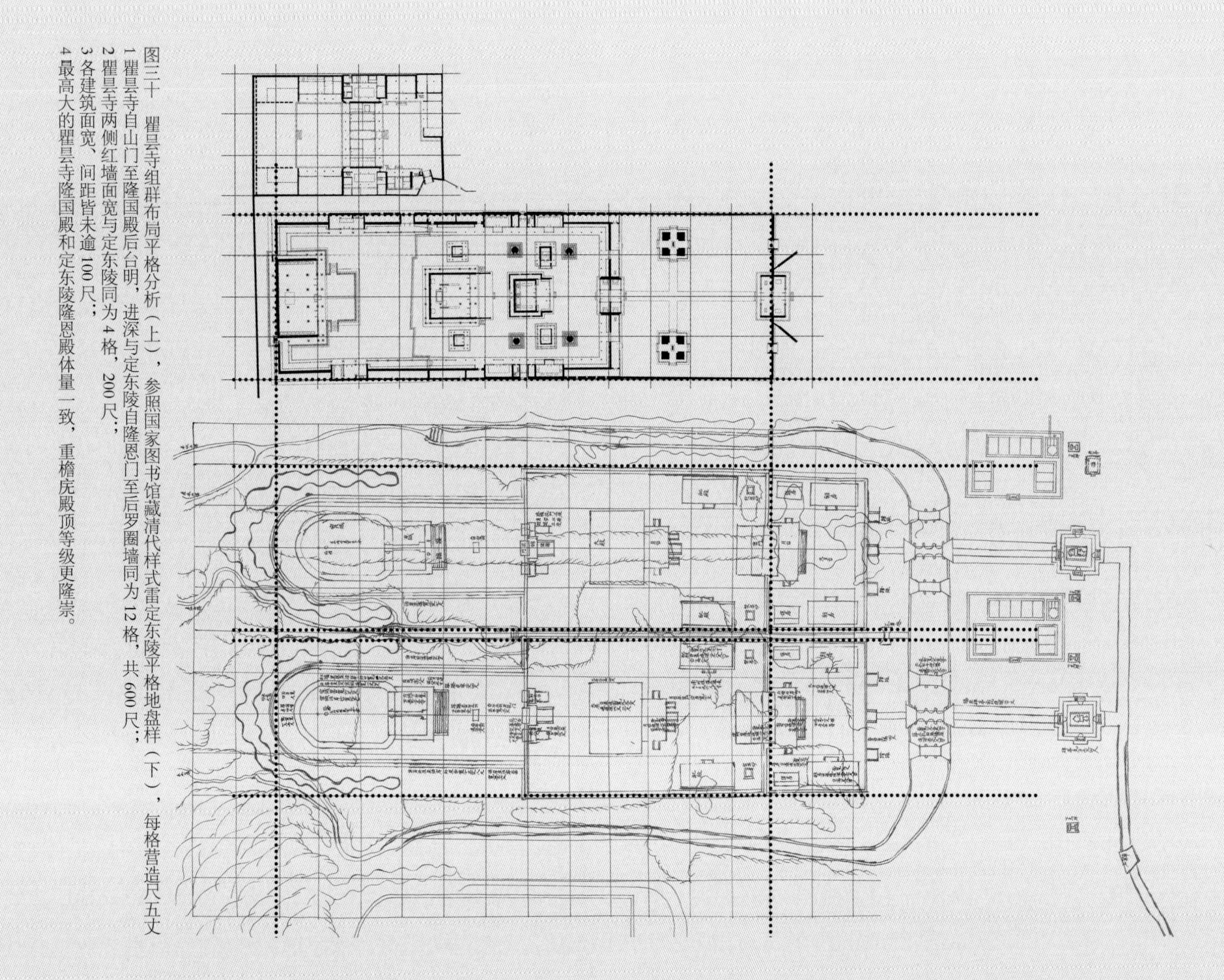

图三十　瞿昙寺组群布局平格分析（上），参照国家图书馆藏清代样式雷定东陵平格地盘样（下），每格营造尺五丈

1 瞿昙寺自山门至隆国殿后台明，进深与定东陵自隆恩门至后罗圈墙同为 12 格，共 600 尺；

2 瞿昙寺两侧红墙面宽与定东陵同为 4 格，200 尺；

3 各建筑面宽、间距皆未逾 100 尺；

4 最高大的瞿昙寺隆国殿和定东陵隆恩殿体量一致，重檐庑殿顶等级更隆崇。

Fig.30 *Ping–ge* analysis on the overall layout of the Qutan Temple complex (upper) Archived *yang–shi–lei* drawing of the *ping–ge* plan for the Ding East Tomb of the Qing Dynasty (lower), the length of each square equals to five *zhang* (equals 50 chi) in terms of the building scale

1 The length from the main gate of the Qutan Temple to the back side of the Longguo Hall equals to that of the Ding East Tomb from the Longen Gate to the backside wall, both reaching 12 squares, *600 chi*.

2 The distance between the two red side walls equals that of the Ding East Tomb, about 4 squares, 200 *chi*.

3 The width of each building and the distance between buildings are all less than 100 *chi*.

4 The tallest building of the Qutan Temple is Longguo Hall, which is about the same volume as the Longen Hall of the Ding East Tomb, both featured with double eaves to suggest their grandeur hierarchy.

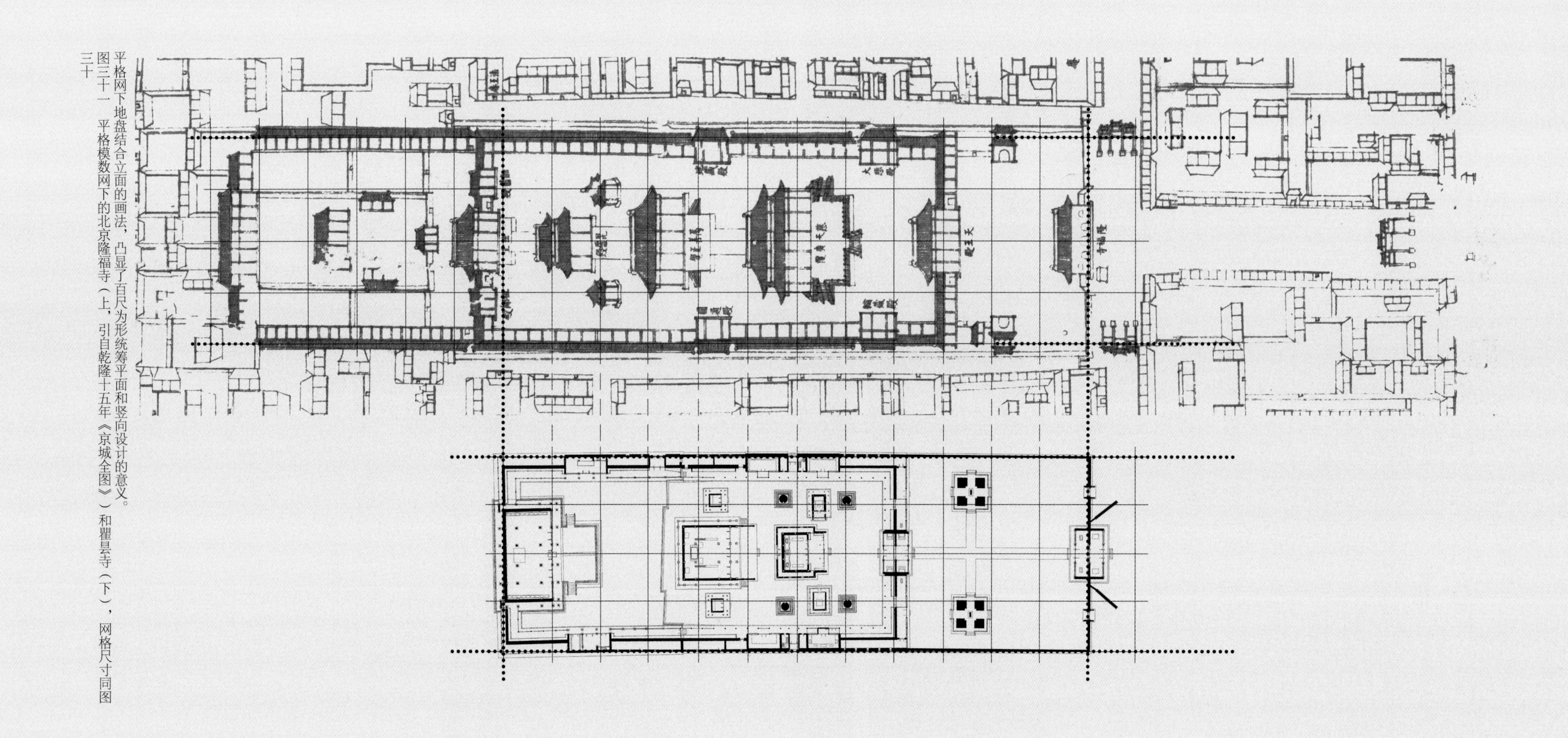

图三十一　平格模数网下的北京隆福寺（上，引自乾隆十五年《京城全图》）和瞿昙寺（下），网格尺寸同图三十

平格网下地盘结合立面的画法，凸显了百尺为形统筹平面和竖向设计的意义。

Fig.31 Longfu Temple (upper) and the Qutan Temple (lower) under the *ping-ge* module network, same network size as in Fig.30

The combination of drawing the plan and building elevation within the same ping-ge network has highlighted the significance of using the 100 chi to control the horizontal and vertical design.

types and methods are summarized in the following table:

Table 1 Types and codes of various buildings in Qutan Temple

Building	year	Foundation	Bay	Depth	Surrounding corridor	Floor	Roof	*Dougong*	Caisson Ceiling
Qutan Hall	1391	Brick-layered foundation above the ground, with stone steps	5	4	Without corridor	one	the gable and hip roof with double eaves	No	Yes
East Side Hall of Qutan Hall	1391	Brick-layered foundation above the ground, with stone steps	3	4	With surrounding corridor	one	the gable and hip roof with single eaves	No	No
West Side Hall of Qutan Hall	1391	Brick-layered foundation above the ground, with stone steps	3	4	With surrounding corridor	one	the gable and hip roof with single eaves	No	Yes
Baoguang Hall	1418	Brick-layered foundation above the ground, with stone steps	5	5	With surrounding corridor	one	the gable and hip roof with double eaves	No	No
East Side Hall of Baoguang Hall	1418	Brick-layered foundation above the ground, with stone steps	3	3	With surrounding corridor	one	the gable and hip roof with single eaves	No	No

（二）建筑规制

瞿昙寺各建筑，除四座砖塔，均采用传统汉式木结构，青砖砌墙，筒板瓦屋面，砖雕脊饰，绘饰彩画和壁画。各建筑式样和做法，略如表一：

表一 瞿昙寺各建筑规制

建筑	年代	台基	开间	进深	周围廊	层数	屋顶	斗栱	藻井
瞿昙殿	一三九一	砖砌台明，覆阶条石	五	四	无	单层	重檐歇山顶	无	有
瞿昙殿东配殿	一三九一	砖砌台明，覆阶条石	三	四	周围廊	单层	单檐歇山顶	无	无
瞿昙殿西配殿	一三九一	砖砌台明，覆阶条石	三	四	周围廊	单层	单檐歇山顶	无	有
宝光殿	一四一八	砖砌台明，覆阶条石	五	五	周围廊	单层	重檐歇山顶	无	无
宝光殿东配殿	一四一八	砖砌台明，覆阶条石	三	三	周围廊	单层	单檐歇山顶	无	无
宝光殿西配殿	一四一八	砖砌台明，覆阶条石	三	三	周围廊	单层	单檐歇山顶	无	无
小鼓楼	一四一八	底层砖砌台明，覆阶条石	三	底层三，二层一	底檐廊，二层周围廊	二层	单檐庑殿顶	无	无
小钟楼	一四一八	底层砖砌台明，覆阶条石	三	底层三，二层一	底檐廊，二层周围廊	二层	单檐庑殿顶	无	无
隆国殿	一四二七	石雕须弥座、望柱栏板	七	五	周围廊	单层	重檐庑殿顶	下檐五踩重昂，上檐七踩单翘重昂	无
大鼓楼	一四二七	底层砖砌台，覆阶条石	三	底层三，二层一	底层檐廊	二层	单檐庑殿顶	下檐平坐五踩重翘；上檐三踩单昂	无
大钟楼	一四二七	底层砖砌台，覆阶条石	三	底层三，二层一	底层檐廊	二层	单檐庑殿顶	下檐平坐五踩重翘；上檐三踩单昂	无
金刚殿	一四二七	砖砌台明，覆阶条石	三	二	无	单层	单檐悬山顶	无	无
厢廊	一四二七	砖砌台明，覆阶条石	五十八	二			单檐两坡顶	中院无，后院一料三升	无
山门	一四二七	砖砌台明，覆阶条石	三	二	无	单层	单檐歇山顶	三踩单昂斗栱	无
东碑亭	一四六六	石雕须弥座、望柱栏板	三	三	无	单层	重檐十字顶	下檐三踩单昂，上檐五踩重昂	无
西碑亭	一四六六	石雕须弥座、望柱栏板	三	三	无	单层	重檐十字顶	下檐三踩单昂，上檐五踩重昂	无

周匝中后院的厢廊即回廊或围廊，自金刚殿两旁各七间，北折四间，各经三世殿、小鼓楼及护法殿、小钟楼檐廊九间，连廊七间，末三间斜升2.6米抵后院；再各接围廊五间，穿大鼓楼、大钟楼檐廊三间，各续回廊四间，又以抄手斜廊两间折向隆国殿，进升2.6米至两山檐廊。大钟鼓楼南，回廊循中柱砌墙，院内侧开敞；外侧隔成房间，小钟鼓楼明间和后院东廊一间各设门外通，西廊中段两间各开小券门，其余皆凿小圆窗。大钟鼓楼北，回廊仍沿中柱砌墙，两侧均敞为双面廊。

厢廊朝向院内的52面墙，曾彩绘佛祖本生故事等，与各殿内满壁藏式佛像及密宗神像相辉映。清代，厢廊少量壁画淋毁重绘；民国大地震后重修期间，囊谦大经堂遭际回禄，皮藏粉本煨灭，24间壁画无缘重摹，仅小钟鼓楼以北28间壁画幸存，明、清各半，备受学界珍视推崇（图三十二）。

文献和考古证明，廊院组群源自商周，唐宋已成庙堂定制，唯存世实物罕见，瞿昙寺廊院堪称遗珍。因效尤皇宫礼制意向，循五行观念，院内鼓楼在东略高，钟楼居西稍低，与汉地伽蓝东钟西鼓惯例相反；后院厢廊，更如北京皇宫朝房，采用一斗三升品字斗栱，凸显了主体建筑的尊崇。

寺院出入口，如山门、后山门（即金刚殿）、小钟鼓楼下明间及后院东廊的大门，都采用了汉地佛寺常用的欢门式样，山门和金刚殿还配以欢窗。

Building	year	Foundation	Bay	Depth	Surrounding corridor	Floor	Roof	*Dougong*	Caisson Ceiling
West Side Hall of Baoguang Hall	1418	Brick-layered foundation above the ground, with stone steps	3	3	With surrounding corridor	one	the gable and hip roof with single eaves	No	No
Small Drum Tower	1418	Brick-layered foundation and ground floor, with stone steps	3	3 on the first floor and 1 on the second floor	eave corridor on the ground floor and surrounding corridor on the second floor	two	Single eaves roof	No	No
Small Bell Tower	1418	Brick-layered foundation and ground floor, with stone steps	3	3 on the first floor and 1 on the second floor	eave corridor on the ground floor and surrounding corridor on the second floor	two	Single eaves roof	No	No
Longguo Hall	1427	stone carving *Ximizuo*, with baluster columns and breast boards	7	5	With surrounding corridor	one	Double eaves roof	Five stepping *dou* with double *ang* under the lower eave; seven stepping *dou* with single *qiao* and double *ang* under the upper eave	No

Building	year	Foundation	Bay	Depth	Surrounding corridor	Floor	Roof	*Dougong*	Caisson Ceiling
Great Drum Tower	1427	Brick-layered foundation and ground floor, with stone steps	3	3 on the first floor and 1 on the second floor	eave corridor on the ground floor	two	Single eaves roof	Five stepping *dou* with double *qiao* under the lower eave; three stepping *dou* with single *ang* under the upper eave	No
Great Bell Tower	1427	Brick-layered foundation and ground floor, with stone steps	3	3 on the first floor and 1 on the second floor	eave corridor on the ground floor	two	Single eaves roof	Five stepping *dou* with double *qiao* under the lower eave; three stepping *dou* with single *ang* under the upper eave	No
Jingang Hall	1427	Brick-layered foundation above the ground, with stone steps	3	2	Without corridor	one	overhanging gable roof with single eaves	No	No

寺内各殿座，包括配殿及大小钟鼓楼，皆采用汉式槅扇门，隆国殿等级类同皇宫，槅心作簇六雪花纹，裙板剔地起雕三幅云；大小钟鼓楼做方棂；其余多为斜格棂子或直棂。殿座外各门，则为双扇或单扇实榻门（即板门）。

凡此，曾深受明初以来日趋森严的礼制限定。

如明洪武二十四年肇建瞿昙殿及配殿（图三十三），均用歇山顶，设藻井，无斗栱，就遵从了洪武三年令：『并不许起造斗栱、彩画梁栋。』其中并未限置藻井。洪武二十六年新规『不许歇山转角、重檐重栱、绘画藻井』，而永乐六年皇帝御赐扩建，仍用歇山，却无藻井；永乐十六年后增华，甚而采用重檐庑殿顶，并起造斗栱，仍无藻井，也没有皇家专用的溜金斗栱。

其中，宣德二年落成的隆国殿、大钟鼓楼、后院厢廊及山门，成化二年告竣的御碑亭，柱头皆安斗栱，与明初北京官式做法略无二致，结构意义外，更凸现了礼制尊卑。权衡斗口大小，隆国殿甚至大于永乐皇帝的长陵祾恩殿；山门、御碑亭、大钟鼓楼及后院厢廊的斗口，也不逊长陵祾恩门，足见朱棣和嗣皇帝对瞿昙寺的青睐（图三十四）。

各建筑屋顶，除厢廊两坡顶、山门两旁角门硬山顶、金刚殿悬山顶，共有九座建筑采用歇山顶，除清代改动的御碑亭重檐十字脊顶外，区别于嗣后官式建筑的，就是坡面舒缓、垂脊内缩即收山普遍较深的特点（图三十五）。

隆国殿和大钟鼓楼等三座建筑，采用高等级的庑殿顶，如同北京同期遗构，正脊外延即推山显著，梁架却省去太平梁和雷公柱，构造大为简化（图三十六）。

如前述，洪武三年，朝廷曾严禁寺观彩画梁栋，却申明：其御赐者……不在禁例。以致瞿昙殿及诸配殿外，各主体建筑木构件彩画均如明初北京官式，惟不作地仗。其中隆国殿、大钟鼓楼和后院厢廊梁枋，用墨、绿色叠晕。隆国殿内檐的旋花心等用黄色，类似点金，级别最高。金刚殿、小钟鼓楼及中院厢廊级别稍逊，枋心不叠晕；凡此，皆为现知明代早期建筑彩画的瑰宝。宝光殿及隆国殿外檐因后世修葺，已趋同厢廊清代壁画表现的彩画式样。乾隆朝，瞿昙殿曾按地方风格添建带斗栱的抱厦，清末到民国，又屡经修缮，彩画形式复杂多样，冷暖色兼用，多属当地手法；但内檐遗存洪武朝彩画，则为现知明代最早木构建筑彩画（图三十七）。

Building	**year**	**Foundation**	**Bay**	**Depth**	**Surrounding corridor**	**Floor**	**Roof**	***Dougong***	**Caisson Ceiling**
Side Corridor	1427	Brick-layered foundation above the ground, with stone steps	58	2			Two slope roof with single eaves	None for those in the middle courtyard; one *dou* and three *sheng* for those in the back courtyard	No
Main Gate	1427	Brick-layered foundation above the ground, with stone steps	3	2	Without corridor	one	the gable and hip roof with single eaves	three stepping *dou* with single *ang*	No
East Stele Pavilion	1466	stone carving Ximizuo, with baluster columns and breast boards	3	3	Without corridor	one	Cross roof with double eaves	Three stepping *dou* with single *ang* on the lower floor; five stepping *dou* with double *ang* on the second floor	No
West Stele Pavilion	1466	stone carving Ximizuo, with baluster columns and breast boards	3	3	Without corridor	one	Cross roof with double eaves	stepping *dou* with single *ang* on the lower floor; five stepping *dou* with double *ang* on the second floor	No

The middle and back yards are enclosed by corridors, starting from the Jingang Hall, seven bays on each side, then four bays northward, passing through the Three Generation Hall, the Small Drum Tower, the Hufa Hall, the Small Bell Tower, featured in a nine-bay eave corridor and a seven-bay ascending side corridor, with the last three bays ascending about 2.6 meters before reaching the back yard. Then, a five-bay corridor is continued on each side, with a three-bay eave corridor passing through the Great Drum Tower and the Great Bell Tower, followed by a four-bay corridor and then another two-bay corridor ascending 2.6 meters towards the Longguo Hall, reaching the latter's eave corridor on the gable ends. South to the Great Drum and Bell Towers, walls are built along the middle column axis, with the inner part opening to the yard, and the outside part being partitioned into rooms. Doors are opened in the middle bays of the Small Drum and Bell Towers, and another one in the east side corridor of the back yard, which are all connected to the outside. Two arched doors are opened in the middle part of the west corridor, and the rest are all opened with small round windows. North to the Great Drum and Bell Towers, the corridors are also built with walls in the axis of middle columns, but with both sides open.

On the fifty-two walls of the side corridors facing towards the yard, once painted with stories of Sakyamuni, responding to all kinds of the Tibetan Buddhist statues and idols of the esoteric doctrine. During the Qing Dynasty, a few of these wall paintings were ruined by rain water and repainted. After the big earthquake took place during the Republic era, and during the repairing process, the Main Assembly Hall was burnt down together with the painting manuscripts, so that the destroyed twenty-four wall paintings were not able to be reproduced, with only twenty-eight paintings north to the Small Drum and Bell Towers survived, half of which were completed during the Ming Dynasty and the other half during the Qing Dynasty, highly treasured by the academic circle (Fig.32) .

Based on the historic literature and archaeology discoveries, the combination of corridor and courtyard system started in the Shang and Zhou dynasties and became formalized by the royal court in the Tang and Song dynasties. But since there are few examples survived till today, the corridor and courtyard composition in the Qutan Temple is considered precious and important. Because it also duplicates the ritual hierarchy of the royal palace, follows the *wuxing* philosophy, the drum towers are always in the east and higher than the bell towers in the west, opposite to those Han-style Buddhist temples, in which the bell tower standing in the east and the drum tower in the west. And the side corridors in the back year, adopts the building codes of those offices for the courtiers, all using the 品 -shaped *dougong* to stress the highness of the main buildings.

在做法上，瞿昙寺各建筑檐柱和角柱内倾，角柱加高，鲜见于嗣后明清官式建筑，却契合宋《营造法式》有关『侧脚』和『生起』规定。瞿昙殿、三世殿及护法殿梁架用叉手，瞿昙殿还用『顺脊串』（图三十八），也见载《营造法式》，却罕见于明清官式建筑。

另一方面，珍贵的是，隆国殿内梁架遗存大量工匠题记，堪称明清官式建筑体系成型的不磨印记。诸如明间、梢间、中金、上金、单步梁、双步梁、五架梁、随梁、行（桁）、方（枋）、同（童）柱等术语（图三十九），明显异于《营造法式》而和清《工程做法则例》同辙，无疑具有弥补明代营造文献缺环的重要价值。

Fig.32 Wall paintings in the Qutan Temple completed during early years of the Ming Dynasty (photo by LUO Wenhua, SU Bai)

1 West side wall paintings in the Qutan Hall during the Hongwu period: Five Dhyani Buddhas (from left to right: Ratnasambhava Buddha in the south, Akshobhya Buddha in the east, Vairocana Buddha in the center, Amoghasiddhi Buddha in the north, Amitayus Buddha in the west) and Master Images

2 West side wall paintings in the Hufa Hall: Images of Guardians (from left to right: Red Devil, Yamantaka, Akshobhyavajra, Kalacakra, Hevajra)

3 North side wall paintings in the Longguo Hall: Trikalea Buddhas and the Guardians (from left to right: Yamantaka, Kassapa Buddha, Sakyamuni Buddha, Maitreya, Kalacakra)

4 South corridor wall paintings in the Great Bell Tower between Yongle and Xuande period: left: the Dragon King inviting Buddha to give speech at the Dragon Palace; right: *Buddha appointing Maitreya* to be the future Buddha

5 South corridor wall paintings in the Great Bell Tower: partial details of the painting *Buddha appointing Maitreya to be the future Buddha*

图三十二 瞿昙寺的明初壁画（罗文华、苏白摄）

5. 大钟楼南廊壁画佛授记弥勒却尽当来度生局部

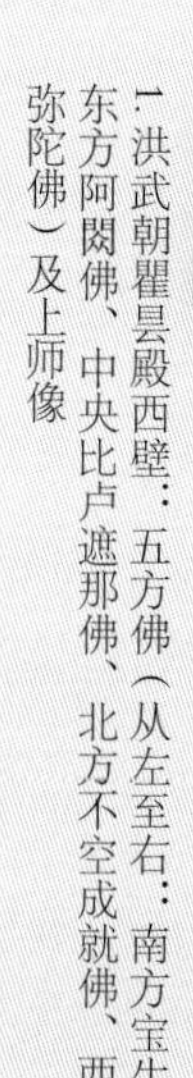

1. 洪武朝瞿昙殿西壁：五方佛（从左至右：南方宝生佛、东方阿閦佛、中央比卢遮那佛、北方不空成就佛、西方阿弥陀佛）及上师像

2. 护法殿西墙：护法像，从左至右：红阎魔敌、大威德金刚、密集不动金刚、时轮金刚、喜金刚

3. 隆国殿北壁：三世佛本尊及护法金刚像，从左至右：大威德金刚、迦叶佛、释迦牟尼佛、弥勒佛、双身时轮金刚

4. 永乐—宣德朝大钟楼南廊壁画。左：龙王迎佛入龙宫说法图；右：佛授记弥勒却尽当来度生

图三十三　瞿昙殿及其西配殿的藻井
1. 瞿昙殿藻井　2. 西配殿的藻井

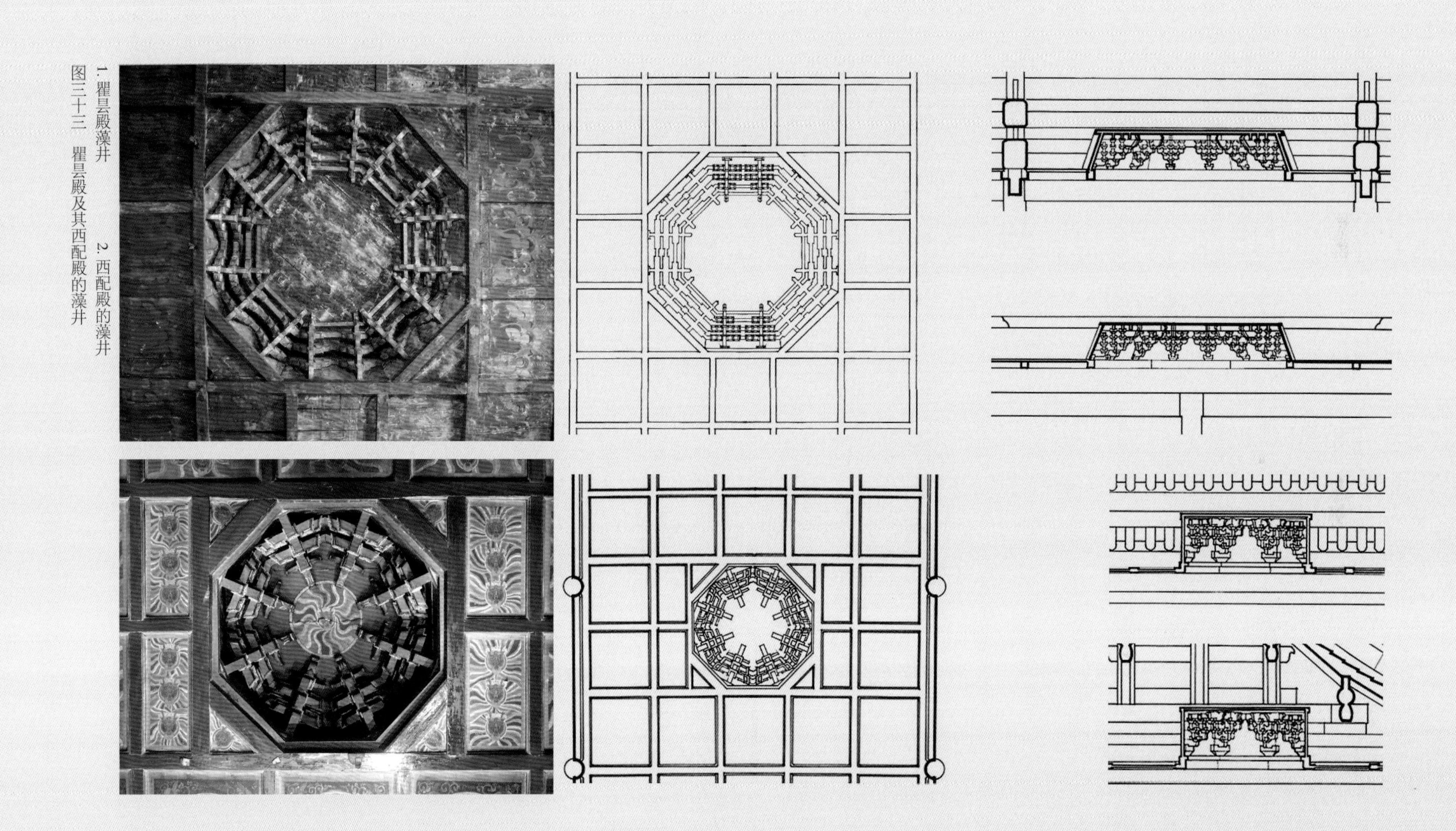

Fig.33 Caisson ceilings in the Qutan Hall and its west side hall
1 Caisson ceilings in the Qutan Hall
2 Caisson ceilings in the west side hall

图三十四 瞿昙寺各式斗栱

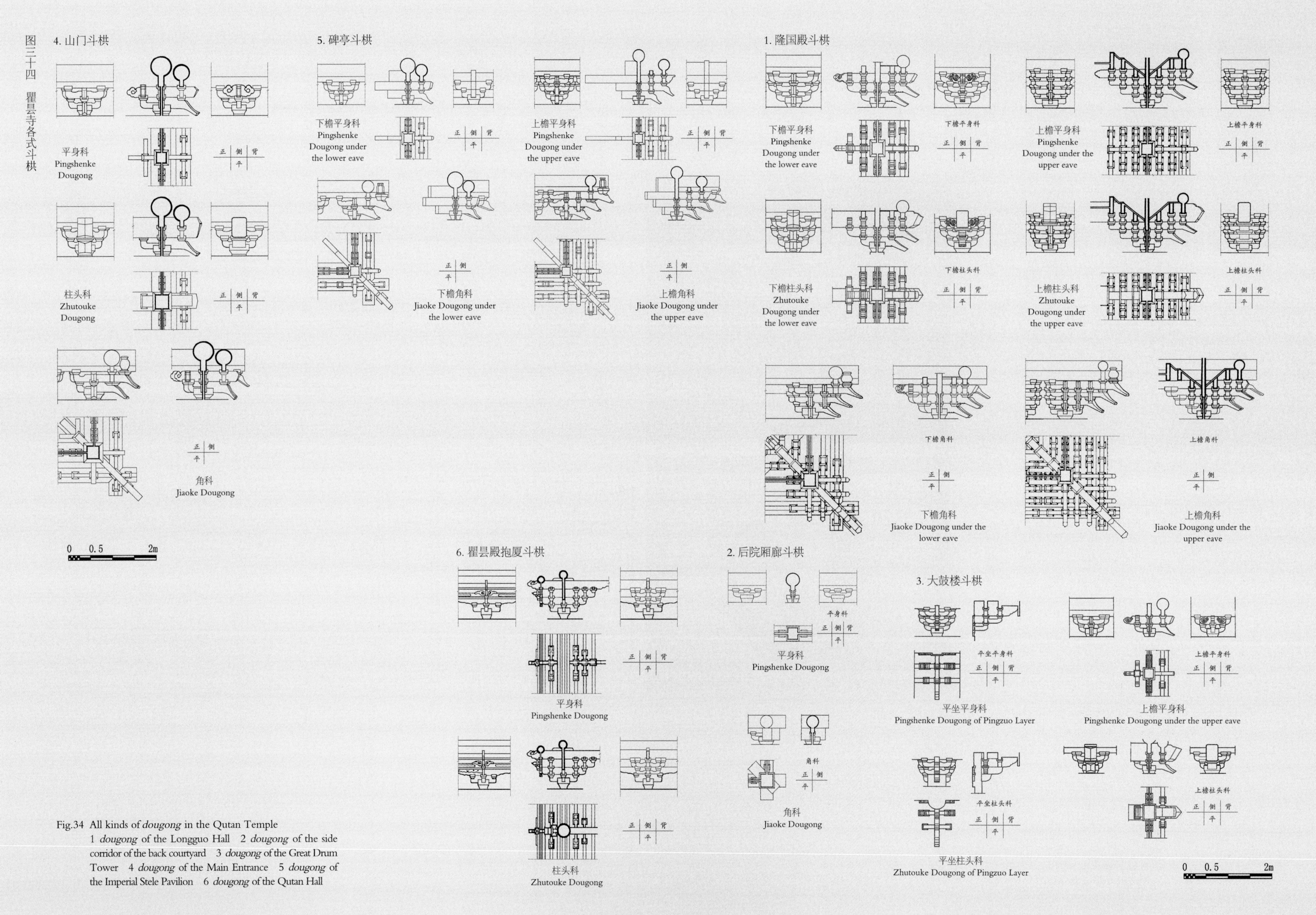

Fig.34 All kinds of *dougong* in the Qutan Temple
1 *dougong* of the Longguo Hall 2 *dougong* of the side corridor of the back courtyard 3 *dougong* of the Great Drum Tower 4 *dougong* of the Main Entrance 5 *dougong* of the Imperial Stele Pavilion 6 *dougong* of the Qutan Hall

图三十五　瞿昙寺各式歇山做法

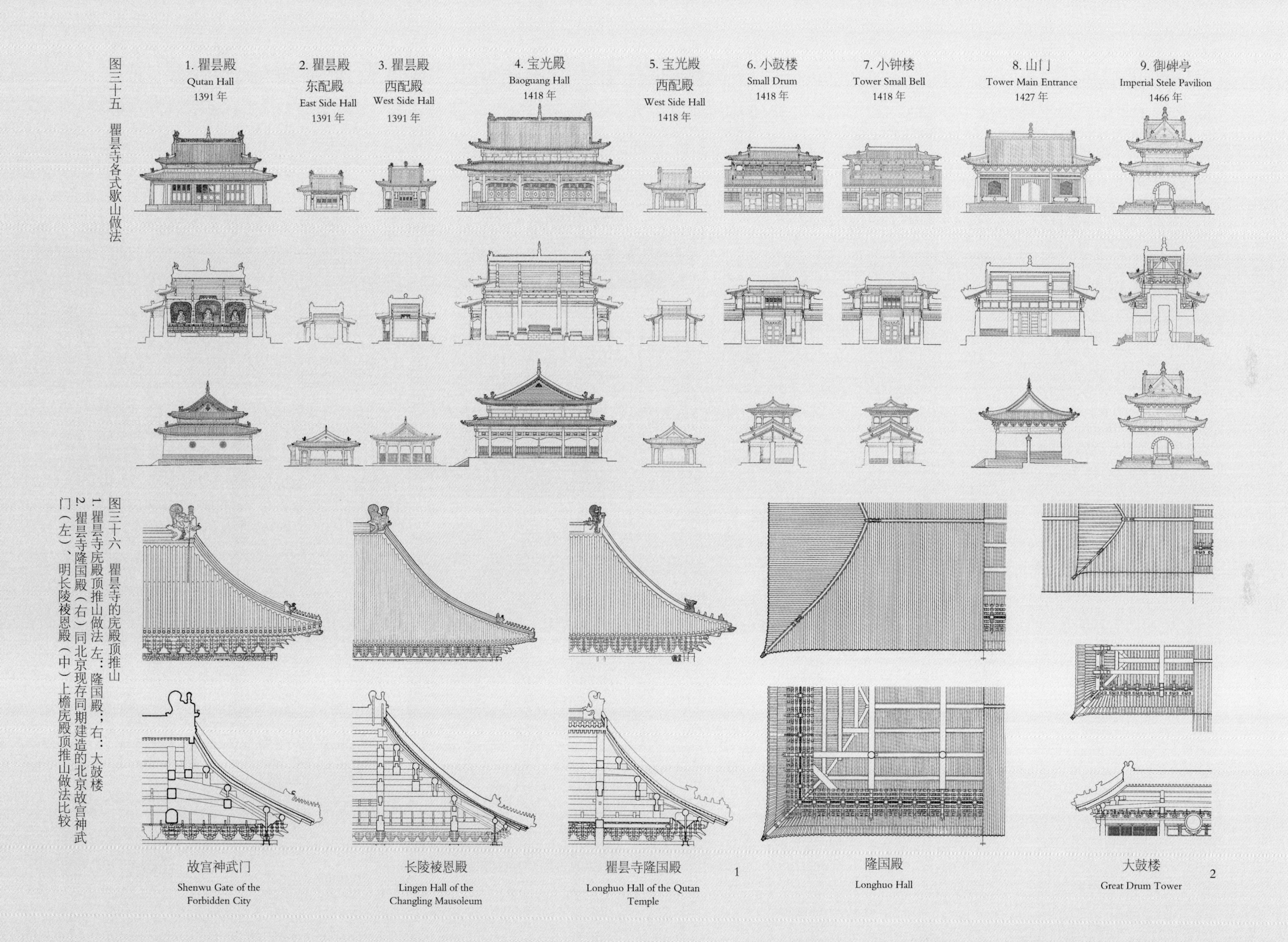

图三十六　瞿昙寺的庑殿顶推山

1. 瞿昙寺庑殿顶推山做法 左：隆国殿；右：大鼓楼

2. 瞿昙寺隆国殿（右）同北京现存同期建造的北京故宫神武门（左）、明长陵祾恩殿（中）上檐庑殿顶推山做法比较

Fig.35 Different gable and hip roofs in the Qutan Temple

Fig.36 Tuishan in the hip roofs in the Qutan Temple

1 Construction method of *tuishan* in the hip roofs of the Qutan Temple left: Longguo Hall; right: the Great Drum Tower

2 Comparisons between the construction methods of *tuishan* in the hip roofs of upper eaves of the Longguo Hall of the Qutan Temple (right), the Shenwu Gate of the Forbidden City (left) and the Linen Hall of the Changling Mausoleum (middle), which were built at the same time and still existing in Beijing

1. 隆国殿明间脊檩彩画
2. 隆国殿明间脊檩彩画局部
3. 隆国殿内梁五架梁及天花彩画
4. 大鼓楼梁架彩画
5. 大鼓楼南廊梁架彩画
6. 大鼓楼下檐彩画复原（吴葱绘）
7. 大钟楼梁架彩画

图三十七 瞿昙寺各式彩画（王其亨、吴葱摄）

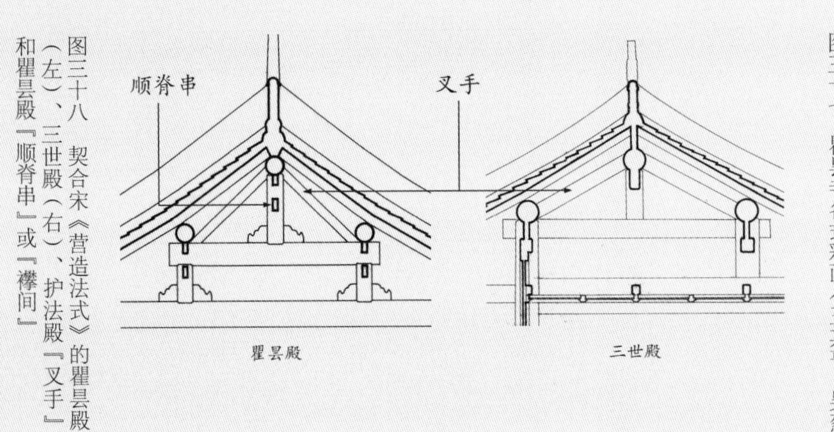

图三十八 契合宋《营造法式》的瞿昙殿（左）、三世殿（右）、护法殿「叉手」和瞿昙殿「顺脊串」或「襻间」

Fig.37 Different kinds of color paintings in the Qutan Temple (photos taken by WANG Qiheng and WU Cong)

1 Color paintings on the middle bay main ridge in the Longguo Hall

2 Details of the color paintings on the middle bay main ridge in the Longguo Hall

3 Color paintings for the 5−purlin beam and the ceiling of the Longguo Hall

4 Color paintings for the beams of the Great Drum Tower

5 Color paintings for the beams of the southern corridors of the Great Drum Tower

6 Restored image of the color paintings for the lower eave of the Great Drum Tower (by WU Cong)

7 Color paintings for beams of the Great Bell Tower

Fig.38 *Chashou* in the Qutan Hall (left), Sanshi Hall (right) and Hufa Hall, and *shunjichuan in* the Qutan Hall, all in comply with *Yingzao fashi*

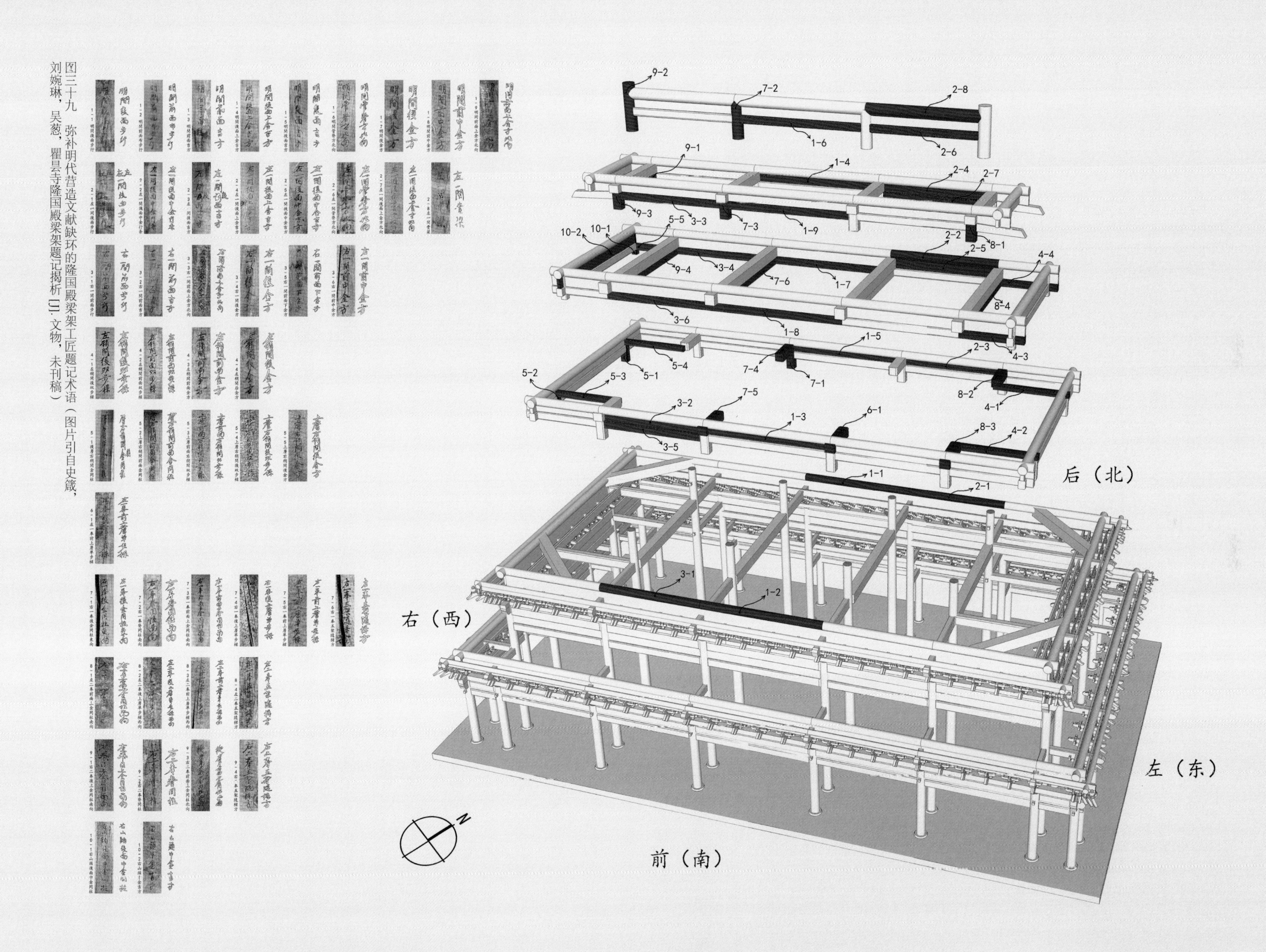

图三十九　弥补明代营造文献缺环的隆国殿梁架工匠题记术语（图片引自史箴，刘婉琳，吴葱，瞿昙寺隆国殿梁架题记揭析[J]. 文物，未刊稿）

Fig.39 Terms in the inscriptions on beams of the Longguo Hall left by craftsmen, making up the lack of construction documentations during the Ming Dynasty (Shi Zhen, Liu Wanlin, Wu Cong, Qutan Si Longguo Dian Liangjia Tiji Jiexi[J], in: Wenwu, Unpublished)

The entrances and exits for the temple, such as the main entrance, the back entrance (also named as the Jingang Hall), the doors in the middle bay of the Small Bell and Drum Towers as well as that of the east side corridor in the back courtyard, have all adopted the style of *huan men* (happy door) that commonly appeared in the Han-style Buddhist temples. In addition, the main entrance and the Jingang Hall have even adopted the style of *huan chuang* (happy window).

While all halls and buildings, including the side halls, Small and Great Bell and Drum Towers, have adopted the Han-style partition doors. Halls like Longguo Hall are treated in the same way as the royal palace, with the middle part of the partition doors hollowed with typical Chinese snowflake pattern and the lower part filled with three-layered clouds relief from the bottom. Square-patterned windows are used for the Bell and Drum Towers, while the rest are often with the oblique square or latticed windows. Doors outside halls are often double-leaf or single-leaf solid doors.

All such building codes were regulated by the ritual system, which had become more and more strict since the early Ming Dynasty.

For example, the Qutan Hall and its side halls (Fig.33) built in 1391 all adopted the gable and hip roof, with caisson ceiling but without *dougong*, following the regulation announced in 1370, which forbidden to build with *dougong* and to paint beams and columns. Yet, there was no limit on the use of caisson ceiling. The new rule announced in 1393 didn't all the use of the gable and hip roof, the double eaves, the multiple *dougong*, the caisson ceiling and the color painting. When the temple was permitted by the emperor to have extensions in 1408, no caisson ceiling was adopted but the gable and hip roof. In 1418 when the temple was further extended, even the double eaved hip roof and *dougong* were adopted, but there were still no caisson ceiling used, and neither the gilded *dougong* specially used by the royal court.

Among them, the Longguo Hall, the Great Bell and Drum Towers, the side corridors of the back courtyard and the main entrance completed in 1427, as well as the Imperial Stele Pavilion completed in 1466, are all featured with *dougong* on top of the columns, which were built in the same way as those in Beijing during the early Ming Dynasty. In addition to the structural meanings, it is also an expression of the norms of etiquette. If we compare the size of the *doukou*, the one of the Longguo Hall is ever larger than that in the Lingen Hall of the Changling Mausoleum, tomb of the Yongle emperor; while the size of the *doukou* for the other buildings are not smaller than the Lingen Gate of the same mausoleum, which indicates the special attention paid to the Qutan Temple by emperor ZHU Di and his successors (Fig.34).

四、瞿昙寺的研究与保护

至当代，1959年设置瞿昙寺文物管理所，由青海省文物管理委员会领导，从此开启了瞿昙寺的研究与保护事业。同年，瞿昙寺被列为青海省重点文物保护单位（图四十）。1960年春，著名建筑史学家张驭寰、杜仙洲勘察该寺，1964年在《文物》第5期发表瞿昙寺研究的开山之作《青海乐都瞿昙寺调查报告》（图四十一），至今被学界仰重。此后直到1980年代，政治动乱，研究停滞，瞿昙寺外围城垣及部分建筑被毁，主体建筑一度被改作粮站。期间，青海省文物主管部门和瞿昙寺管理所马骏等为保护瞿昙寺奔走呼吁，才得较完整地保存至今。

Fig.40 Landmark stone tablets of the Qutan Temple as cultural relics protection
left: Landmark stone tablet of the Qutan Temple as the province-level key unit for cultural relics protection, 1964
right: Landmark stone tablet of the Qutan Temple as the state-level key unit for cultural relics protection, 1982
Fig.41 Investigation Report on the Qutan Temple in Ledu, Qinghai, by ZHANG Yuhuan, DU Xianzhou, 1964

图四十 瞿昙寺被列为文物保护单位标志牌
左：1964年《青海省重点文物保护单位瞿昙寺》标志碑
右：1982年《全国重点文物保护单位瞿昙寺》标志碑

乐都在青海省西宁市东，湟水北岸，自古称作"湟中"。从汉代开始在这里設县，于公元五世紀南凉禿髮烏孤曾經作为都城，明初置碾伯卫，洪武十九年移卫于西宁后，乐都遂改为右所①。这里在古代曾是通往西域的交通要道，与內地接触比較頻繁，是海东經济文化荟萃之区，因此，至今还保留着不少明代寺院。

1960 年春我們調查青海民居和塔儿寺时，路过乐都，听当地群众讲这里有一座明代大寺院，叫做瞿曇寺，規模宏大，建筑壮丽，因而引起了我們的注意。順便作了初步勘查，因时間所限，对各殿結构未作詳細測繪，玆就略查所得簡述如下。

寺建在乐都城南 40 里瞿曇堡城中，南向。城前崗巒起伏，后有高山屏障，滾滾湟水流經其間，景色清幽。

城堡略呈方形，分內外二城。外城为居民宅巷，內城为瞿曇寺和僧侶房舍。城墻用黄土夯筑，城門甕城建筑在河岸高地上，曲折通进，形势頗險固（图三）。

一、历史沿革

瞿曇寺，据文献記載，始建于明太祖洪武年間，第一代开山僧是三罗喇嘛，寺額曰瞿曇②，永乐間重建佛殿③，并立碑"諭示"軍民人等，不許干扰喇嘛的宗教生活，或侵占寺院的田产山林，由于受到封建王朝的保护，从此，瞿曇寺的規模和声望便逐步建立了起来。至宣德二年（公元 1427），宣宗朱瞻基为了追念他的祖宗，又在寺的后部修建了一座大殿——隆国殿④，于寺前树立《御制碑》二通，并賜給一批銅制的瓶炉香案等物，至使寺院体制益臻完备，达于极盛时期⑤。一直到明末天启、崇禎年間，未見損坏或修繕的記載⑥。

① 乾隆十二年《西宁府新志》卷十五。
② 永乐六年《大明皇帝勅諭碑》："曩者喇嘛三罗蔵[illegible]居，寺額曰瞿曇。令其在班丹藏卜、端約藏卜演如来之教法……以嗣承其叔三罗之宗教……今特令住持瞿曇寺。"
③ 洪熙元年《御制碑》："太宗文皇帝……乃于瞿曇寺重作奉佛之殿……祖宗之誠宪率不忘，重惟玆寺，太祖高皇帝肇之于前，太宗皇帝紹之于后。"
④ 宣德二年《御制瞿曇寺后殿碑》："……于是即西宁之境，肇建梵宇，賜名曰瞿曇寺，以居其徒，……玆于瞿曇寺继作后殿"。
⑤ 乾隆《西宁府新志》："碾伯县瞿曇寺，在城南四十里，明永乐初勅建，授国师一，禪师三，賜地甚广，……殿宇雄丽，有御制碑文，洪熙、宣德中复賜御制二碑文，瓶炉香案皆宣德佳制也"。
⑥ 瞿曇寺前院左廊悬有天启、崇禎間所制匾額数方。

·46·　　1964年

图四十一 1964年张驭寰、杜仙洲《青海乐都瞿昙寺调查报告》

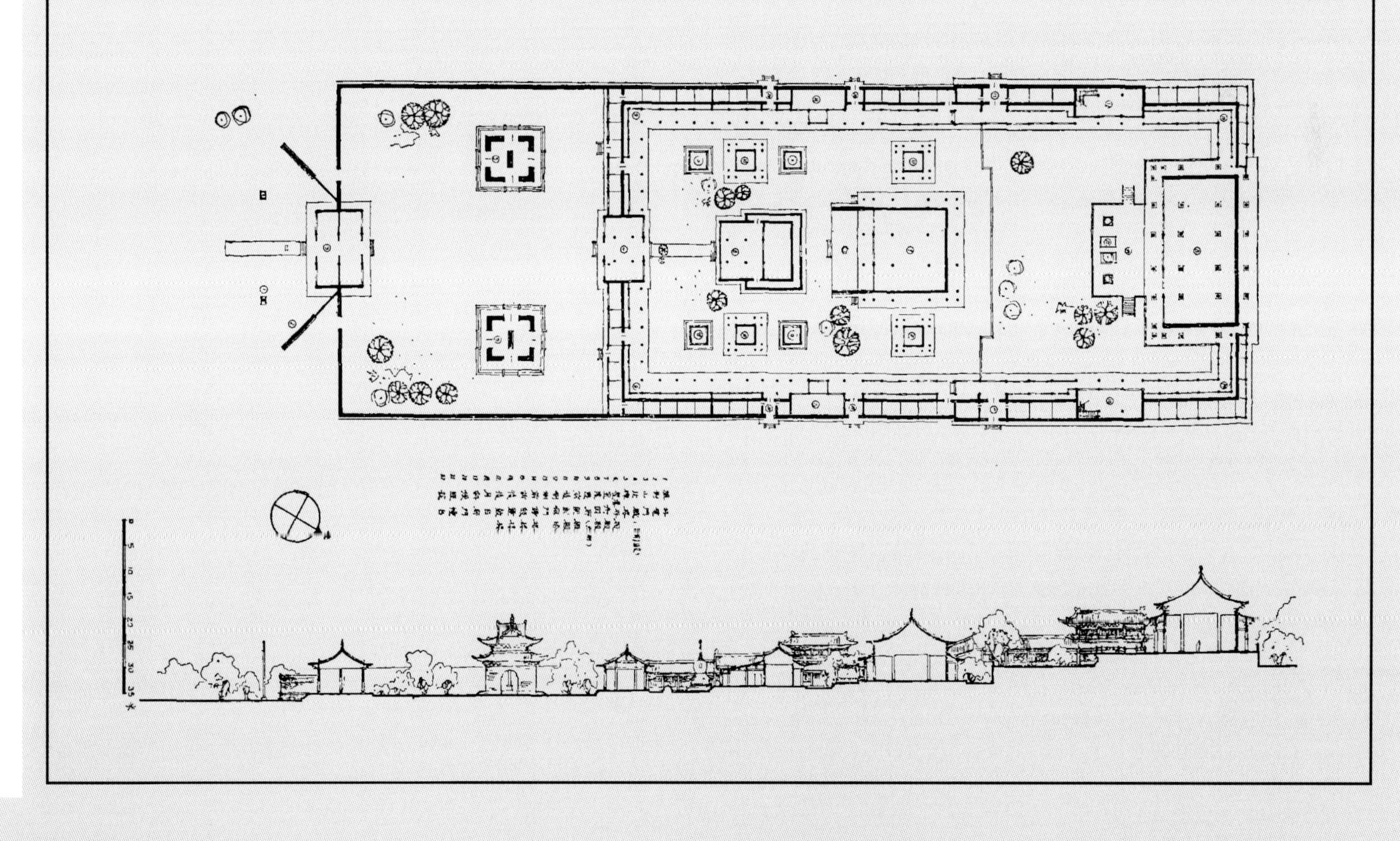

改革开放后，青海省学者谢佐、芈一之、赵生琛等深入爬疏藏、汉文史籍，方志和碑文等，结合现场调查，相继发表《青海乐都瞿昙寺考略》《瞿昙寺补考》《瞿昙寺及其在明代西宁地区的地位》《青海乐都瞿昙寺文物调查记》等论文，出版《瞿昙寺》等著作（图四十二），从历史、政治、宗教、艺术等视角，系统梳理了瞿昙寺的历史及其文物价值。

基于这些工作，1982年瞿昙寺被国务院列为第二批全国重点文物保护单位（图四十）。1984年瞿昙寺文物管理所恢复后，旋即编辑出版《瞿昙寺》（图四十二）。1985年，青海省文物管理处委托杰出古建筑专家、中国文物研究所祁英涛所长主持设计，瞿昙殿落架大修（图四十三），在明间脊枋发现了洪武二十四年九月初六日上梁典礼的墨书题记，为考证瞿昙寺的创建年代提供了确凿证据。

Other than the two-slope top of the side corridors, gabled roof of the corner gates beside the main entrance, and the suspension roof of the Jingang Hall, there are nine buildings in the Qutan Temple that have adopted the gable and hip roof style. Except for the double-eaved crossing ridge roof of the Imperial Stele Pavilion altered in the Qing Dynasty, the main features that differ from the followed royal buildings lie in its rather gentle roof slope and the indentation of its vertical ridges (Fig.35).

The Longguo Hall, the Great Bell and Drum Towers have adopted the hip roof style, similar to those built in the royal capital during the same period of time. With its main ridge pushed much outward while the Taiping beams and Leigong columns are left out, their construction structures were much simplified (Fig.36).

As mentioned before, the royal court had once forbidden to color paint beams and columns in temples, yet excluding those bestowed by the emperor. Therefore, all major buildings in the Qutan Temple have their beams color painted like those in the capital, except for the Qutan Hall and its side halls. The only difference is that the former has no base layer. Among them, black and green colors are overlapped and blended for the beams and tiebeams of the Longguo Hall, the Great Bell and Drum Towers, and the side corridors in the back courtyard. Yellow color is used to paint the spinning heart of the inner eave of the Longguo Hall, similar to the method of gold pointing, indicating its the highest status within the complex. No overlapping and blending colors are used in the less important buildings such as the Jingang Hall, the Small Bell and Drum Towers and the side corridors in the middle courtyard. All these are referred as treasures of the color paintings found on building structures in the early Ming Dynasty. Paintings in the later renovations on the Baoguang Hall and the outer eave of the Longguo Hall are more similar to those expressed in the frscoes of the Qing Dynasty. When the Qutan Hall was extended with the front attached hall during the Qianlong period, it was built with local style, which is featured with *dougong*. After several repairing works between the late Qing and Republic era, its color paintings style had become complicated and varied, with both warm and cold colors, following the common local craftsmanship. But the color paintings on the inner eave belong to the Hongwu period remain to be the earliest existing wood structure color painting of the Ming Dynasty (Fig.37) .

In terms of construction methods, eave columns and corner columns of the buildings in the Qutan Temple were built with the same way: all slightly leaning towards the inside part of the building, and the corner columns heightened, which were rarely seen in the latter royal buildings in Ming and Qing Dynasties. However, they are much in comply with the construction methods, the so-called *cejiao* and *shengqi*, as regulated in the *Yingzao fashi* (*Rules for Structural Carpentry*) compiled during the Song Dynasty. In addition, the inverted V-shaped brace (*chashou* in Chinese) found in the Qutan Hall, the Sanshi Hall and the Hufa Hall, as well as the *shunjichuan* (a series of wood structure elements under the ridge rafters) found in the Qutan Hall (Fig.38), are also in comply with rules in the *Yingzao fashi*, while rarely seen in the royal buildings of the Ming and Qing Dynasties.

On the other hand, it is so rare to find many inscriptions on the upper eave beams in the Longguo Hall, left by its craftsmen. These are indeed precious and never-die hallmarks to record how the royal building structures had become systematized during the Ming and Qing Dynasties. For example, terms like *mingjian*, *shaojian*, *zhongjin*, *shangjin*, *danbuliang*, *shuangbuliang*, *wujialiang*, *suiliang*, *hang*, *fang*, *tongzhu*, etc (Fig.39), differed from those recorded in *Yingzao fashi* of the Song Dynasty , but exactly the same as those recorded in *Gongcheng zuofa zeli (Examples of Engineering Practices)* of the Qing Dynasty, which indicates its important value to make up the lack of the construction documentation of the Ming Dynasty.

4. Research and Protection of the Qutan Temple

Up to the contemporary time, the research and protection work around the Qutan Temple is marked by the establishment of the Qutan Temple Ancient Building Management Station in 1959, under the leadership of the Qinghai Province Heritage Management Committee. The temple was listed as the provincial relics protection site in the same year (Fig.40). In spring 1960, famous architectural historians ZHANG Yuhuan, DU Xianzhou investigated and surveyed this temple and published the investigation report on the periodical of *Cultural Relics*, No.5 in 1964, the earliest contemporary academic research achievement about the Qutan Temple (Fig.41). Since then, research on the temple had been stagnated until the 1980s, during which the exterior walls outside the temple and some buildings were destroyed, and the main buildings had been converted into a grain supply station. It is the effort made by MA Jun, the chief of the Qutan Temple Management Station and the related provincial government who continued to appeal for the protection of the Qutan Temple, which help to protect the temple in a rather complete shape and form till today.

After the Reform and Open Policy was initiated, local scholars such as XIE Zuo, MI Yizhi and ZHAO Shengchen began to search and sort all kinds of local records and epigraphs, together with their on-site investigations, and published several papers and books (Fig.42), which helped to sort out the histories and values of the Qutan Temple in the historical, political, religious and

1993年2月，经国家文物局专家组推荐，青海省文化厅委托天津大学建筑系承担瞿昙寺维修设计，期望结合高校的教学与科研优势促进文物建筑的保护研究。为此，该系建筑历史研究所由王其亨教授主持，组建了建筑学、建筑结构、岩土地基等多专业的协作团队，聘请单士元、罗哲文、冯建逵、于倬云、杜仙洲、傅连兴、郭旃、晋宏逵、傅清远等文物建筑权威和专家为顾问，经过缜密的调查测绘，深入研究相关历史，获得大量新发现，从而高质量地完成了修缮设计。1995年5月国家文物局划拨上千万元专项经费，由青海省文化厅苏生秀（格桑本）副厅长担纲，文物处李智信处长、考古研究所刘溥所长等主持施工，延请敦煌研究院资深文物建筑专家孙儒僩为顾问，开始了历时5年的瞿昙寺大规模维修。

在瞿昙寺修缮设计和施工中，『不改变文物原状』的理念得以严格实施，按照真实性、完整性的原则，以最小干预、可逆性措施解决实际问题。针对最大的病害原因——当地三级湿陷性黄土遇水沉降的特性，项目组聘请天津大学土力学专家吴家琦教授为顾问，处理寺院因排水不利致使地基下沉的险情，圆满完成相关建筑基础的加固，恢复并完善了旧有的地面海墁和排水、道路等系统。全部屋面苫背，则参照清代官式建筑的改性做法；不得不更换的砖瓦和石雕以及木构件等，则严格遵循原式样、原材料、原工艺的传统做法。其中，王其亨教授等还发现大量明代题记，尤其是隆国殿梁架上数十款工匠墨书题记，具有弥补有明一代营造文献缺环的重要价值。修缮后的瞿昙寺各建筑，经过20多年冻胀循环的严酷考验，至今未发生地基沉降、墙体和屋面开裂渗漏病害，实现了国家文物局专家组的愿景，成为名副其实的文物建筑修缮工程的『样板』。

此后，天津大学建筑历史与理论研究所大规模开展甘青地区传统建筑及其保护研究，厘清了清代以来瞿昙寺建筑修缮中融入甘青地区所谓『河州做法』的历史。瞿昙寺系统测绘及研究的部分成果，先后被东南大学建筑学院潘谷西教授主编的《中国古代建筑史·元明卷》、青海省文化厅编著的《瞿昙寺》以及郭华瑜著《明代官式建筑大木作研究》等专著吸收。

artistic perspectives in a systematic way.

Based on such works, the Qutan Temple was listed as one of the second patch of the state-level key units of cultural relics protection in 1982 (Fig.40). In 1984, the Qutan Temple Heritage Management Station was restored, who soon published the book of *Qutan Temple* (Fig.42). When the Qutan Hall went through a major repair in 1985 (Fig.43), under the guidance of QI Yingtao, the chief of the Chinese Relics Research Institution, an ink inscription was found on the middle part of the ridge tiebeam, written on the day of beam raising ceremony dated on September 6th, 1391, which no doubt provided a solid evidence for the starting year of the Qutan Temple.

In February 1993, under the recommendation of the expert group of State Cultural Relics Bureau, Department of Architecture of Tianjin University was authorized by the Qinghai Provincial Department of Culture to provide the repairing proposal for the Qutan Temple, in the hope to take the educational and researching advantages of the higher education institutions and to boost the local research on historical relics and architecture. Hence, Prof. WANG Qiheng, the chief of the architectural history and theory research office from the Department of Architecture, became the person in charge of the project, who organized an interdisciplinary research team from architecture, building structure, rock and soil foundation, inviting experts such as SHAN Shiyuan, LUO Zhewen, FENG Jiankui、YU Zhuoyun, DU Xianzhou, FU Lianxing, GUO Zhan, JIN Hongkui and FU Qingyuan to be the project consultants. Based on careful building survey and investigation, as well as researches on the history and architecture of the Qutan Temple, a high-quality repairing design was completed. In May 1995, with an over ten million special funds granted by the State Cultural Relics Bureau, a five-year long large-scale repairing and maintenance work on the Qutan Temple was started, led by SU Shengxiu（Kelsang Ben）, the deputy director of the Qinghai Provincial Department of Culture, managed by LIU Pu and LI Zhixin from the Qinghai Province Institute of Archaeology, and with the consultant work provided by SUN Ruxian, an expert on relics and ancient architecture from the Dunhuang Academy.

During the design and construction process of this repairing and maintenance work of the Qutan Temple, the principle of "without changing the original state of cultural relics" had been strictly engaged, and all practical problems were solved in comply with the measures of minimum intervention and reversibility. Facing the challenge of local third class collapsible loess that would suffer water settlement, the project team engaged prof. WU Jiaxun, an expert in soil mechanics from Tianjin University, to be the special consultant, who helped to successfully strengthen

图四十二 瞿昙寺相关研究著作
左：1982 年谢佐《瞿昙寺》 右：1985 瞿昙寺文物管理所《瞿昙寺》

图四十三 1985 年祁英涛、张剑波等《瞿昙寺维修工程》

图四十四 2015 年格桑本主编《藏族美术集成·绘画艺术·壁画·青海卷》

Fig.42 Research publications of Qutan Temple
left: XIE Zuo, *Qutan Temple*, 1982 right: Qutan Heritage Management Station, *Qutan Temple*, 1985

Fig.43 QI Yingtao, ZHANG Jianbo, *Repairing and Maintenance Work on the Qutan Temple*, 1985

Fig.44 *Collection of Tibetan Fine Arts, Art of Painting, Frescoes, Qinghai Volume*, edited by Kelsang Ben, 2015

与此同时，瞿昙寺规模巨大的壁画遗存，也引起国内外美术界高度关注，展开了系统深入的研究，对各殿内遗存的佛像及藏传佛教密宗诸神像、厢廊汉式壁画的来龙去脉，以及艺术价值的定位，都有了前所未有的突破，拓展并深化了瞿昙寺深厚历史文化内涵的当代认知。其中，基于对瞿昙寺壁画的完整记录、辨识与评价，所形成的集大成的代表性成果，就是2015年国家出版基金重点资助的《藏族美术集成》第一册『瞿昙寺壁画专卷』（图四十四）。这些努力和成果，为未来瞿昙寺的文物建筑保护和研究，以及展示和旅游开发，打下了良好基础。

the foundation from subsiding that might be caused by the temple's drainage disadvantage, restoring the flat brick pavement, drainage system and the road system. The base of all roofs were following traditional method and craftsmanship, referring to the modified techniques adopted in the imperial architecture of the Qing Dynasty. Bricks, tiles, stone sculptures and wood elements, which had to be replaced, were all redone in the traditional methods, with the same style, materials and techniques. And during this process, many inscriptions of the Ming Dynasty, especially those left by craftsmen on the beams of the Longguo Hall, all found by Prof. WANG Qiheng, have made up to the lack of construction documentations of the Ming Dynasty. The repaired buildings of the Qutan Temple have gone through the test of more than twenty years' frost, with no sufferings in terms of foundation settlement, cracking or leaking of its walls and roofs. It has thus successfully meet the expectation of the expert team of the State Cultural Relics Bureau, becoming a "show case" that is worthy of its name in the field of repairing works of cultural relics.

Afterwards, a series of academic papers have been completed by the architectural history and theory research office in Tianjin University, who has continued to conduct the research on how the traditional building methods and craftsmanship in the Gansu and Qinghai area, the so-called "hezhou practice" had been integrated into the repairing works for the Qutan Temple since the Qing Dynasty for the protection purpose. Part of the survey and research work were absorbed and quoted in several monographs, such as *History of Ancient Chinese Architecture - Yuan and Ming Volume*, edited by Prof. PAN Guxi, *Qutan Si*, edited by the Qinghai Province Department of Culture, and *Research on the Carpentry Work in the Imperial Buildings of the Ming Dynasty*, written by GUO Huayu.

At the same time, the abundant murals in the Qutan Temple has drawn increasing attentions from scholars in the field of art history both inside and outside China in recent years. Systematic researches have been carried out on the remaining Buddhist sculptures and images, the ins and outs of those Han-style wall paintings in the side corridors, and as well as their artistic evaluations. Such researches have made some unprecedented breakthrough, which have extended and deepened contemporary understandings of the rich historic and cultural connotations of the Qutan Temple. Among them, the special volume devoted to the frescoes of the Qutan Temple, No.1 in the Collection of the Tibetan Fine Arts (Fig.44), key founding by the State Publishing Founding in 2015, is the representative achievement, based on the complete documentation, identification and evaluation of the frescoes in the Qutan Temple. All these efforts and outputs have built a fine foundation for not only the protection and study on the historic monuments and buildings, but also for their public presentation and tourism development.

图版

Figure

瞿昙寺组群
Qutan Temple Complex

前院 Front Courtyard
1. 山门 Main Gate
2. 垂花门 Festooned Door
3. 御碑亭 Stele Pavilion

中院 Middle Courtyard
4. 金刚殿 Jingang Hall
5. 瞿昙殿 Qutan Hall
6. 香趣塔 Scent Tower
7. 瞿昙殿东配殿 East Side Hall of Qutan Hall
8. 瞿昙殿西配殿 West Side Hall of Qutan Hall
9. 小鼓楼 Small Drum Tower
10. 三世殿 Sanshi Hall
11. 小钟楼 Small Bell Tower
12. 护法殿 Hufa Hall
13. 宝光殿 Baoguang Hall
14. 宝光殿东配殿 East Side Hall of Baoguang Hall
15. 宝光殿西配殿 West Side Hall of Baoguang Hall

后院 Back Courtyard
16. 大鼓楼 Great Drum Tower
17. 大钟楼 Great Bell Tower
18. 隆国殿 Longguo Hall

囊谦 Nangqian
19. 囊谦垂花门 Festooned Door of Nangqian
20. 囊谦倒座 Inverted Room of Nangqian
21. 囊谦过厅 Passage Hall of Nangqian
22. 囊谦前东楼 Qiandonglou of Nangqian
23. 囊谦前西楼 Qianxilou of Nangqian
24. 囊谦下转楼 Xiazhuanlou of Nangqian
25. 囊谦后东楼 Houdonglou of Nangqian
26. 囊谦后西楼 Houxilou of Nangqian

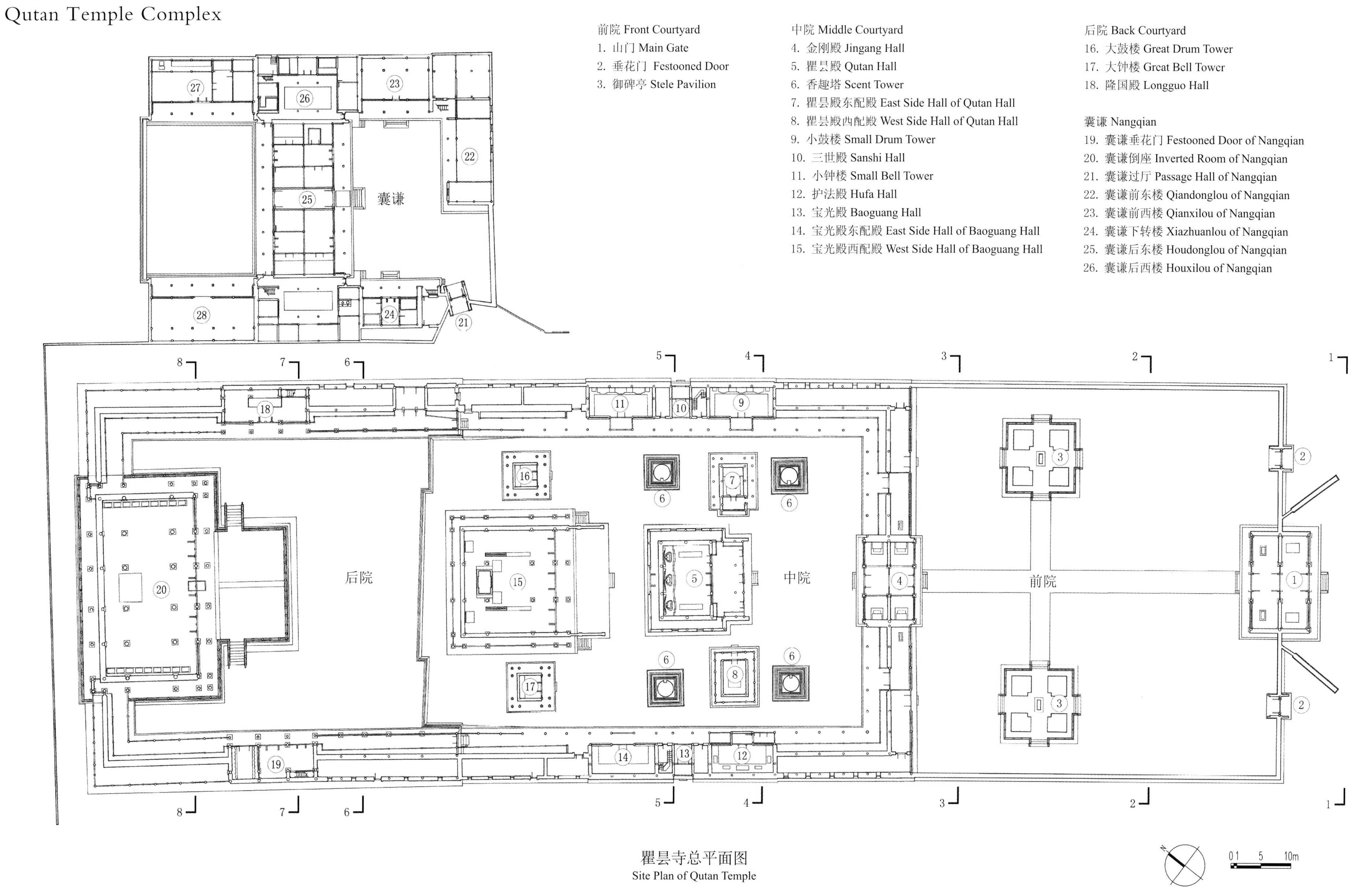

瞿昙寺总平面图
Site Plan of Qutan Temple

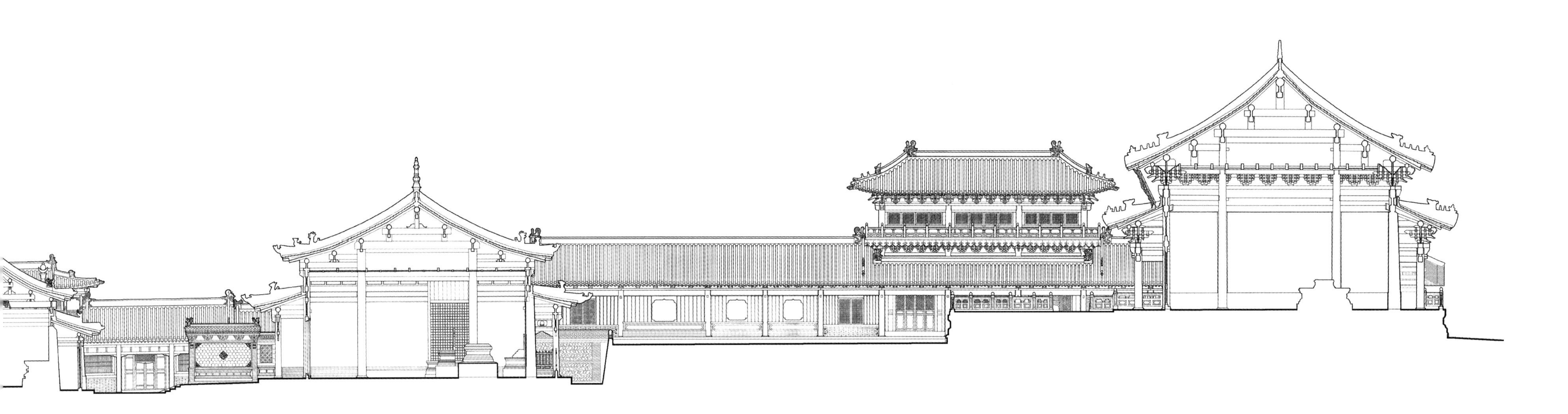

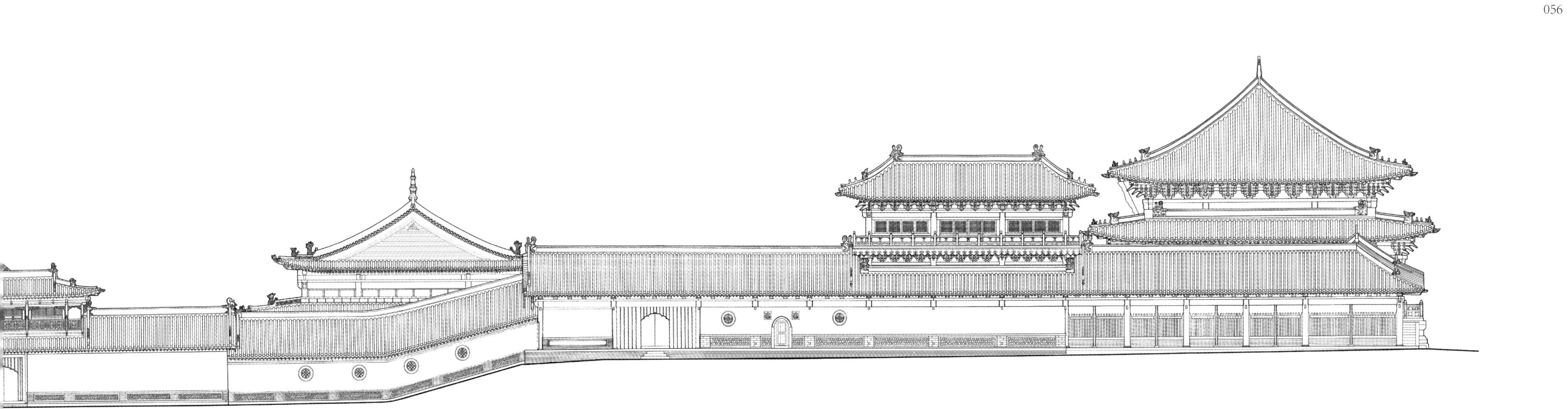

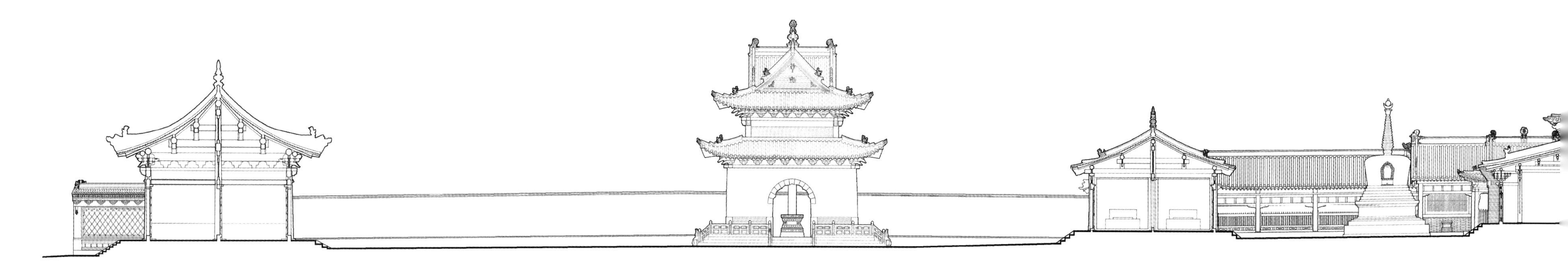

瞿昙寺组群剖面图
Cross-section of Qutan Temple

瞿昙寺组群东立面图
East Elevation of Qutan Temple

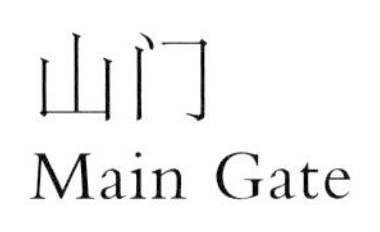

山门
Main Gate

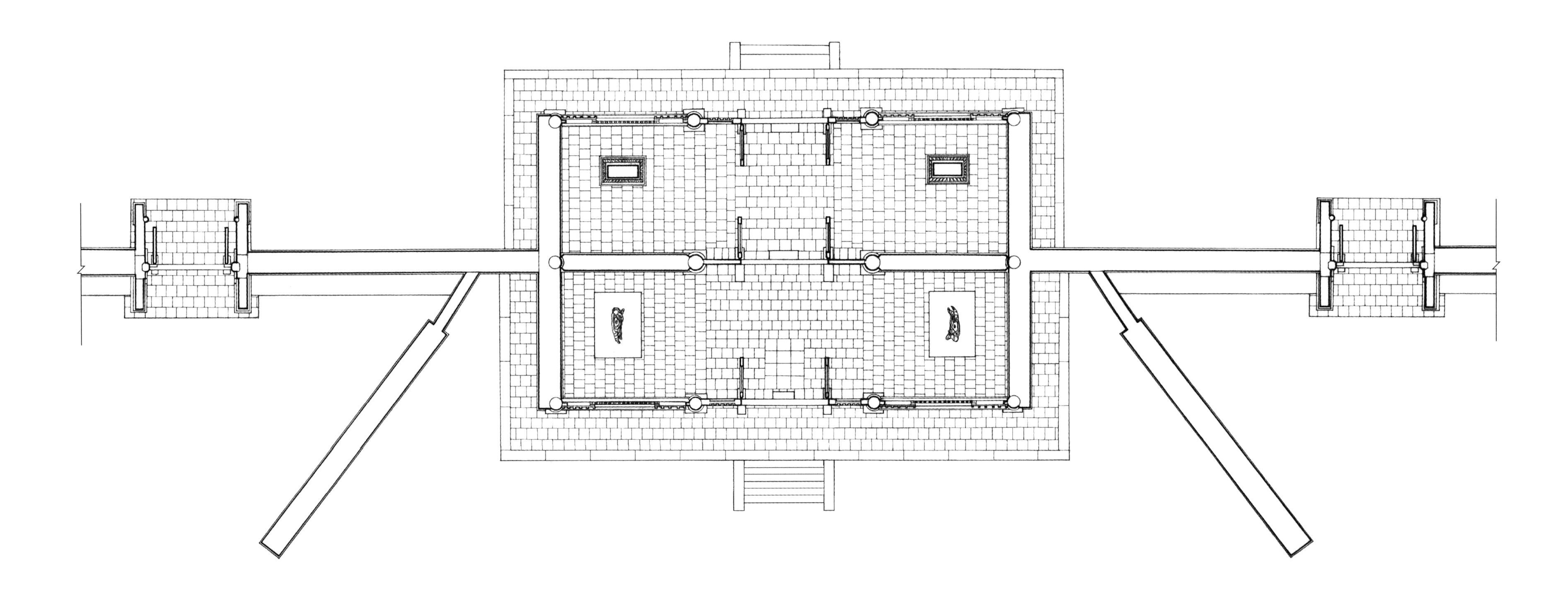

山门垂花门平面图
Plan of Main Gate and Festooned Door

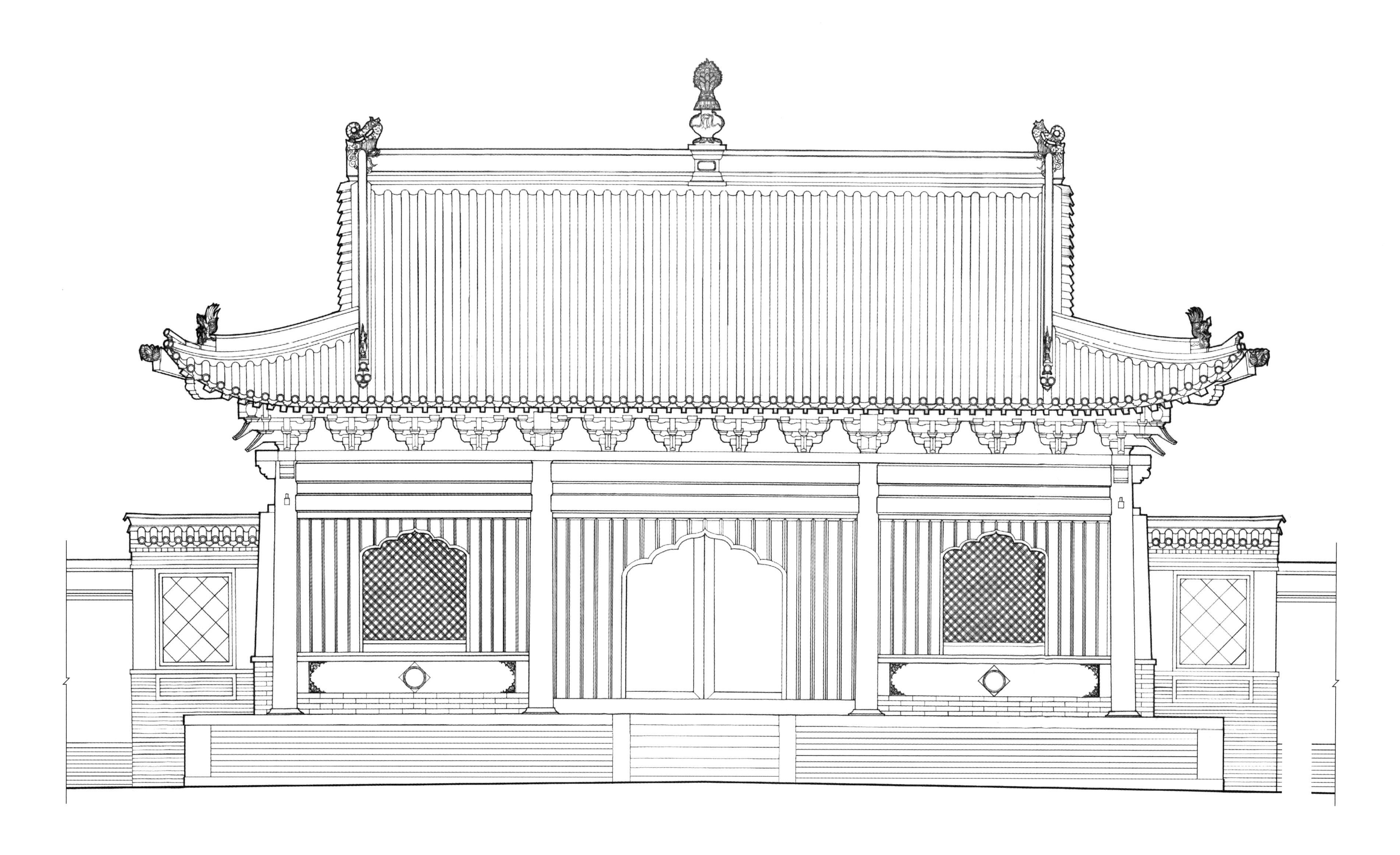

山门南立面图
South Elevation of Main Gate

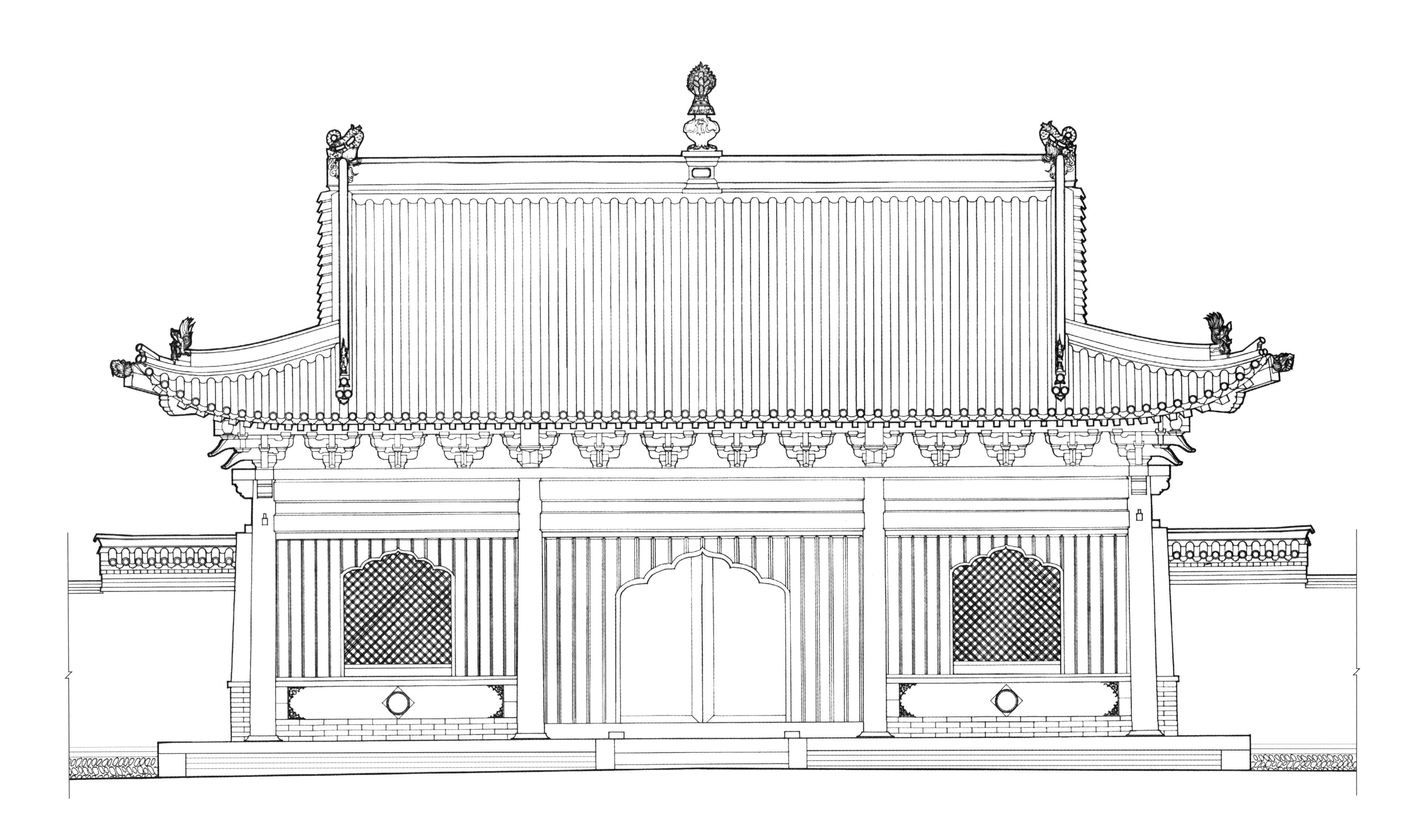

山门北立面图
North Elevation of Main Gate

山门东立面图
East Elevation of Main Gate

山门明间剖面图
Mingjian Cross-section of Main Gate

0 0.5 2m

山门次间剖面图
Cijian Cross-section of Main Gate

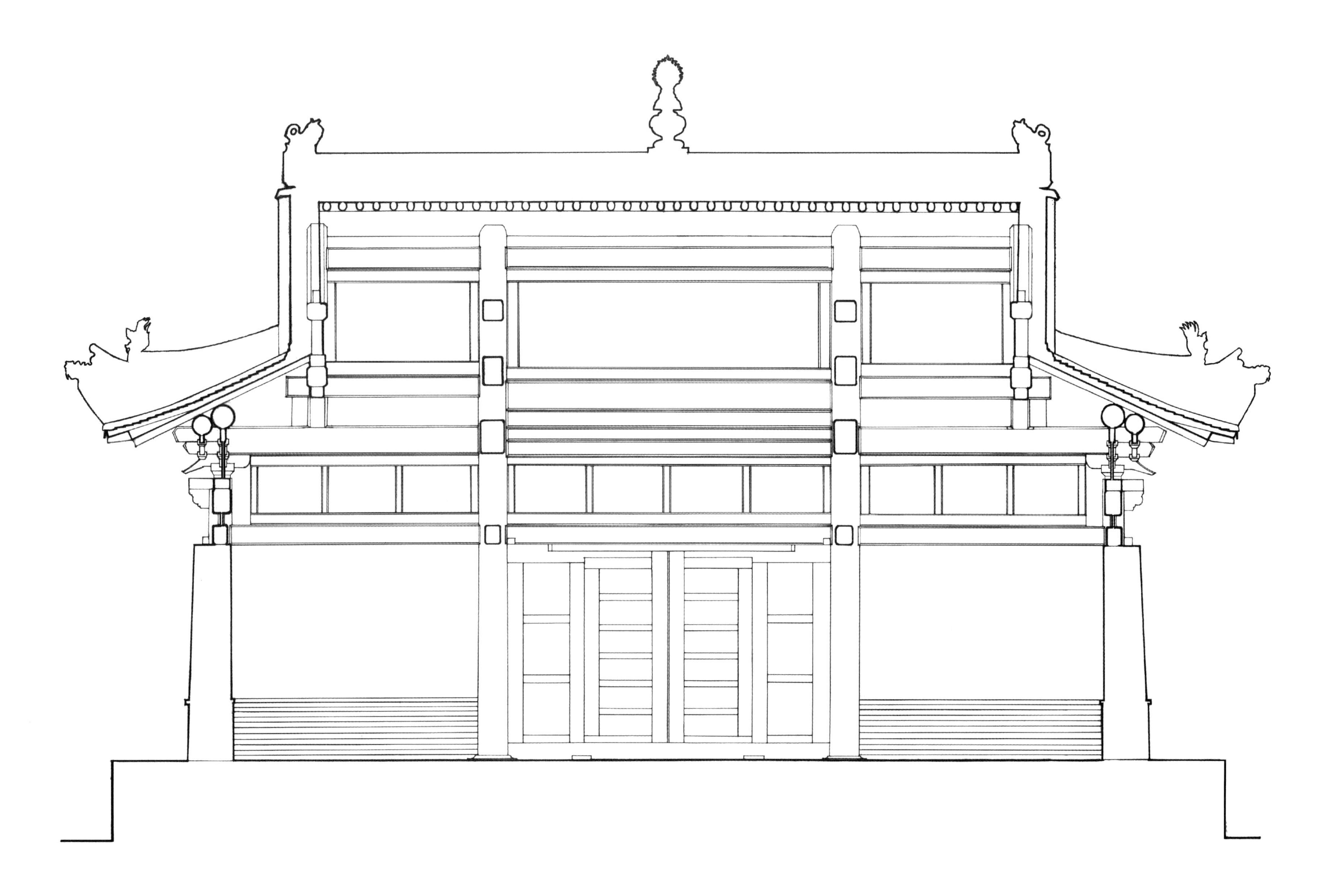

山门纵剖面图
Longitudinal Section of Main Gate

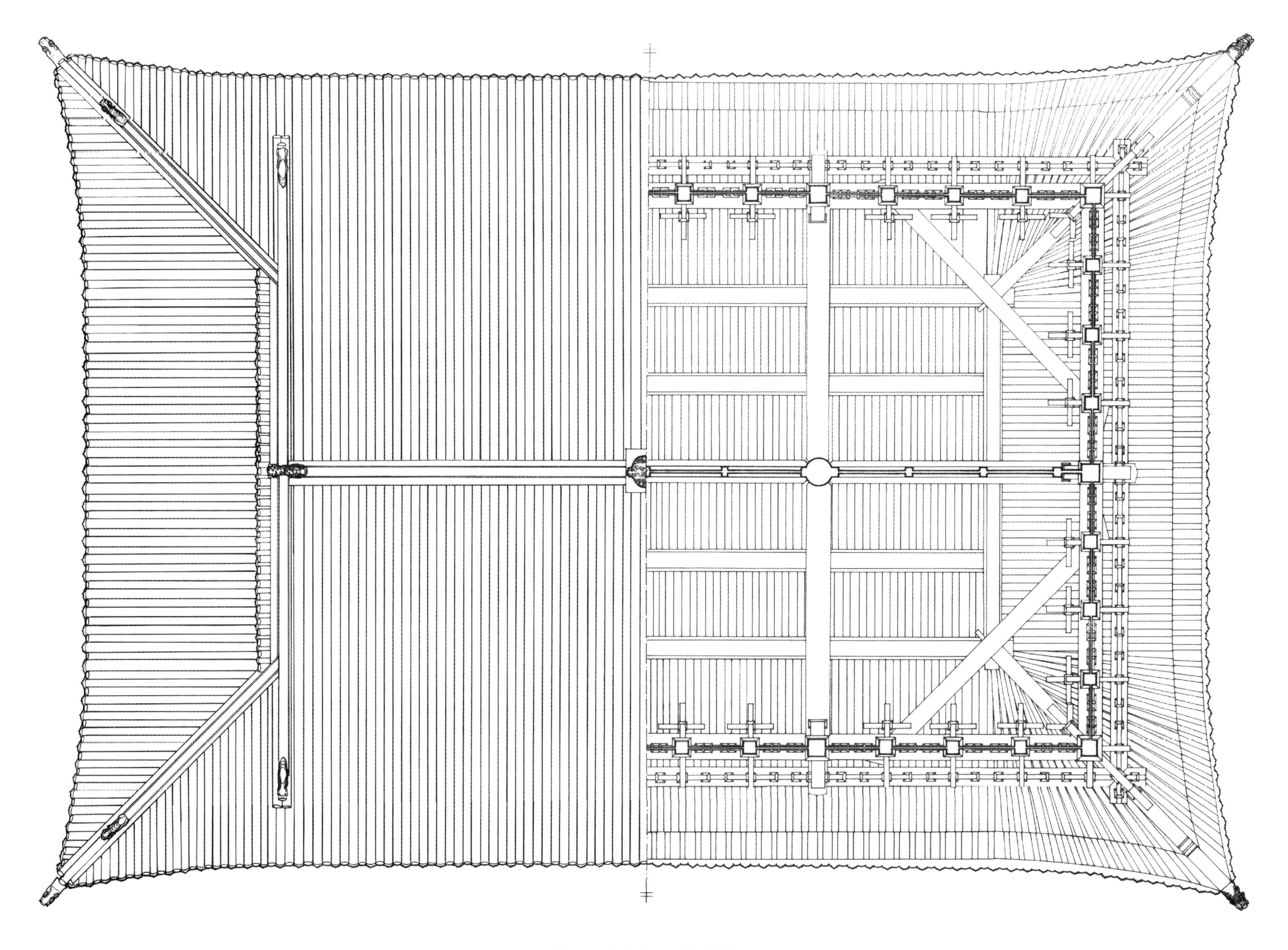

山门屋顶平面及梁架仰视图
Roof Plan and Framework Plan of Main Gate

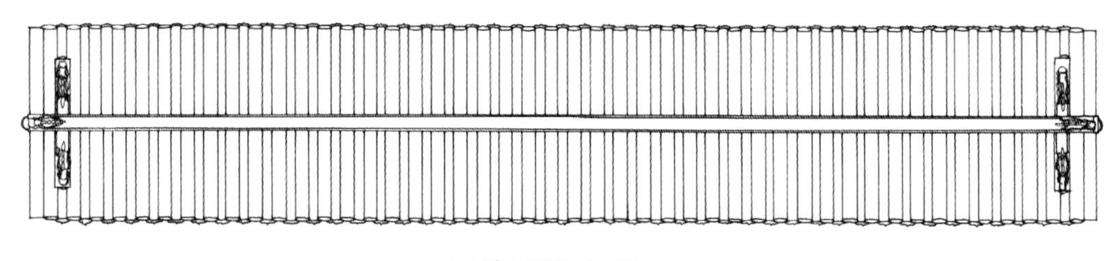

影壁屋顶平面图
Roof Plan of Screen Wall

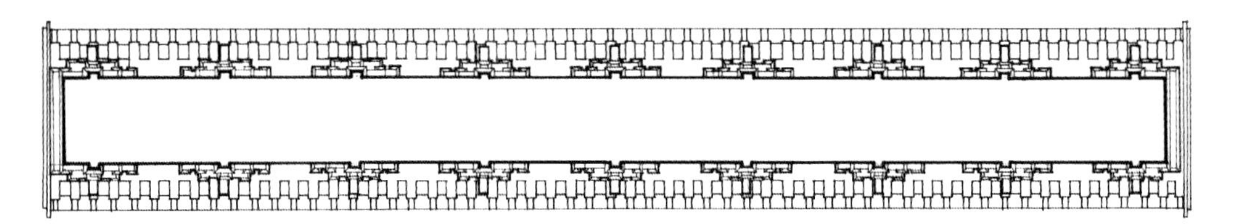

影壁梁架仰视图
Framework Plan of Screen Wall

影壁剖面图
Cross-section of Screen Wall

影壁东立面图
East Elevation of Screen Wall

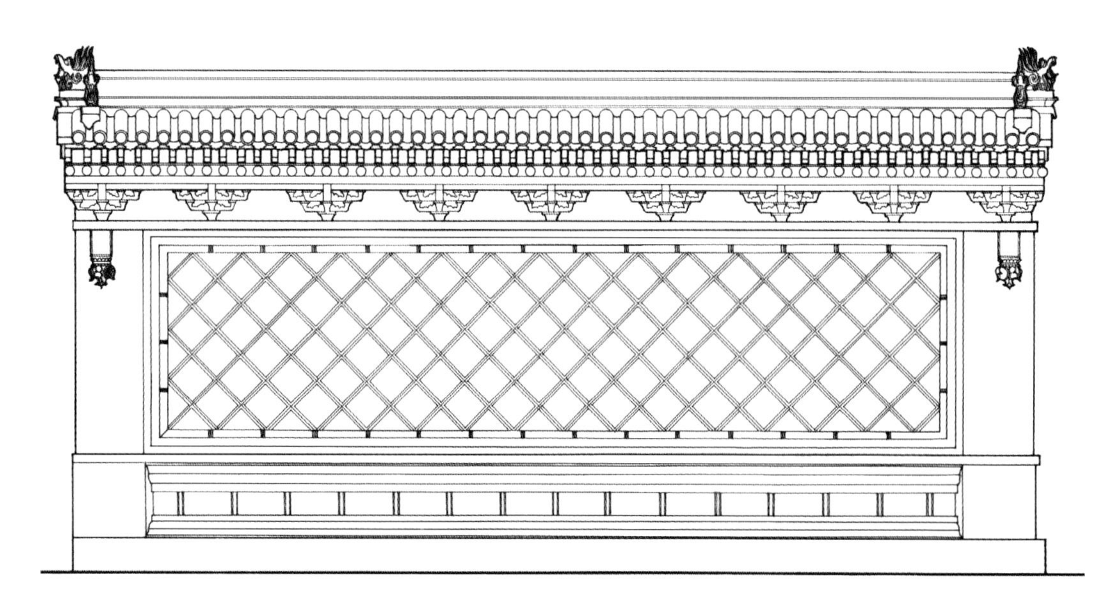

影壁西立面图
West Elevation of Screen Wall

0 0.5 1 2m

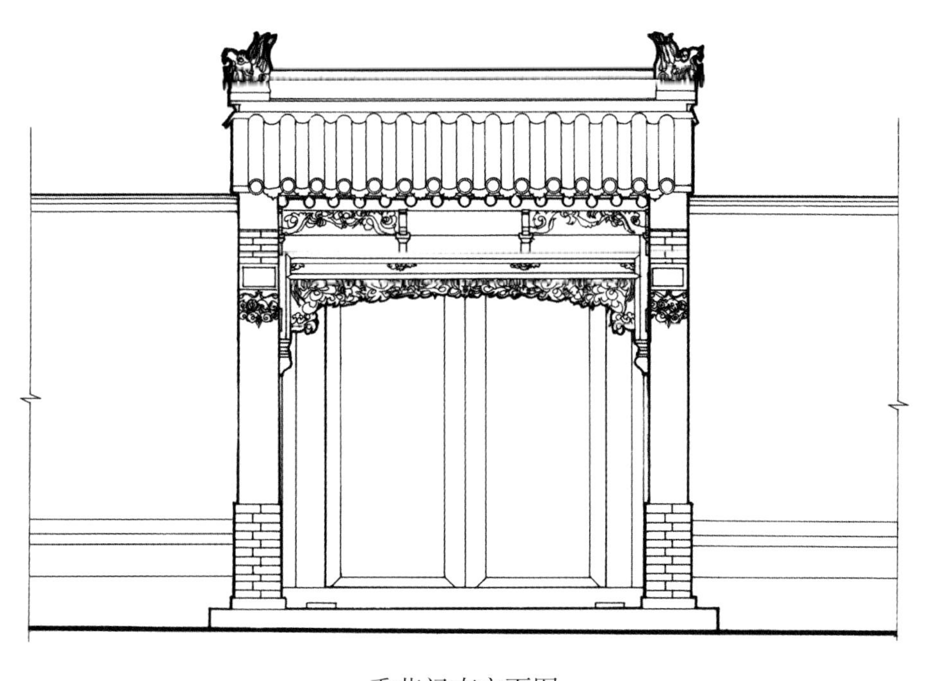

垂花门南立面图
South Elevation of Festooned Door

垂花门北立面图
North Elevation of Festooned Door

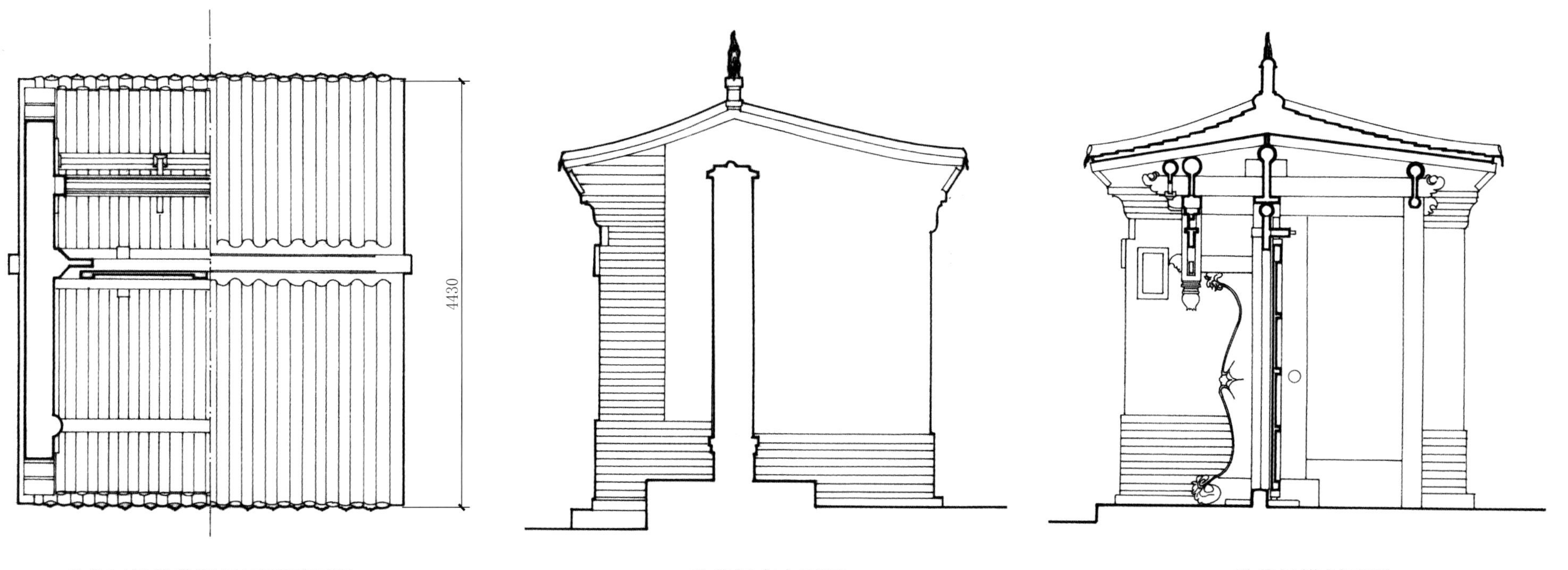

垂花门梁架仰视及屋顶平面图
Framework Plan and Roof Plan of Festooned Door

垂花门东立面图
East Elevation of Festooned Door

垂花门横剖面图
Cross-section of Festooned Door

御碑亭
Stele Pavilion

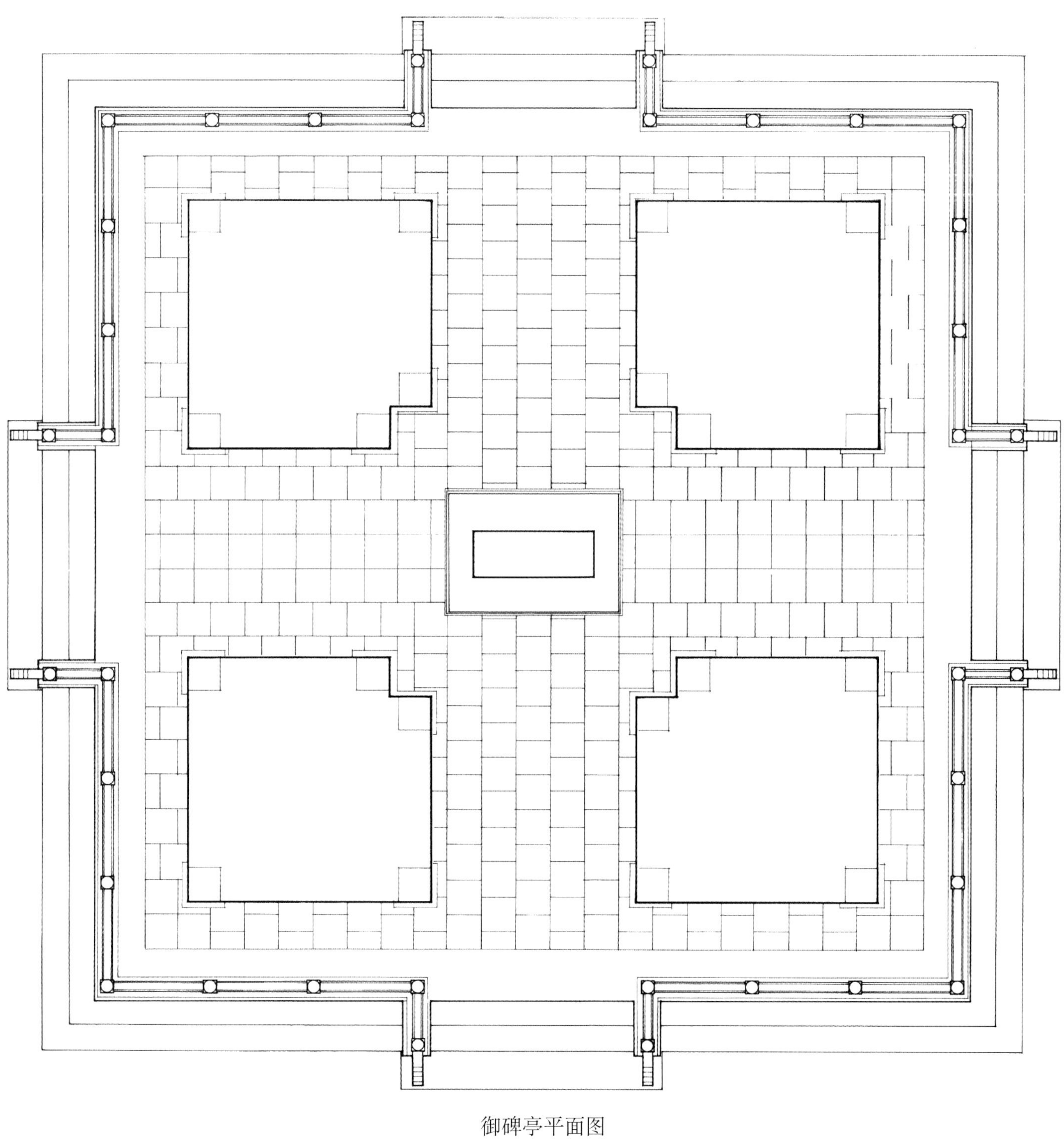

御碑亭平面图
Plan of Stele Pavilion

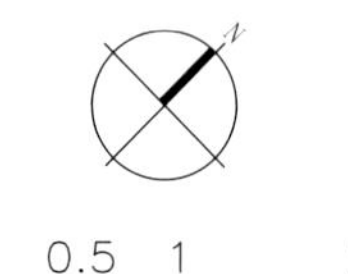

0 0.5 1 2m

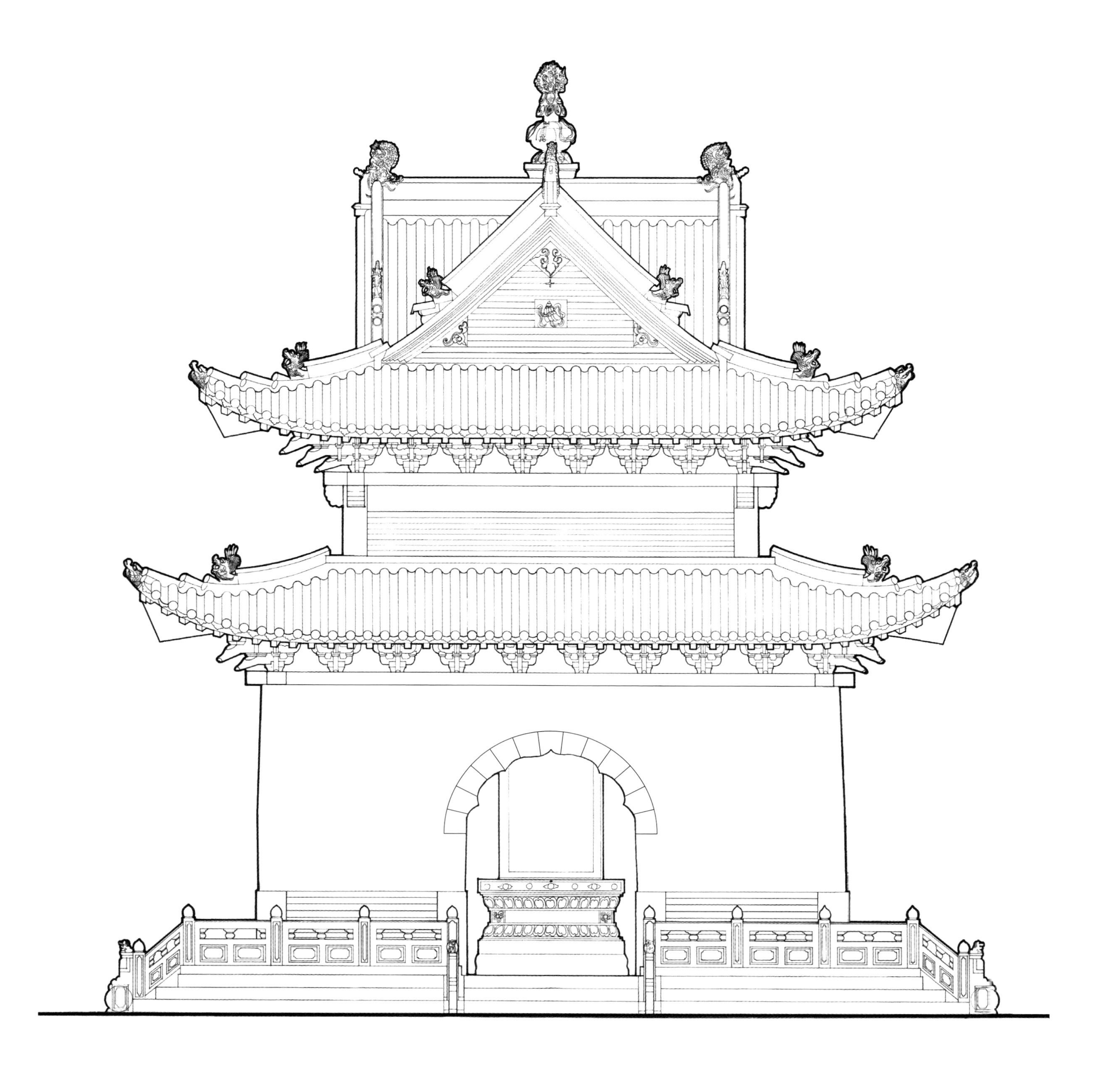

御碑亭立面图
Elevation of Stele Pavilion

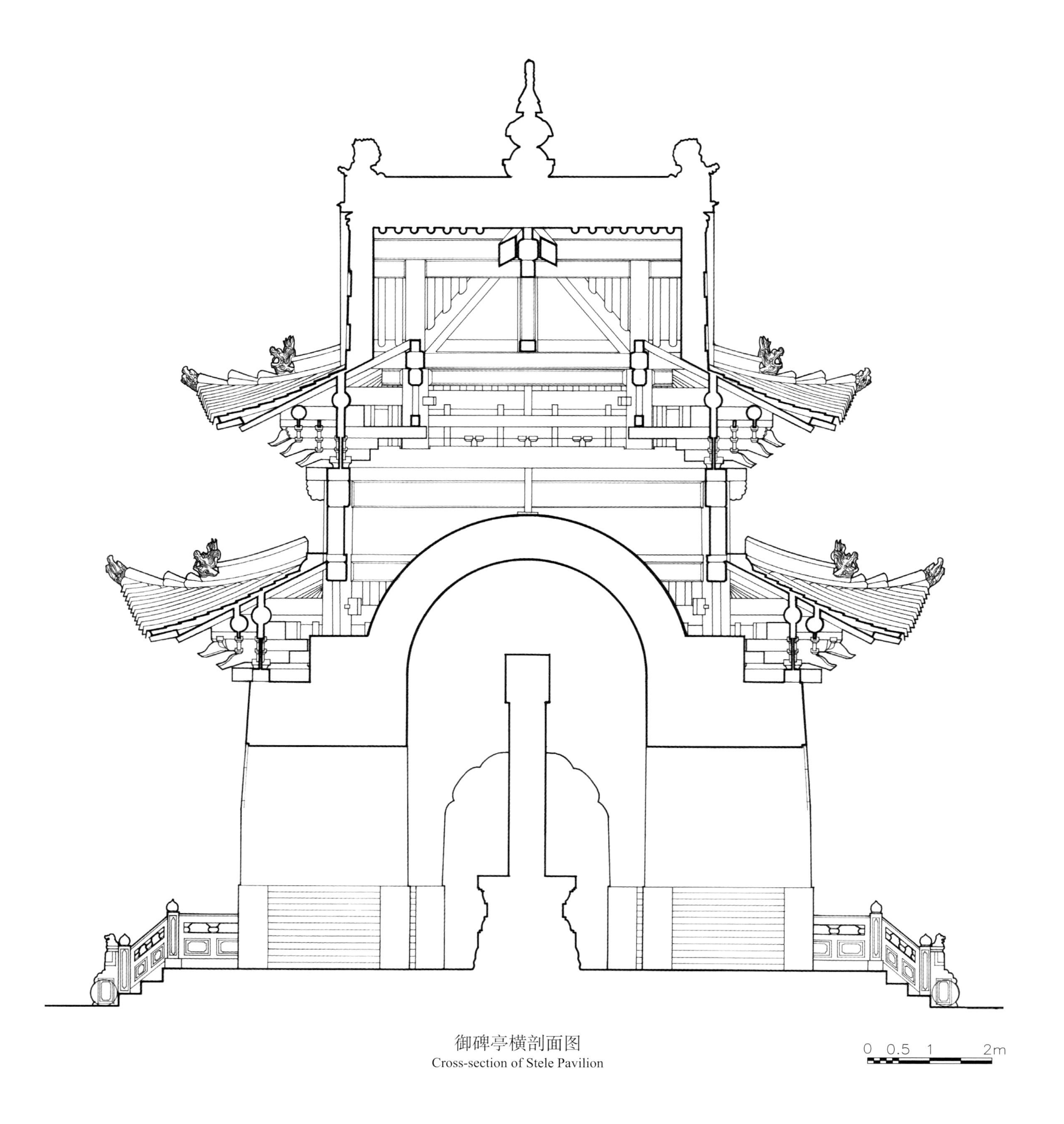

御碑亭横剖面图
Cross-section of Stele Pavilion

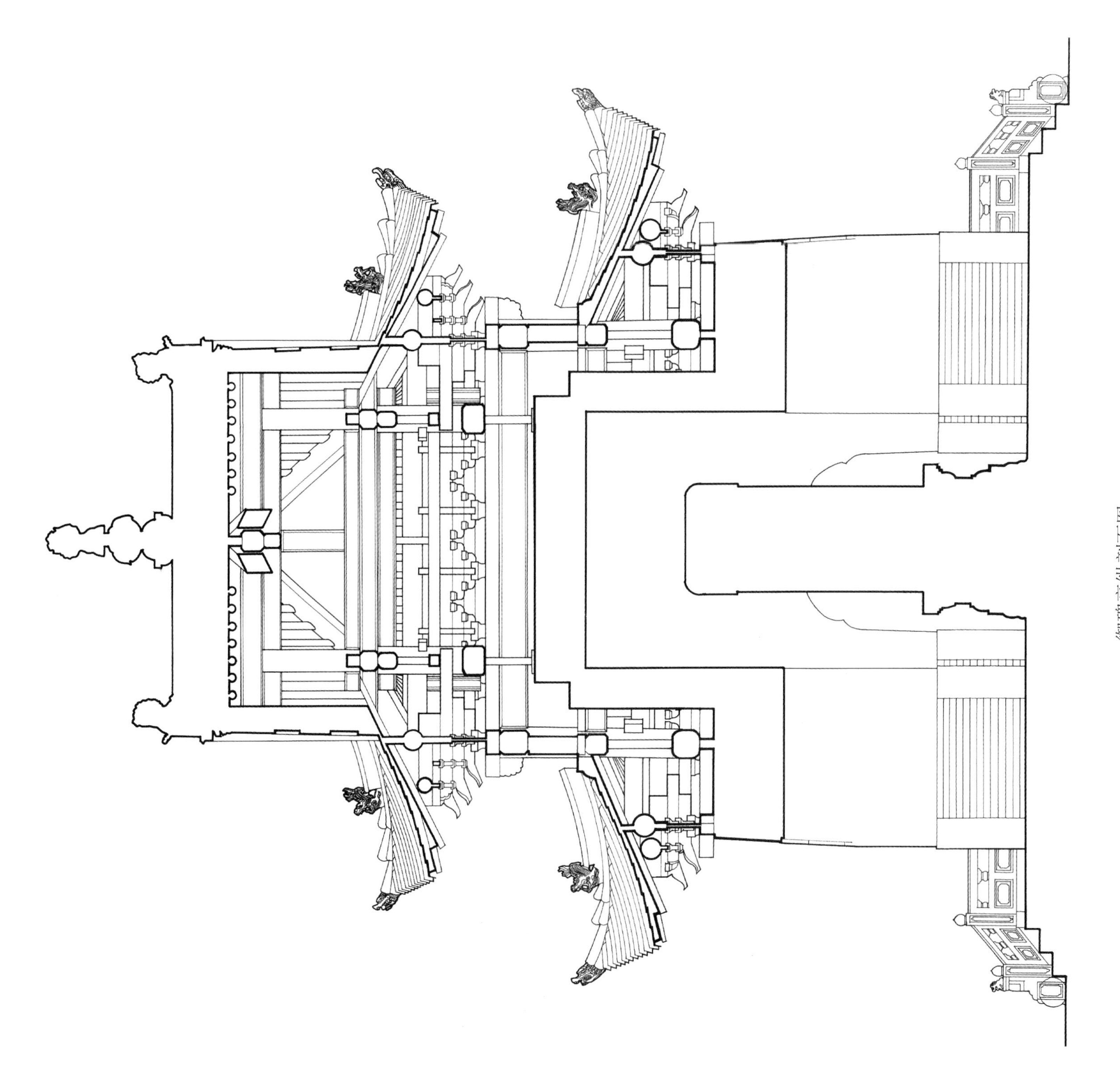

御碑亭纵剖面图
Longitudinal Section of Stele Pavilion

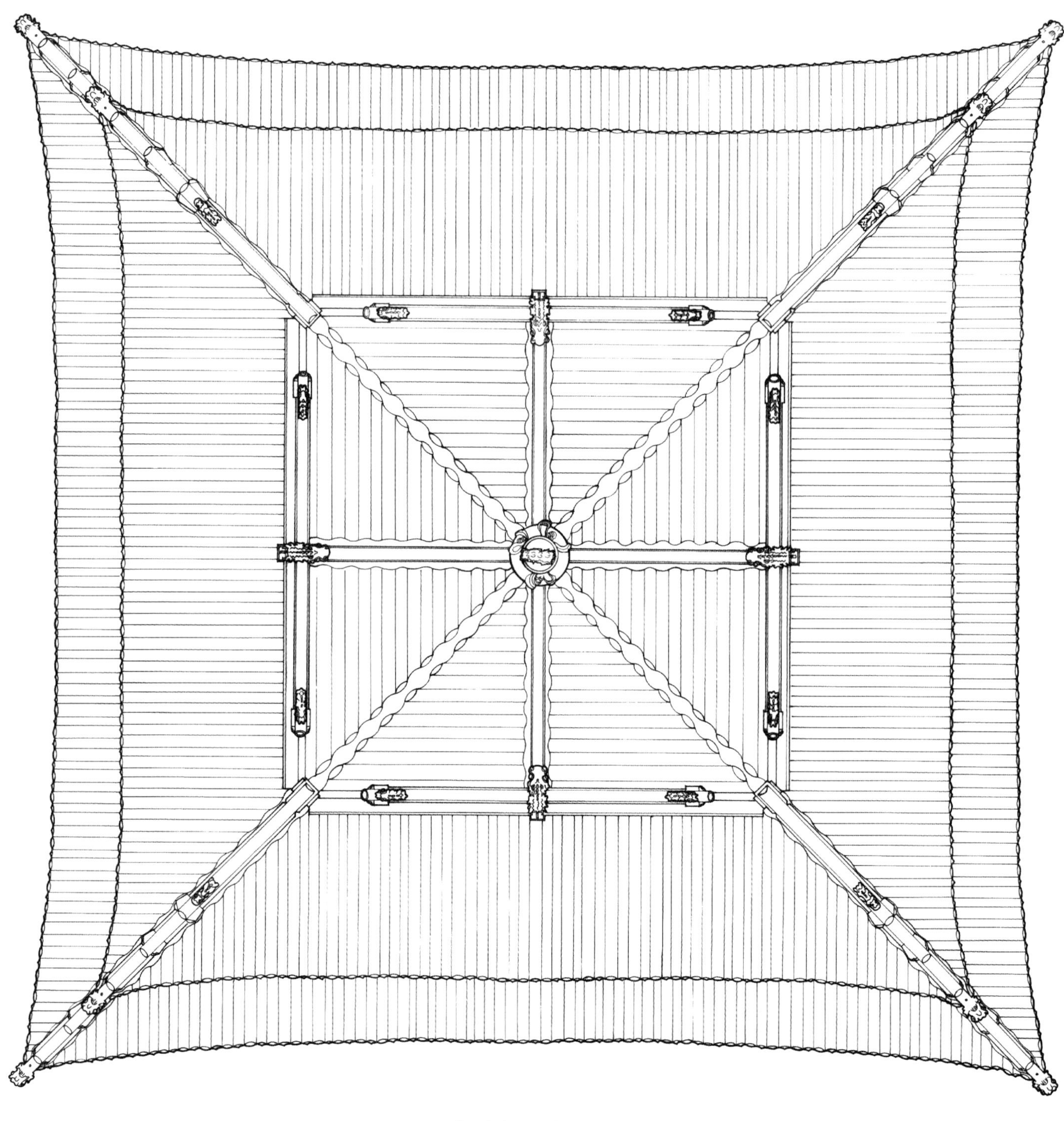

御碑亭屋顶平面图
Roof Plan of Stele Pavilion

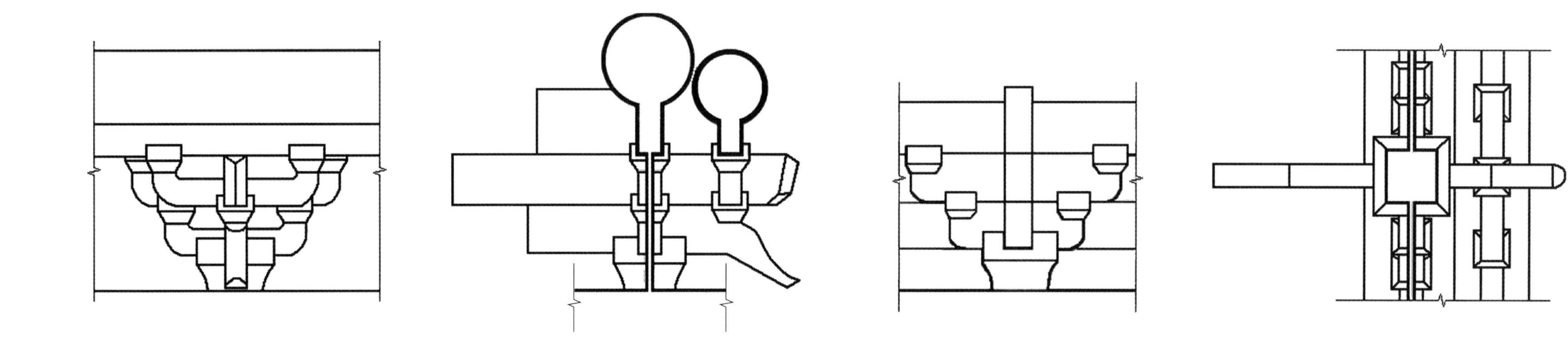

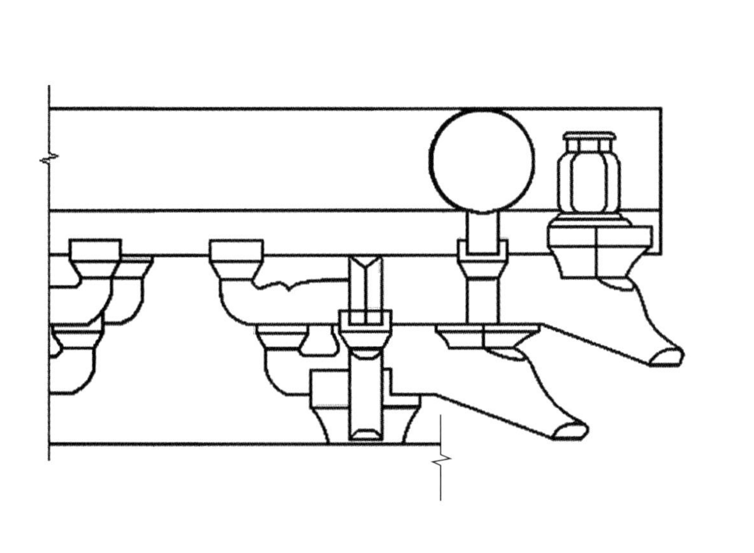

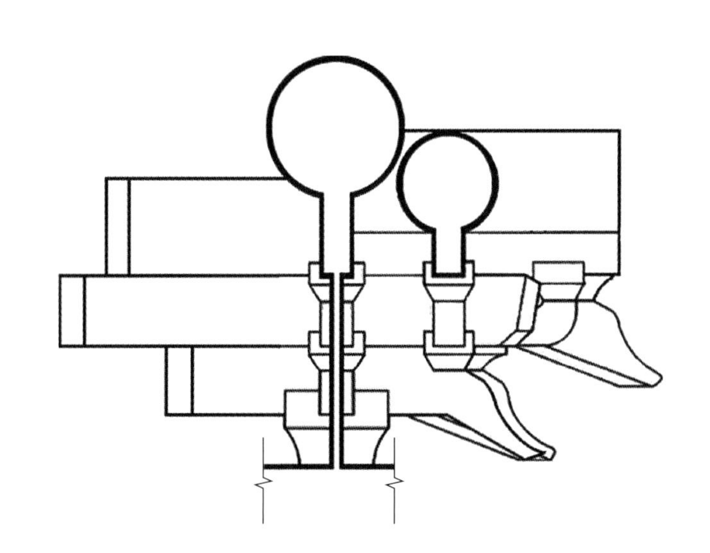

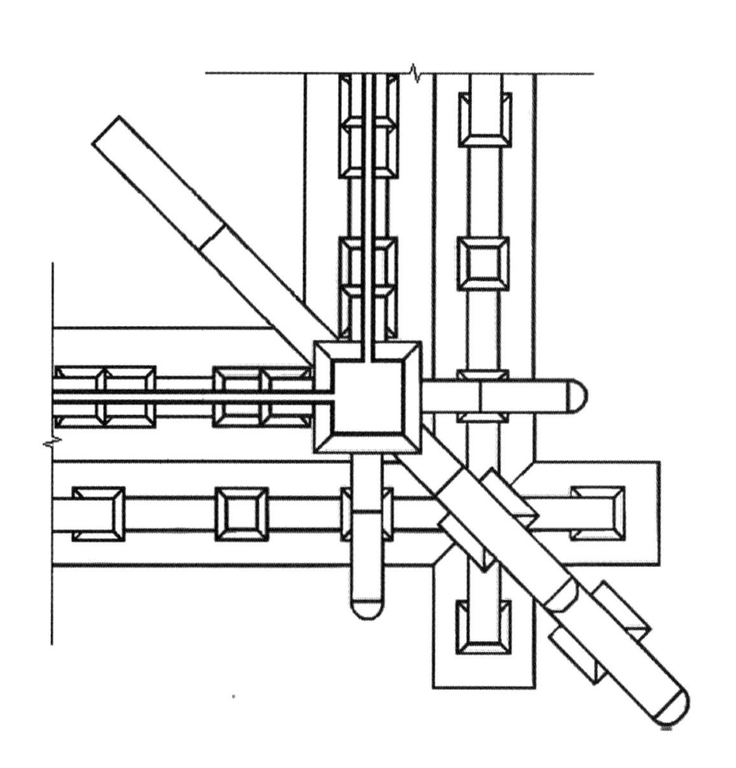

御碑亭下檐斗栱大样图
Dougong Under the Lower Eave of Stele Pavilion

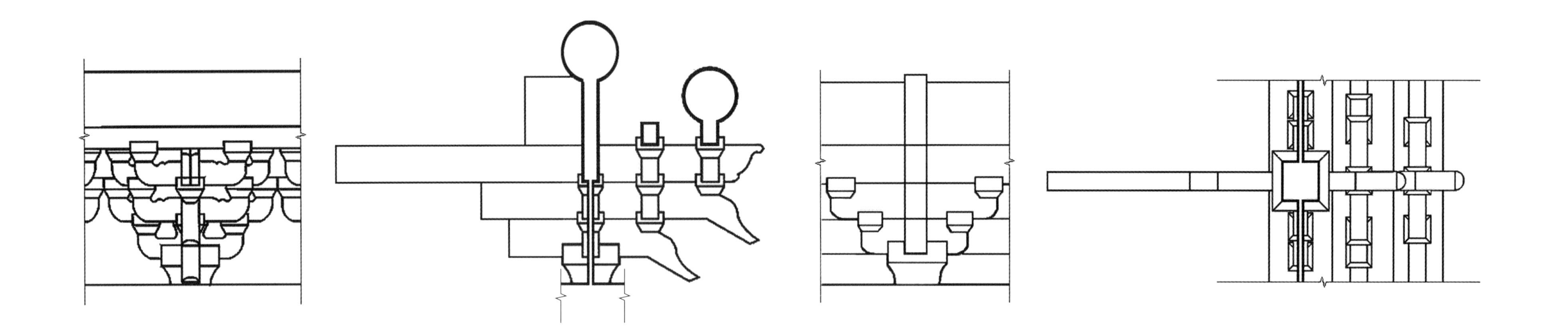

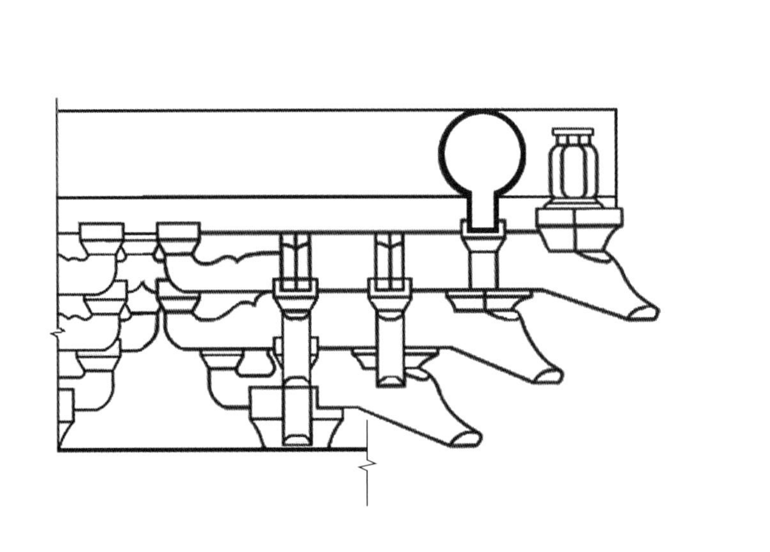

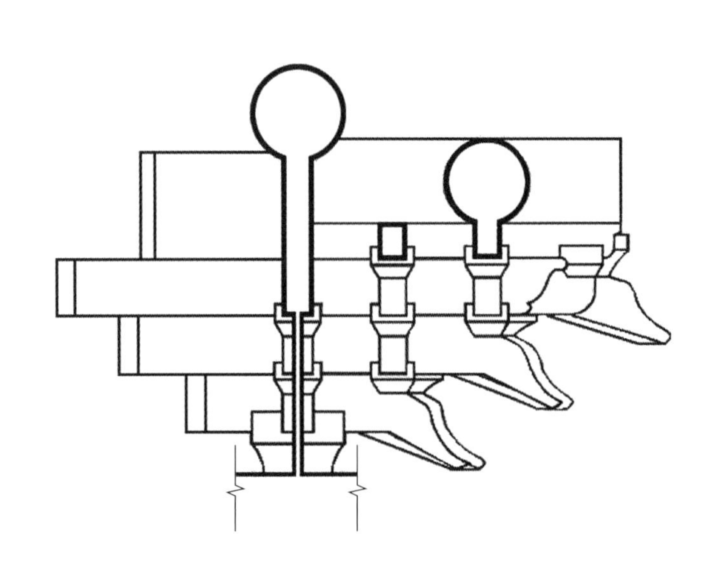

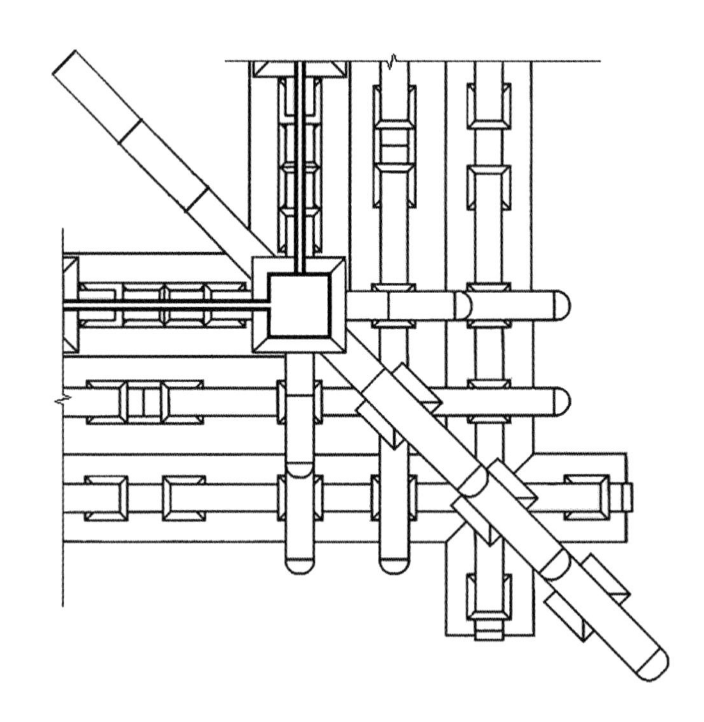

御碑亭上檐斗栱大样图

Dougong Under the Upper Eave of Stele Pavilion

金刚殿
Jingang Hall

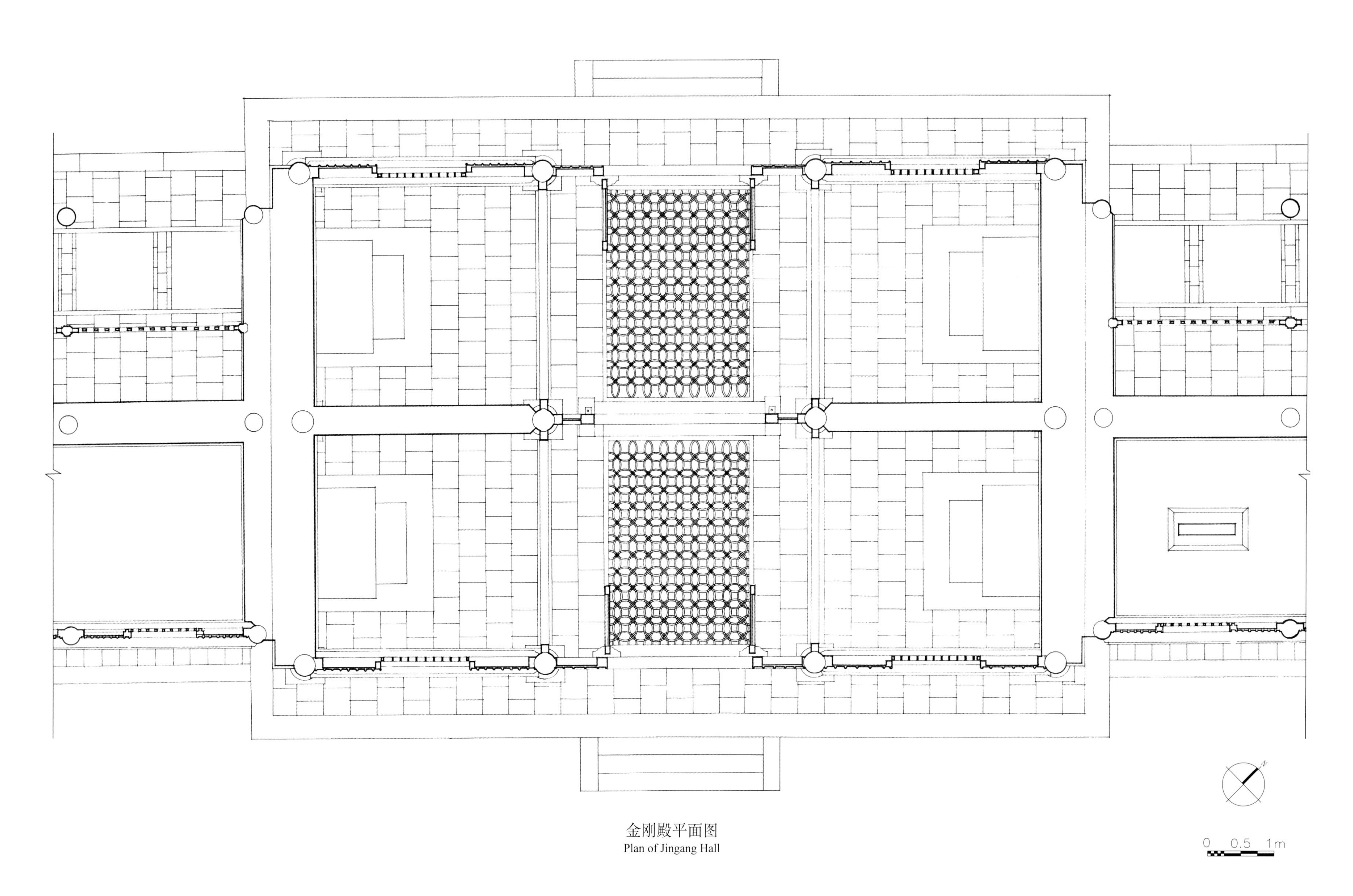

金刚殿平面图
Plan of Jingang Hall

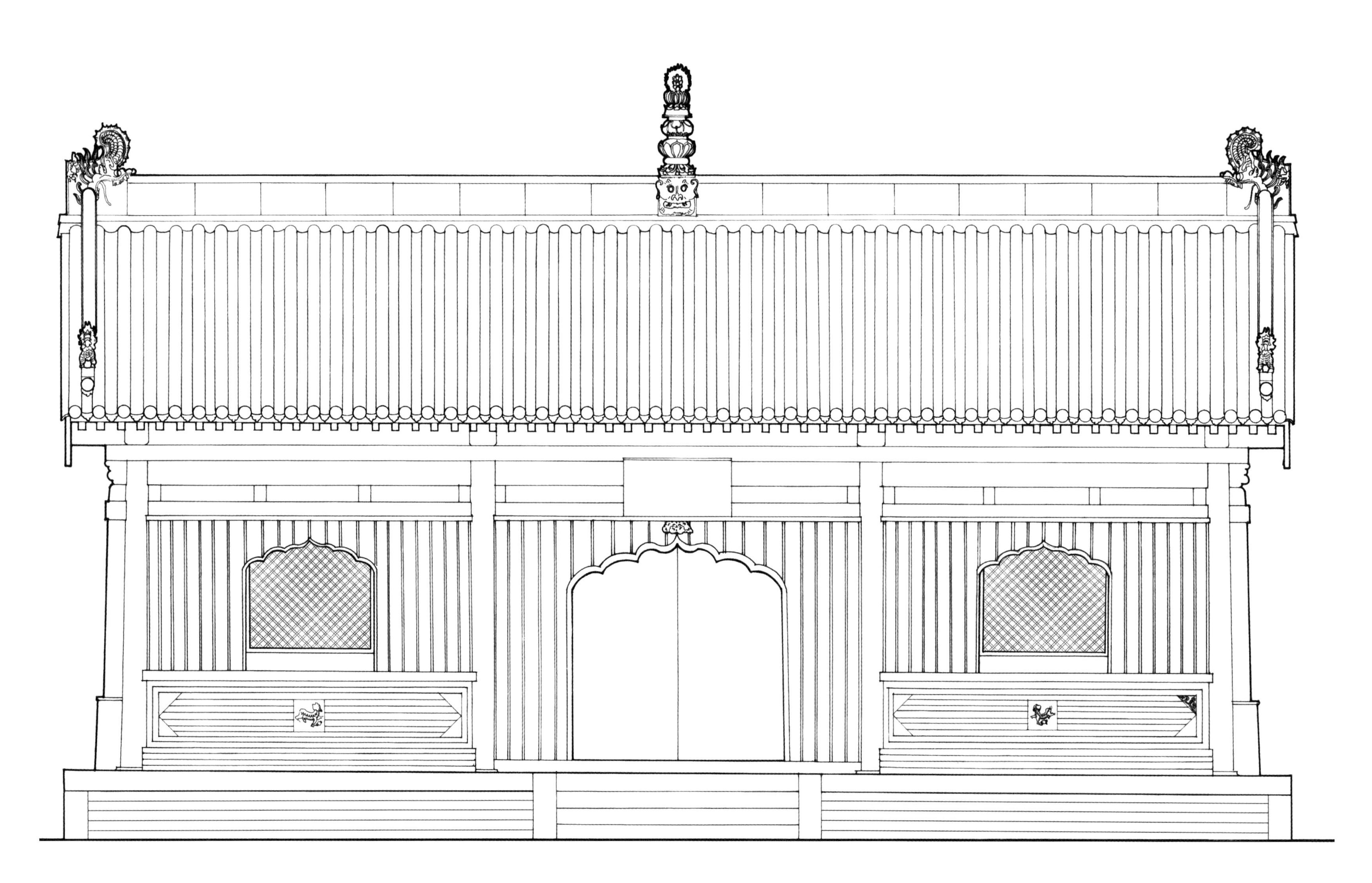

金刚殿南立面图
South Elevation of Jingang Hall

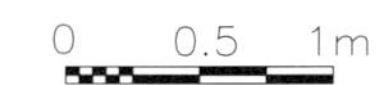

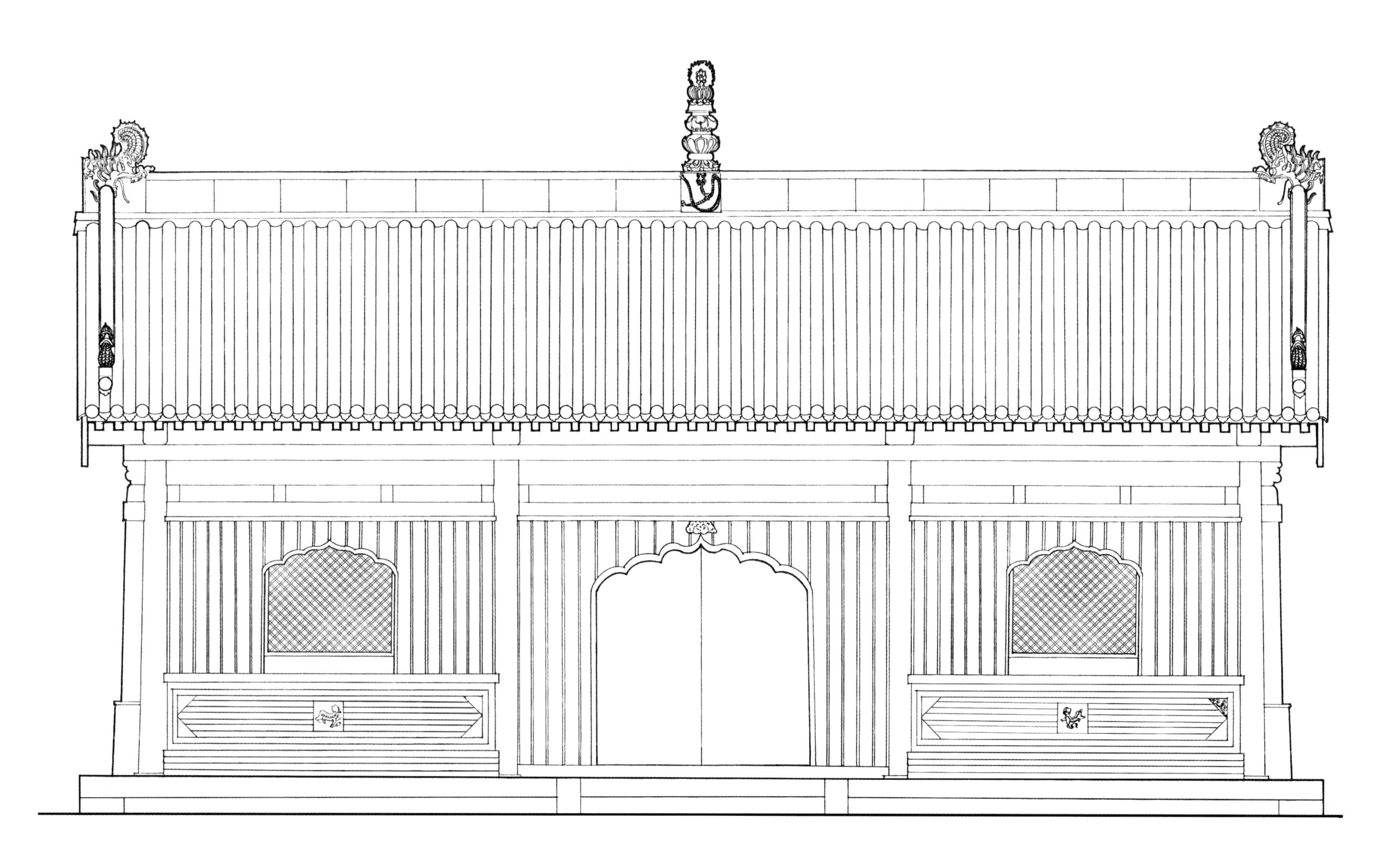

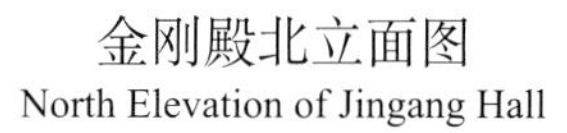

金刚殿北立面图
North Elevation of Jingang Hall

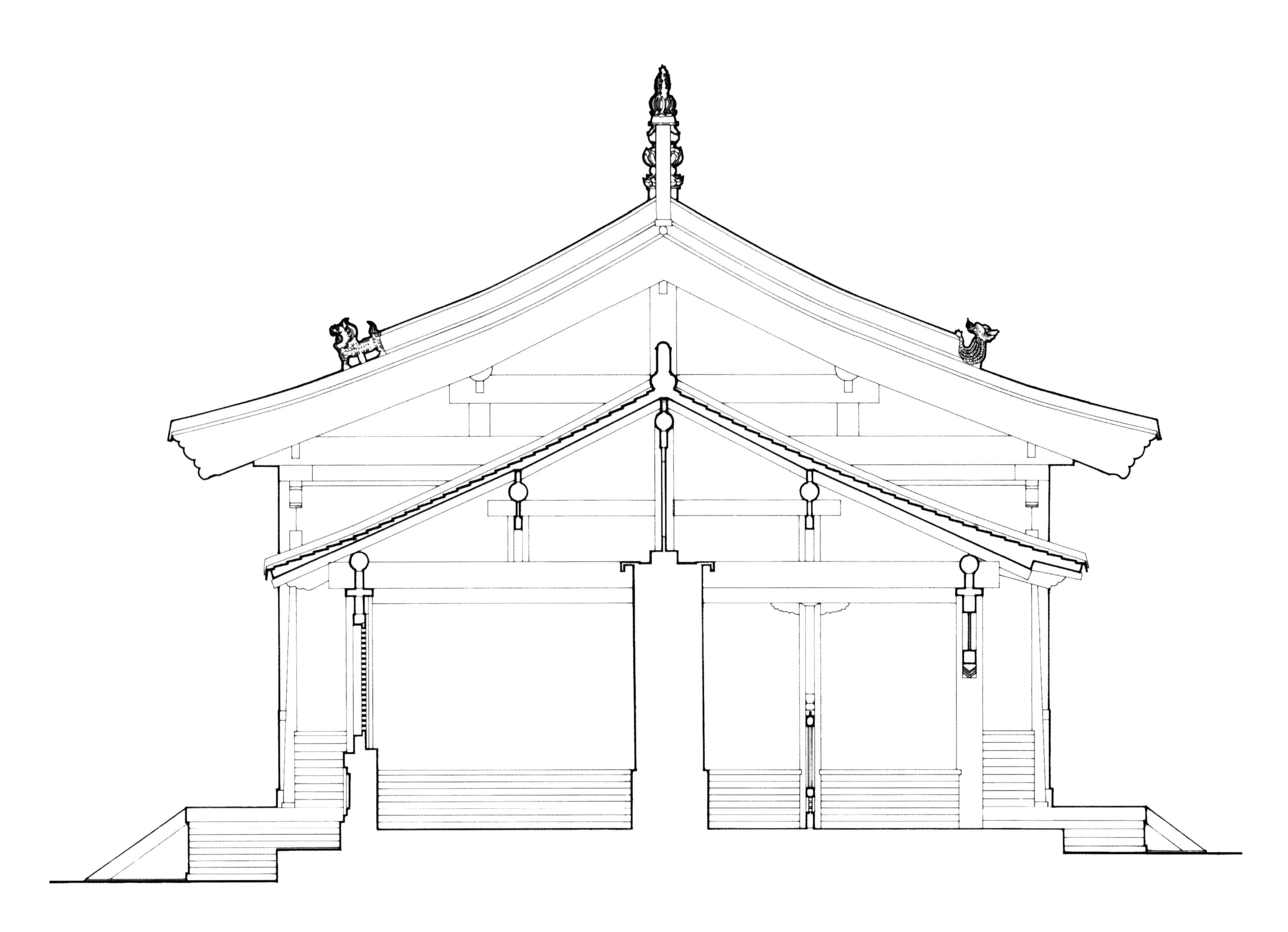

金刚殿东立面图
East Elevation of Jingang Hall

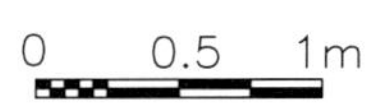

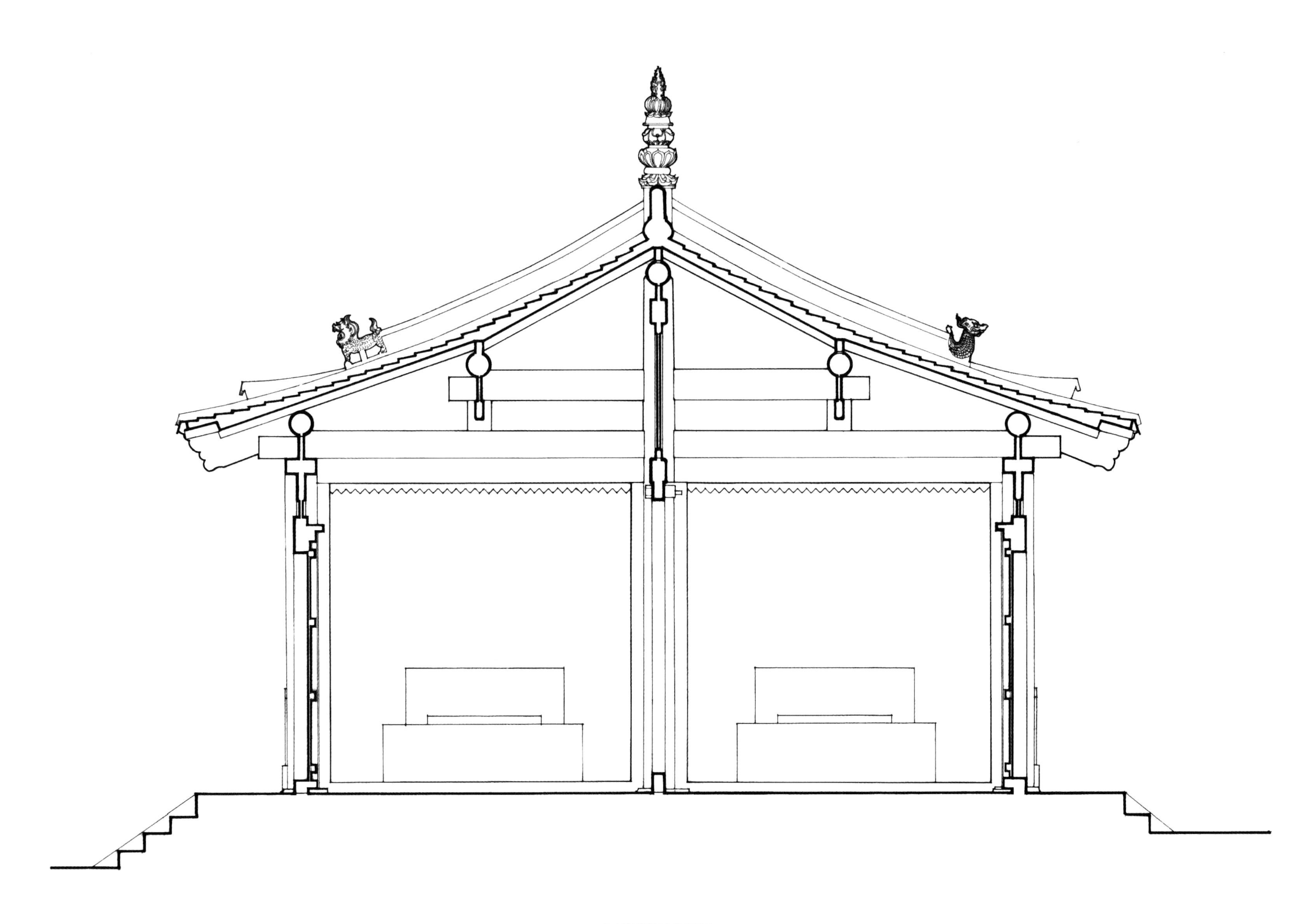

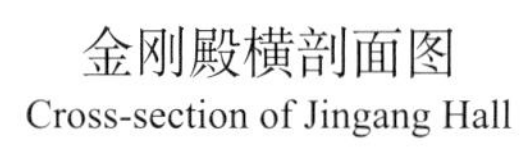
金刚殿横剖面图
Cross-section of Jingang Hall

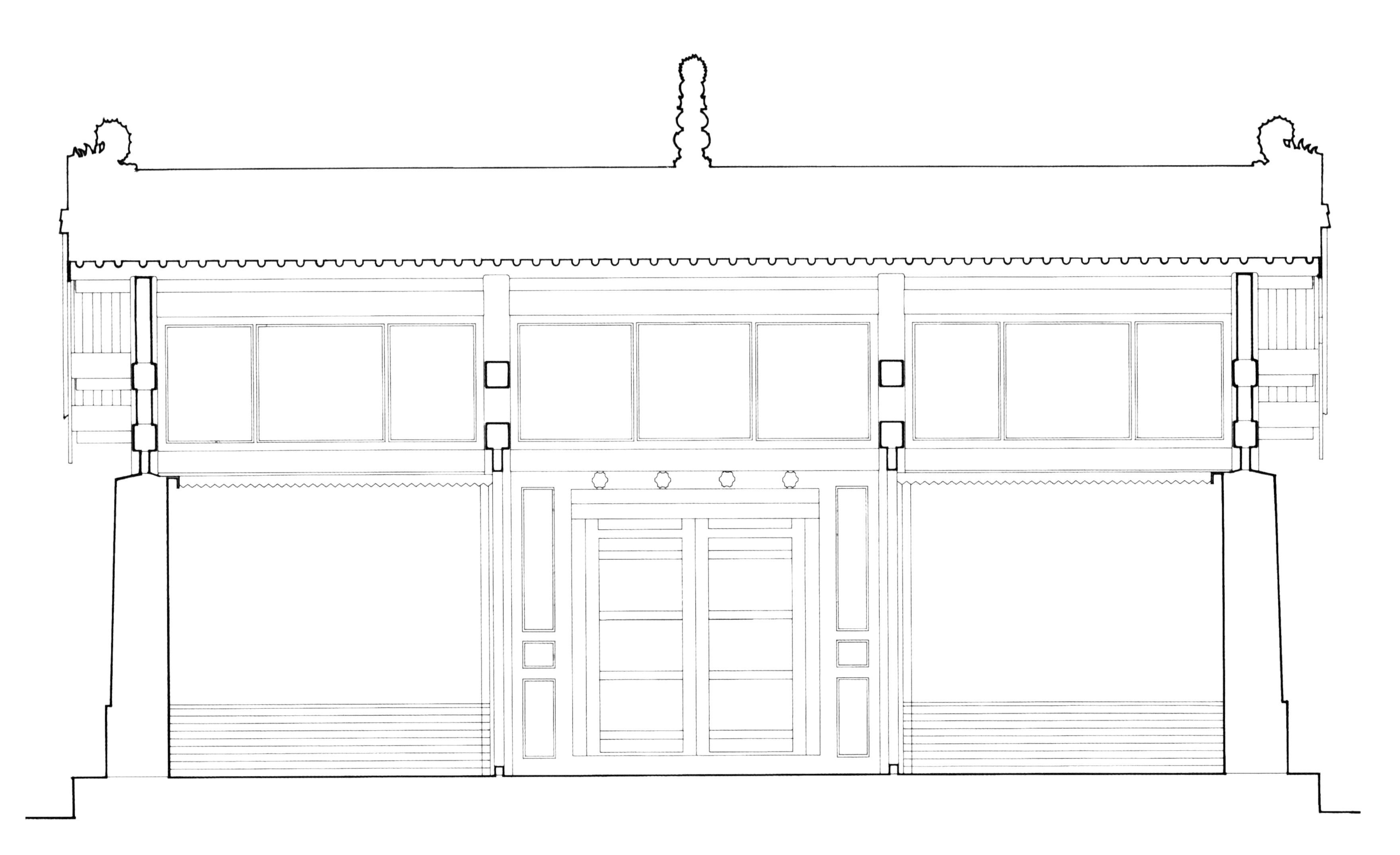

金刚殿纵剖面图

Longitudinal Section of Jingang Hall

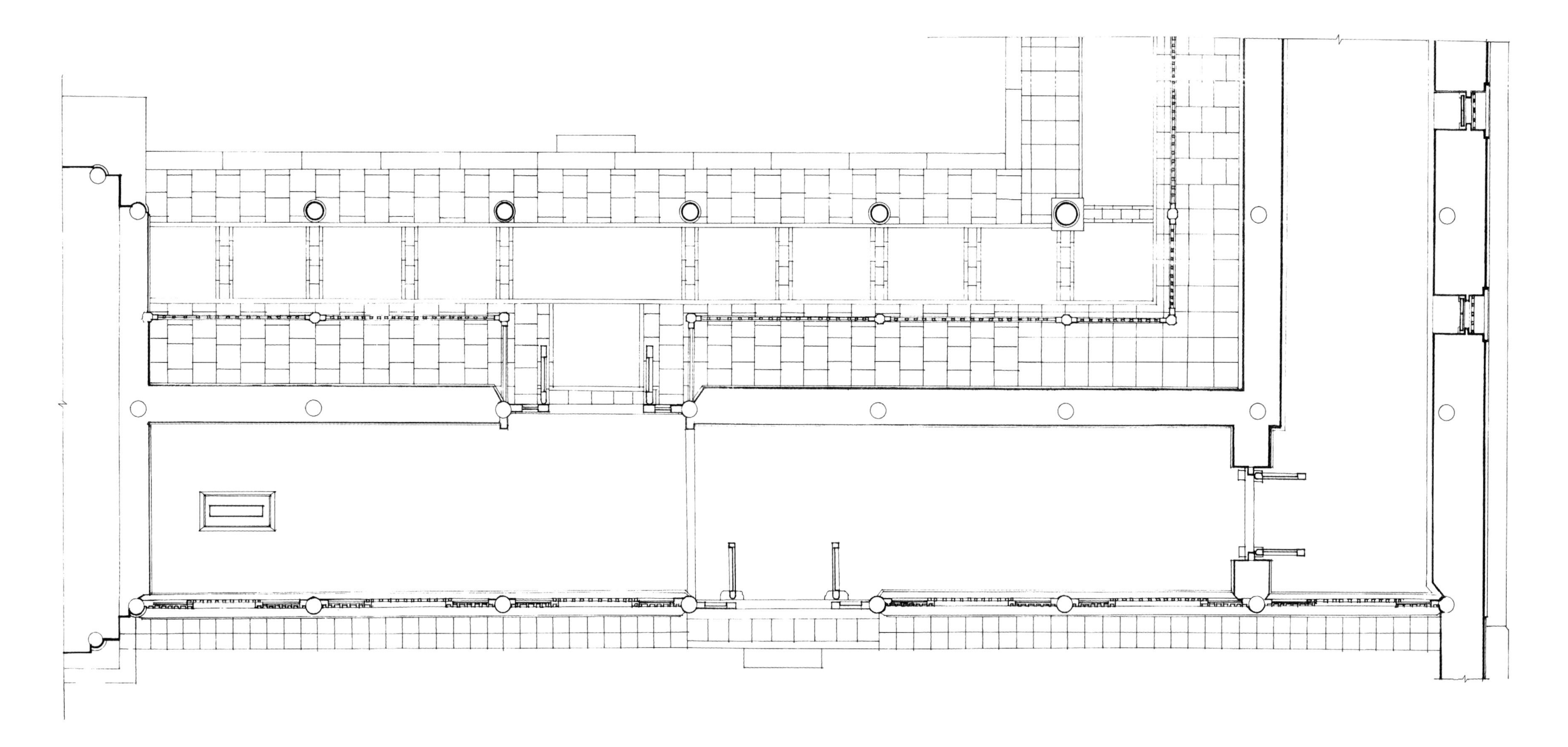

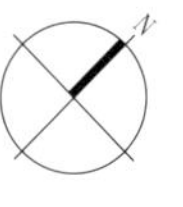

金刚殿东廊平面图
Plan of East Corridor of Jingang Hall

0 0.5 1 2m

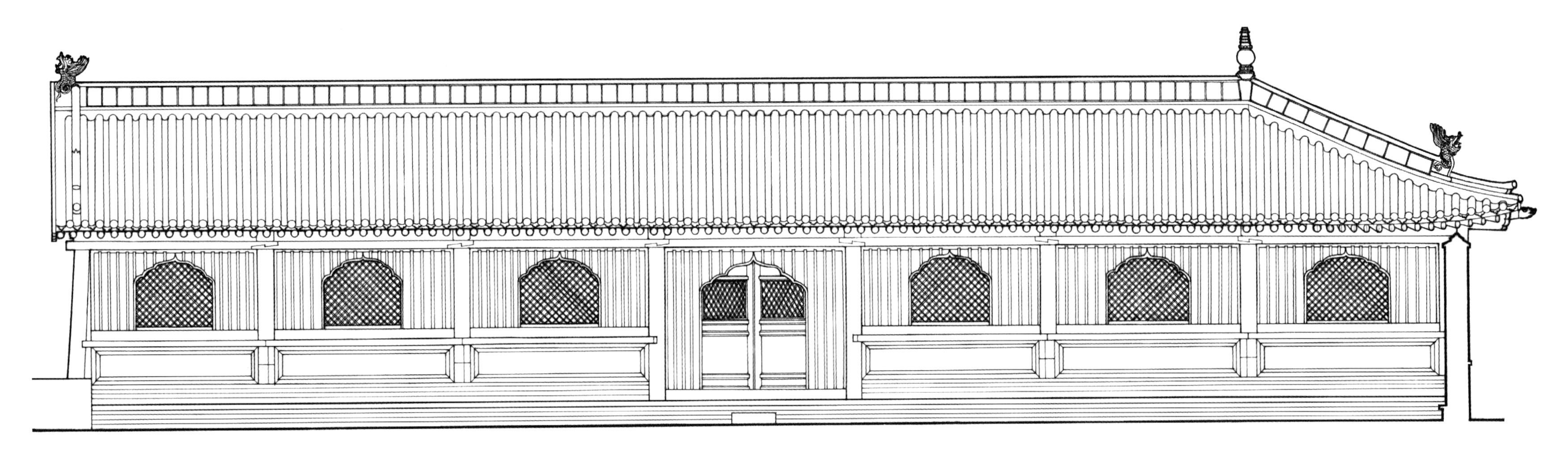

金刚殿东廊南立面图
South Elevation of East Corridor of Jingang Hall

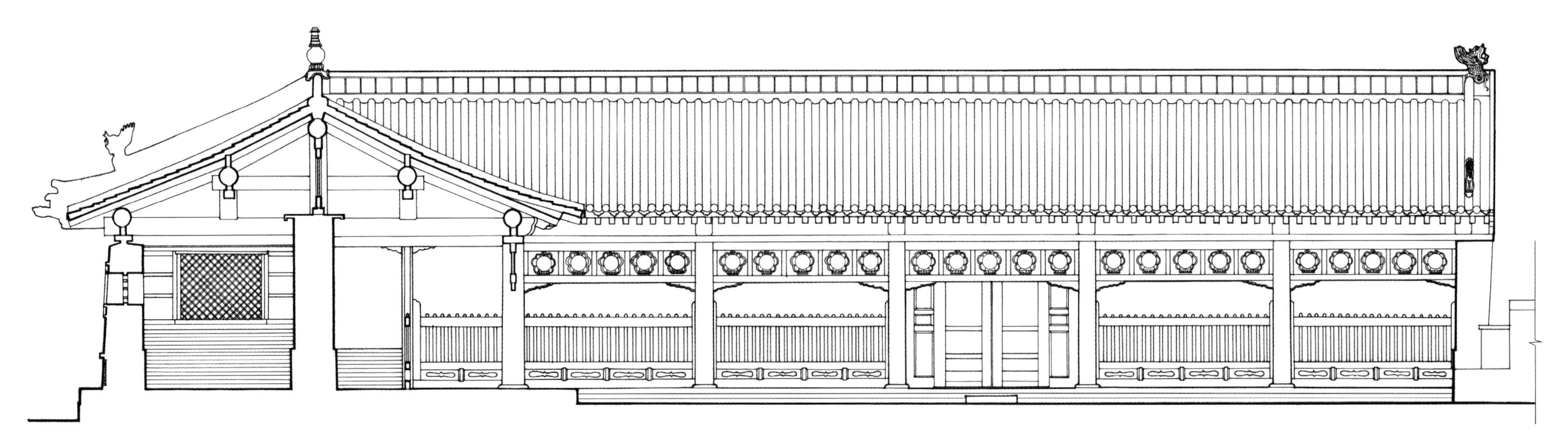

金刚殿东廊北立面图
North Elevation of East Corridor of Jingang Hall

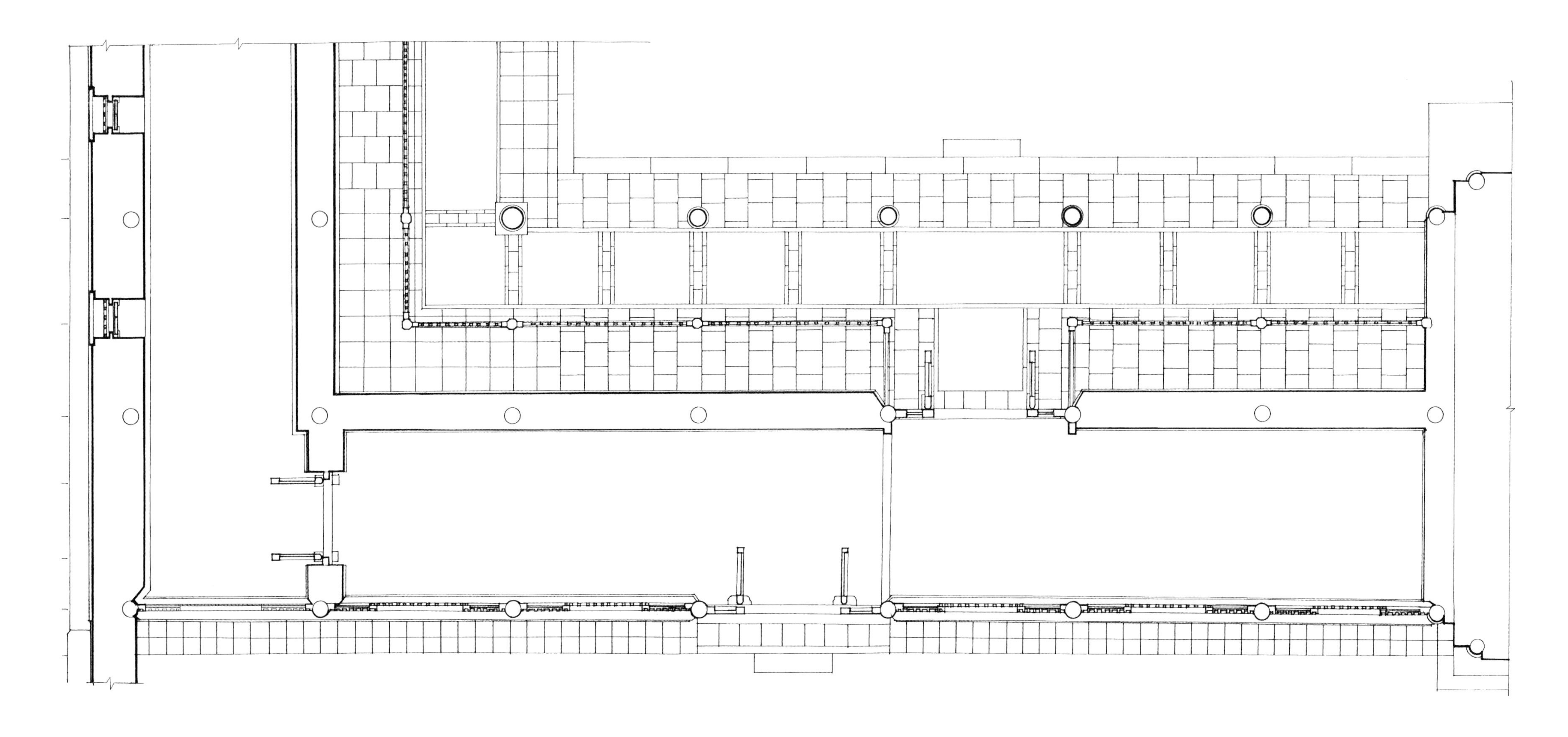

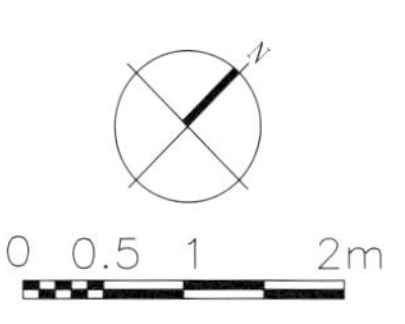

金刚殿西廊平面图
Plan of West Corridor of Jingang Hall

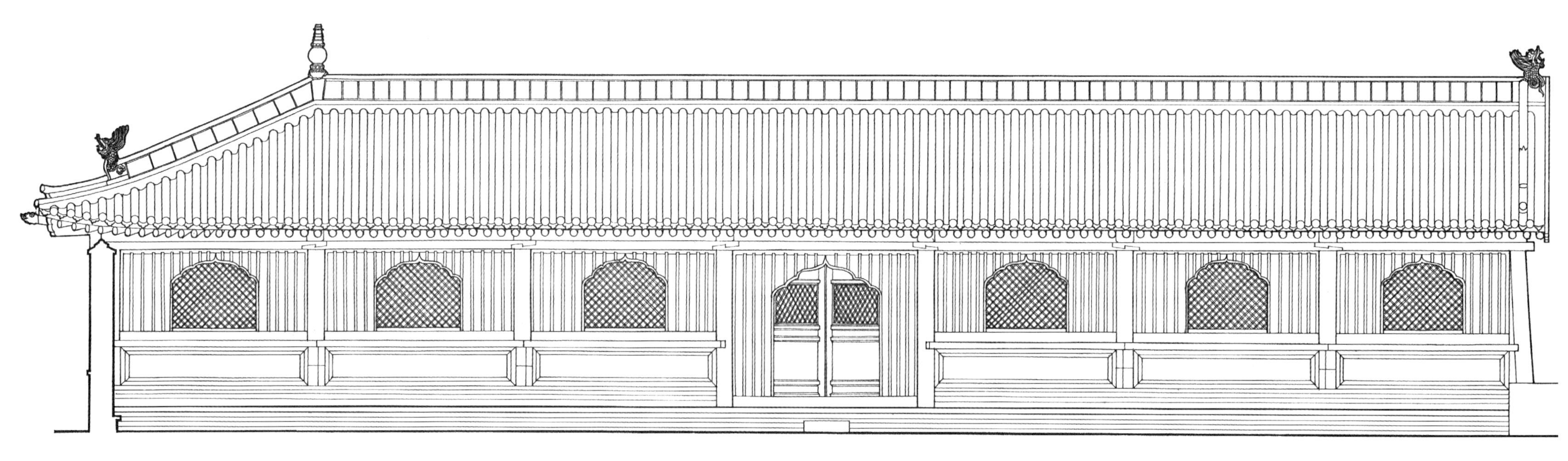

金刚殿西廊南立面图
South Elevation of West Corridor of Jingang Hall

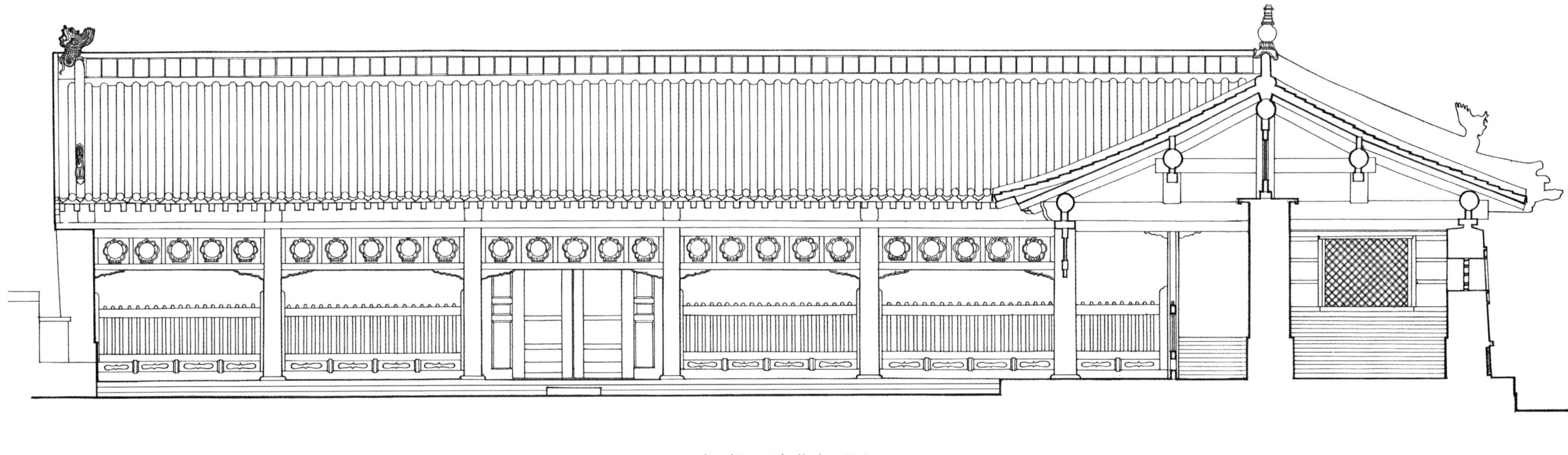

金刚殿西廊北立面图
North Elevation of West Corridor of Jingang Hall

香趣塔
Scent Tower

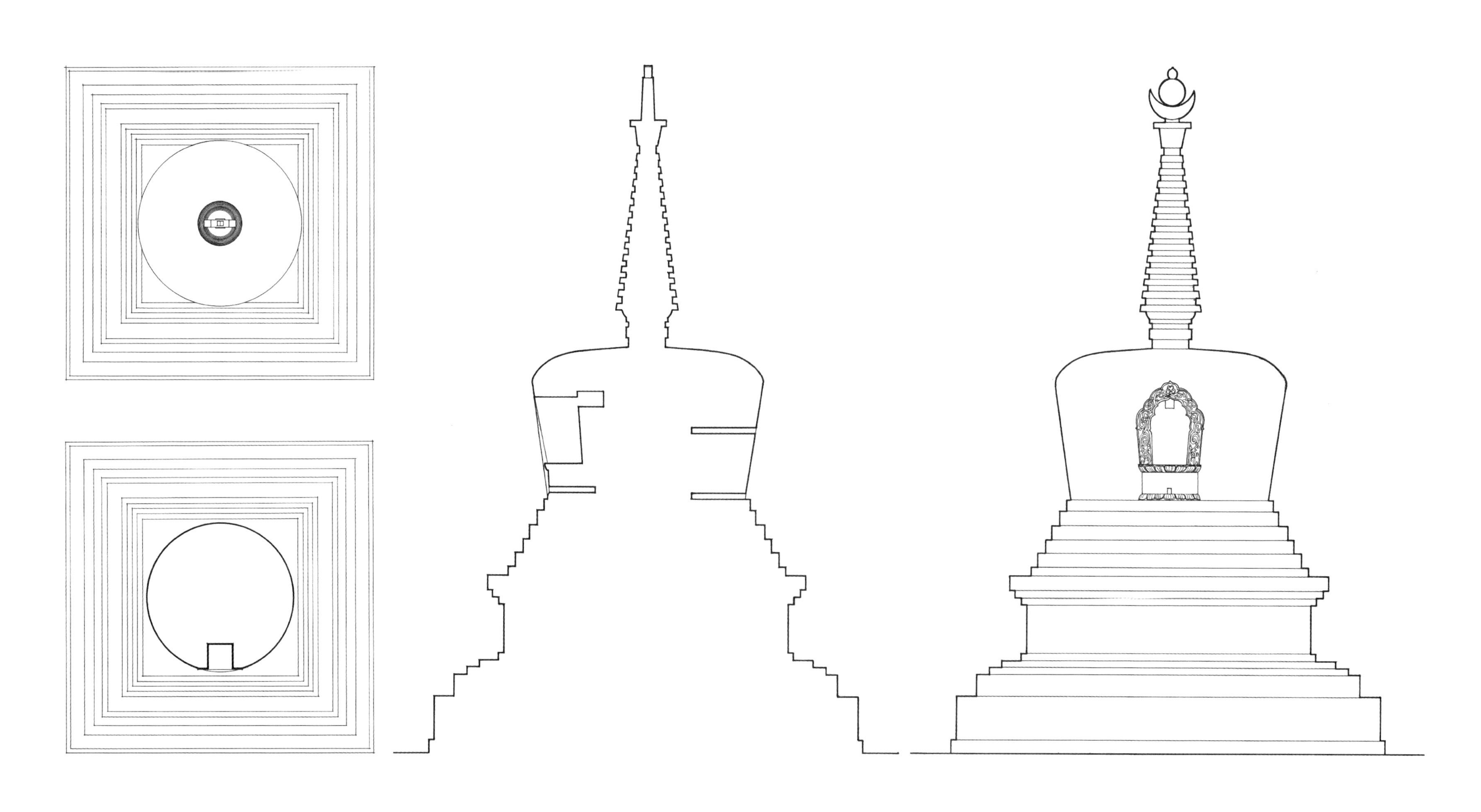

香趣塔平面图
Plan of Scent Tower

香趣塔剖面图
Section of Scent Tower

香趣塔立面图
Elevation of Scent Tower

瞿昙殿
Qutan Hall

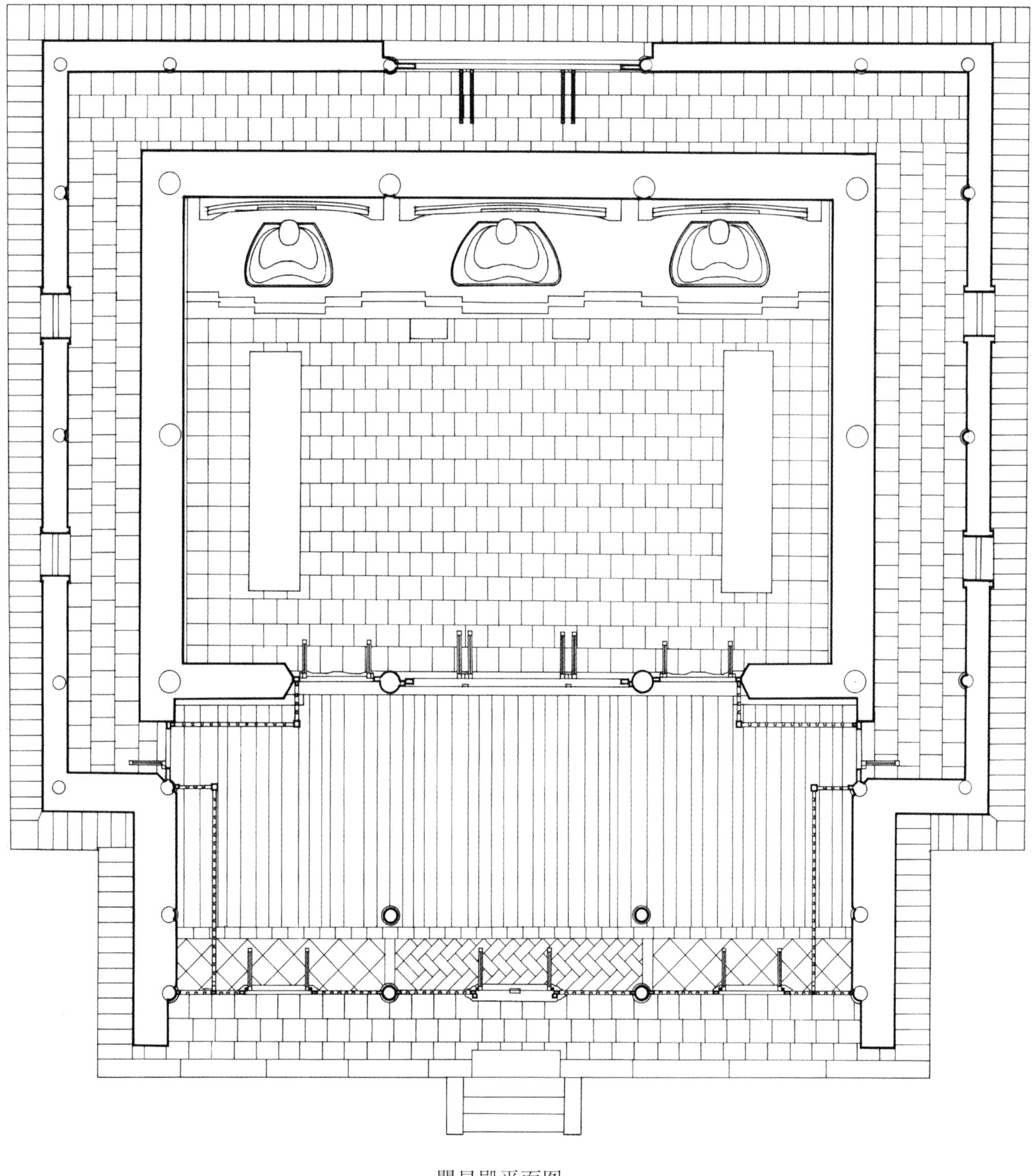

瞿昙殿平面图
Plan of Qutan Hall

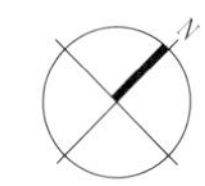

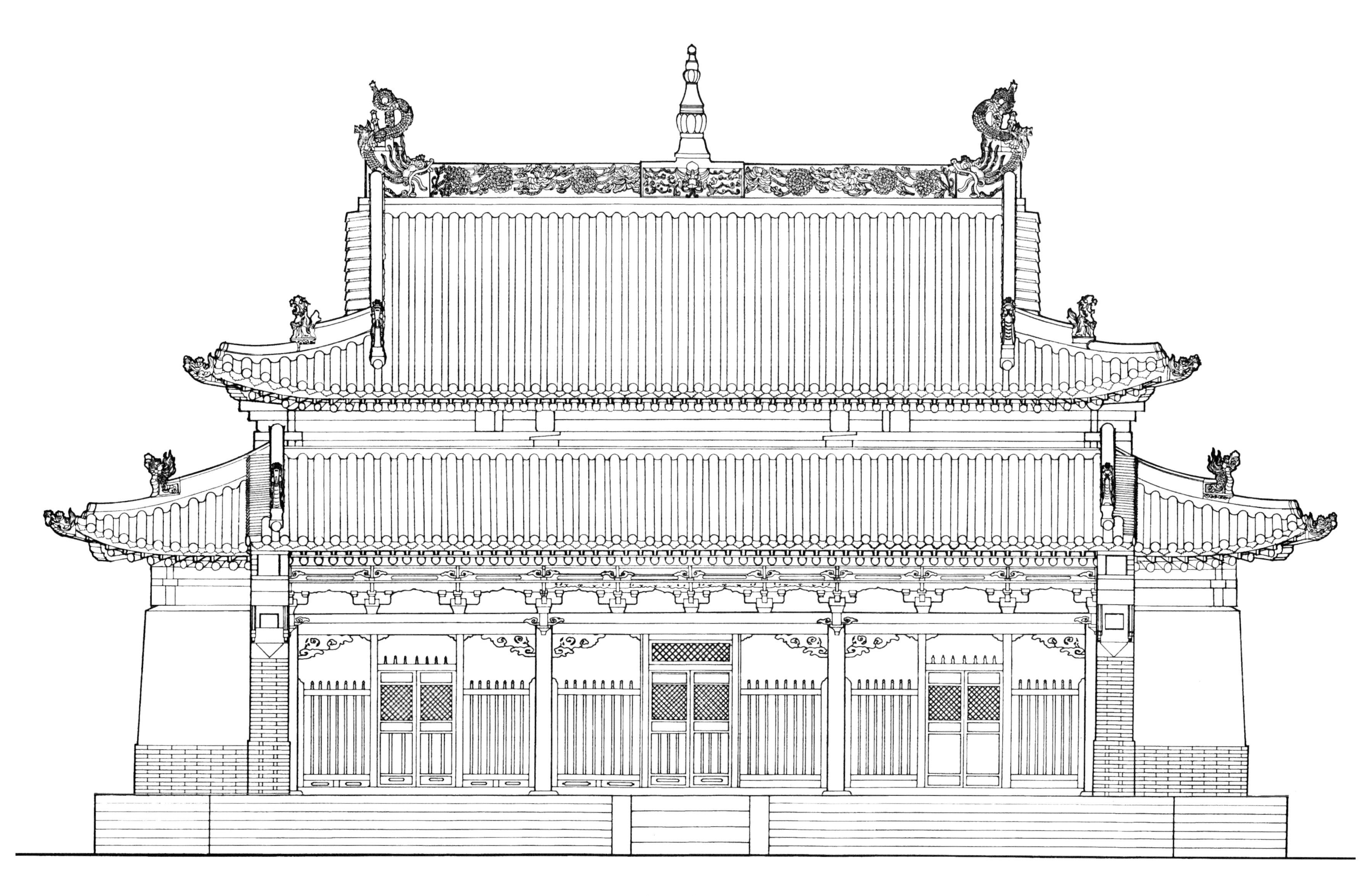

瞿昙殿南立面图
South Elevation of Qutan Hall

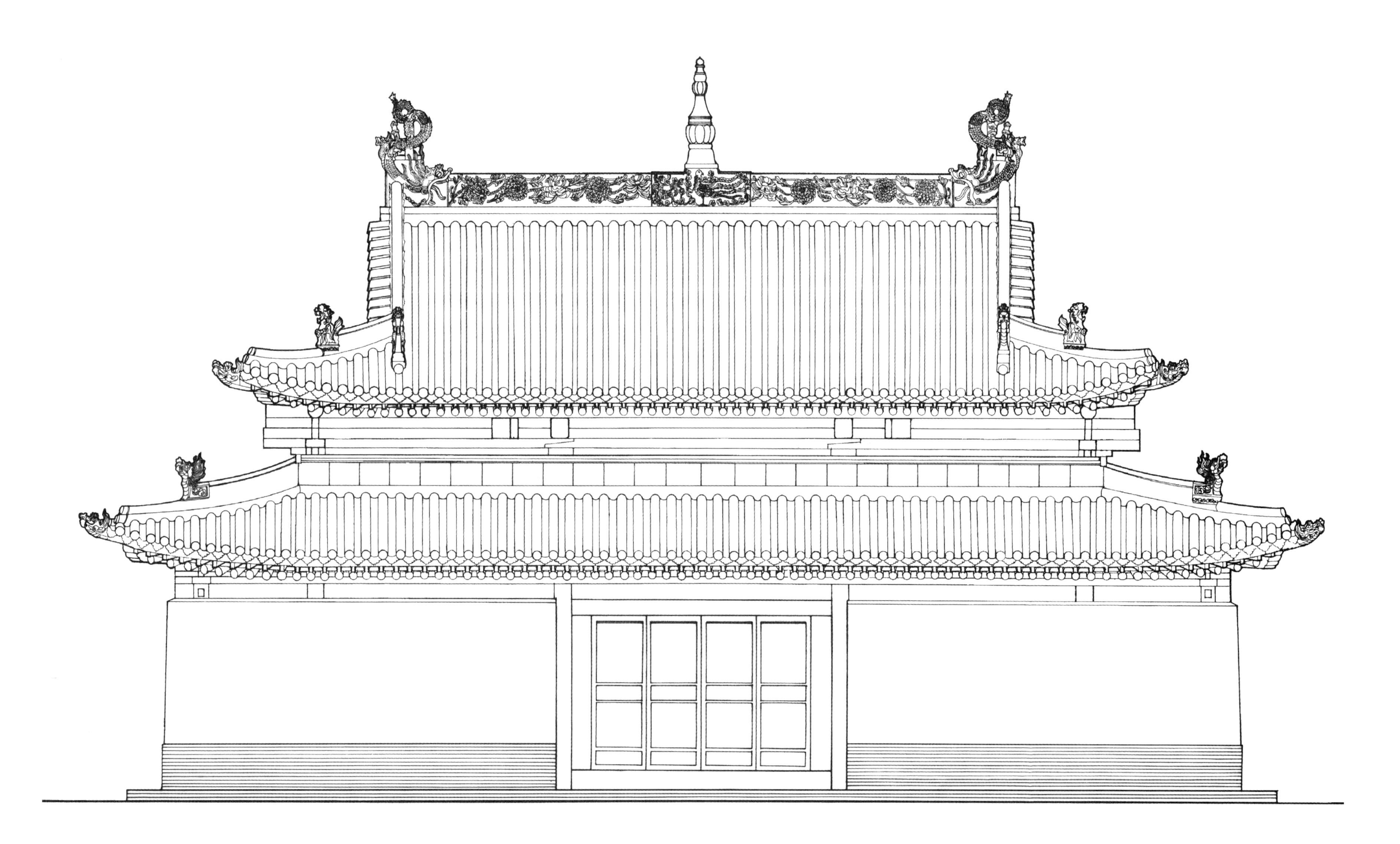

瞿昙殿北立面图
North Elevation of Qutan Hall

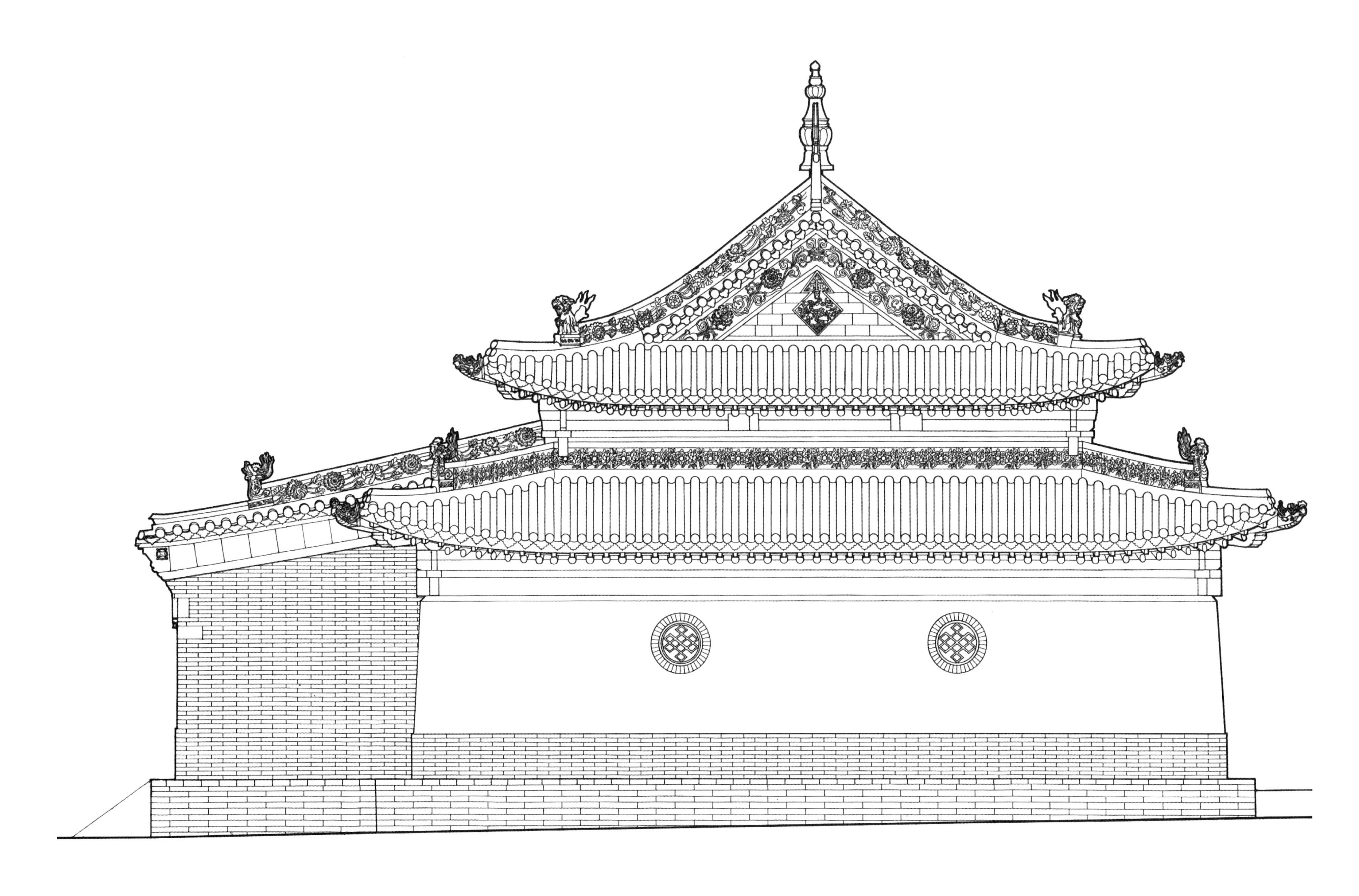

瞿昙殿东立面图
East Elevation of Qutan Hall

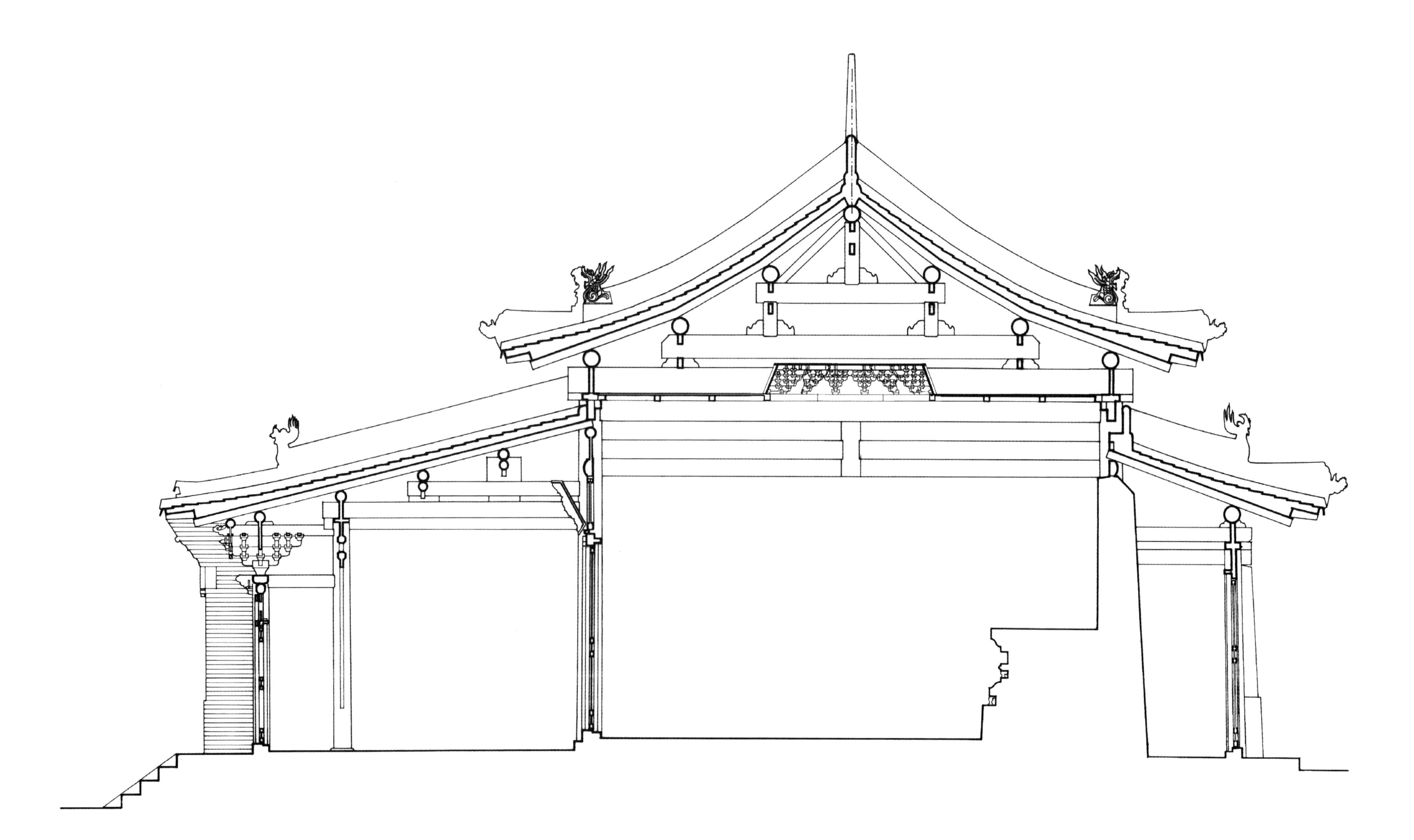

瞿昙殿明间剖面图
Mingjian Cross-section of Qutan Hall

0 0.5 1 2m

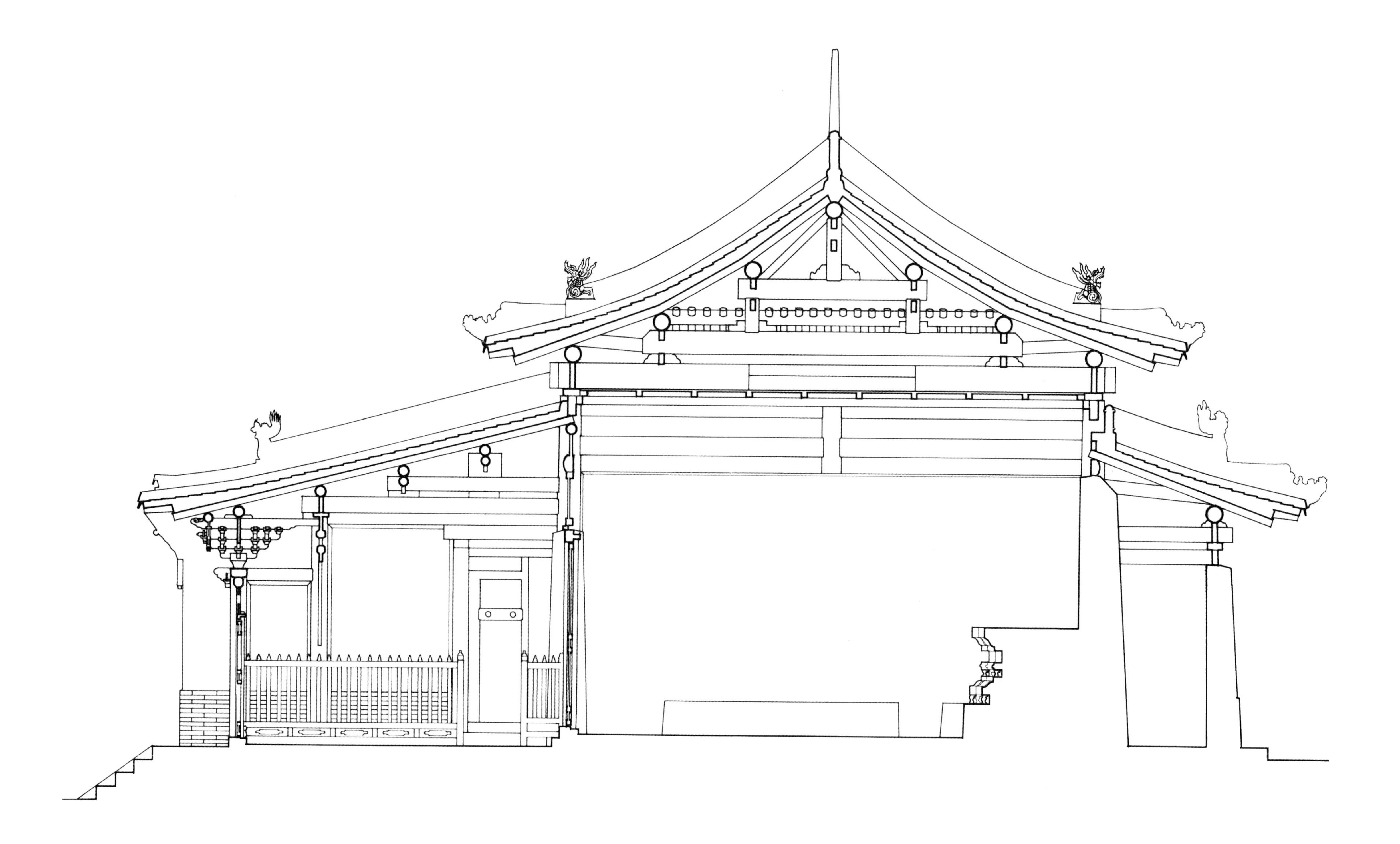

瞿昙殿次间剖面图
Cijian Cross-section of Qutan Hall

瞿昙殿纵剖面图
Longitudinal Section of Qutan Hall

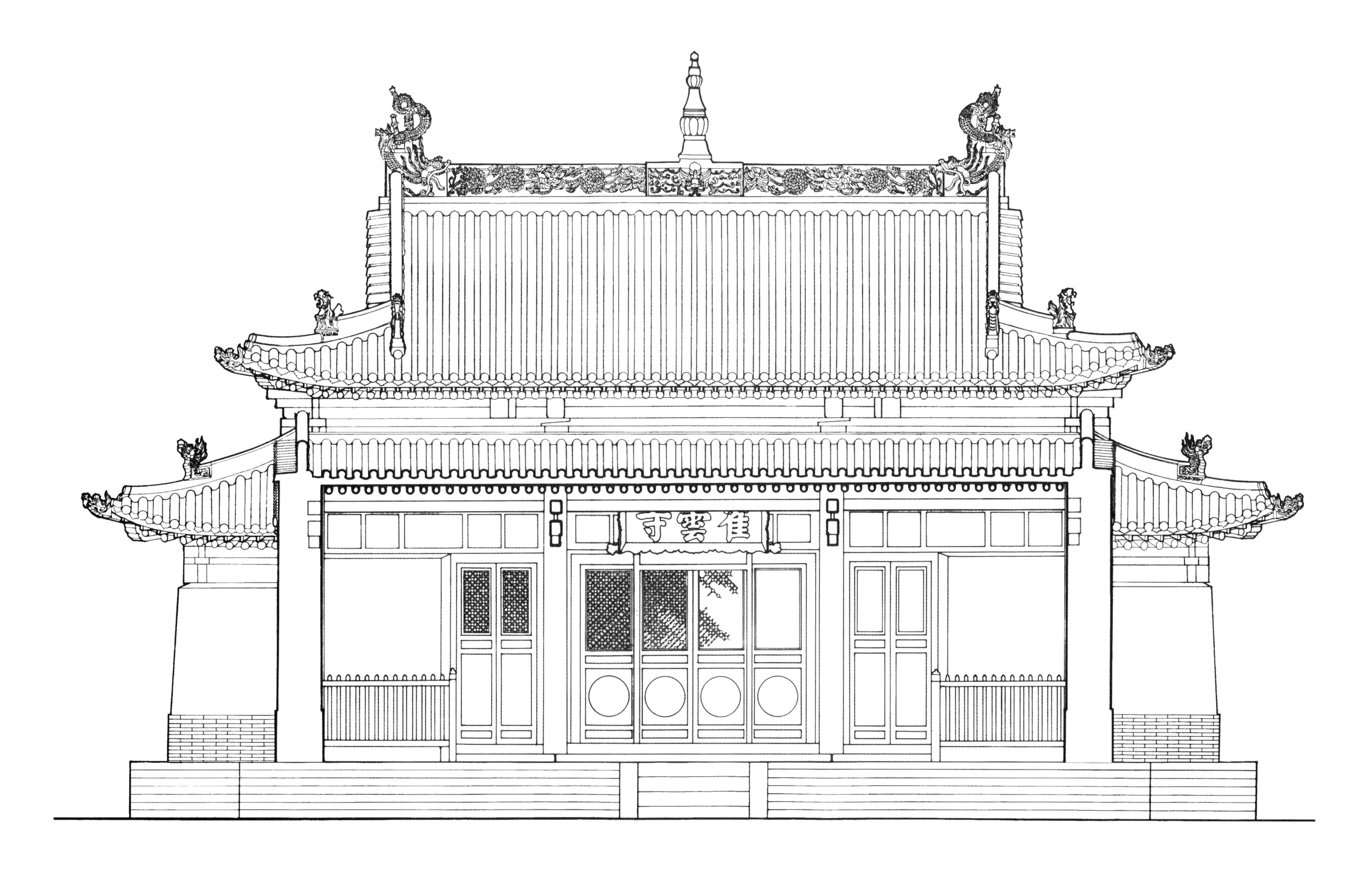

瞿昙殿抱厦纵剖面图
Longitudinal Section of Baosha of Qutan Hall

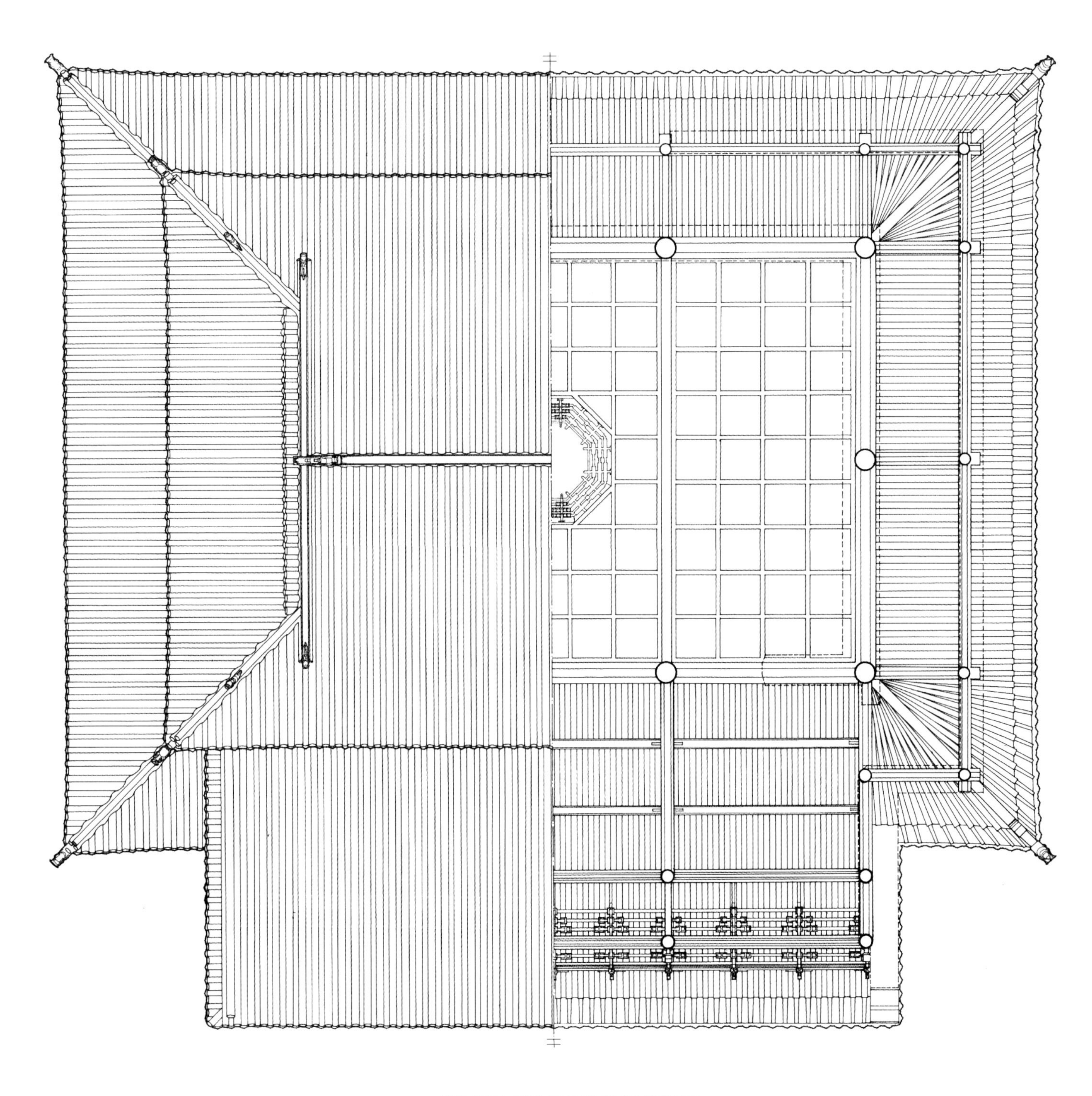

瞿昙殿屋顶平面及梁架仰视图
Roof Plan and Framework Plan of Qutan Hall

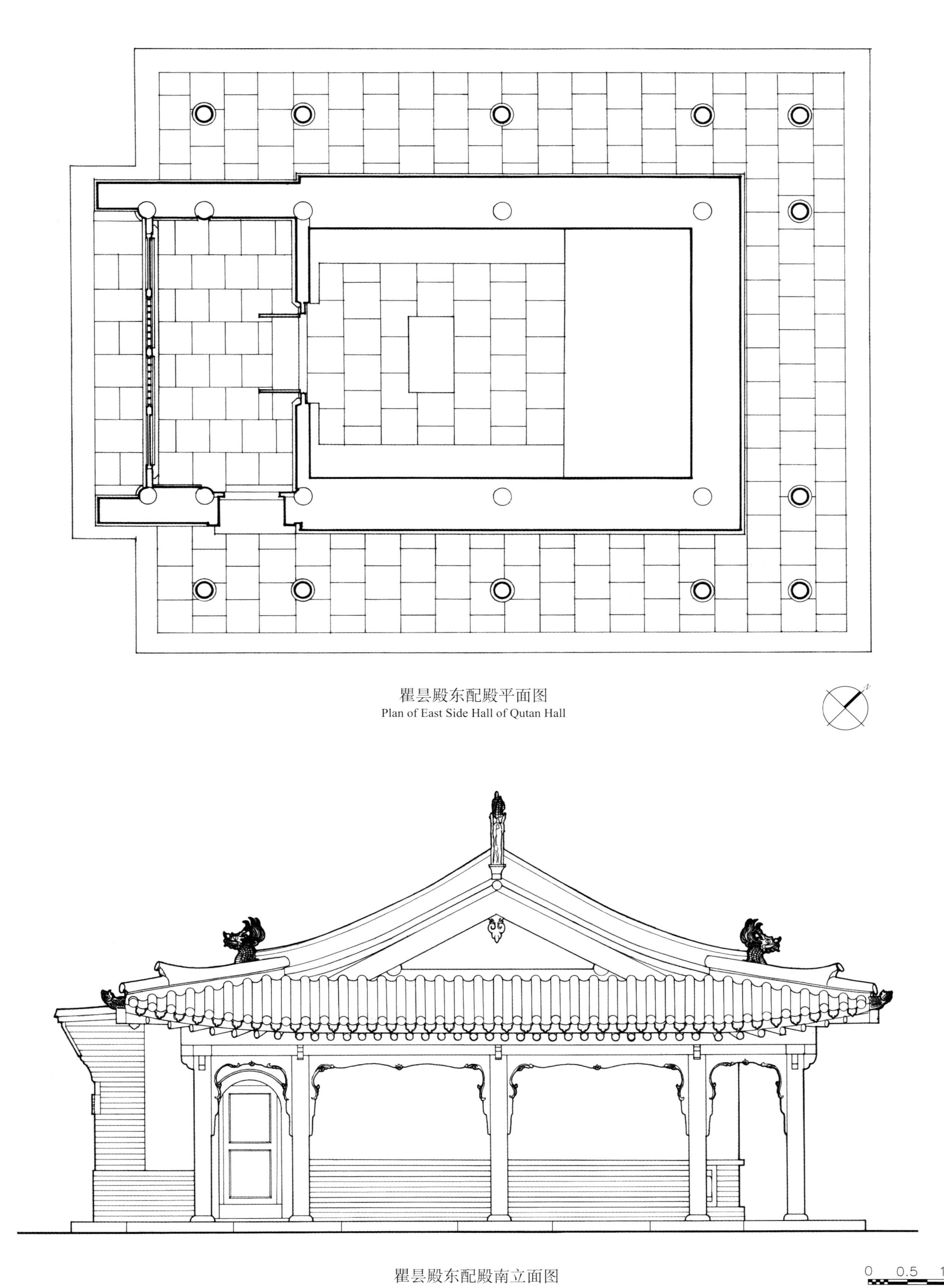

瞿昙殿东配殿平面图
Plan of East Side Hall of Qutan Hall

瞿昙殿东配殿南立面图
South Elevation of East Side Hall of Qutan Hall

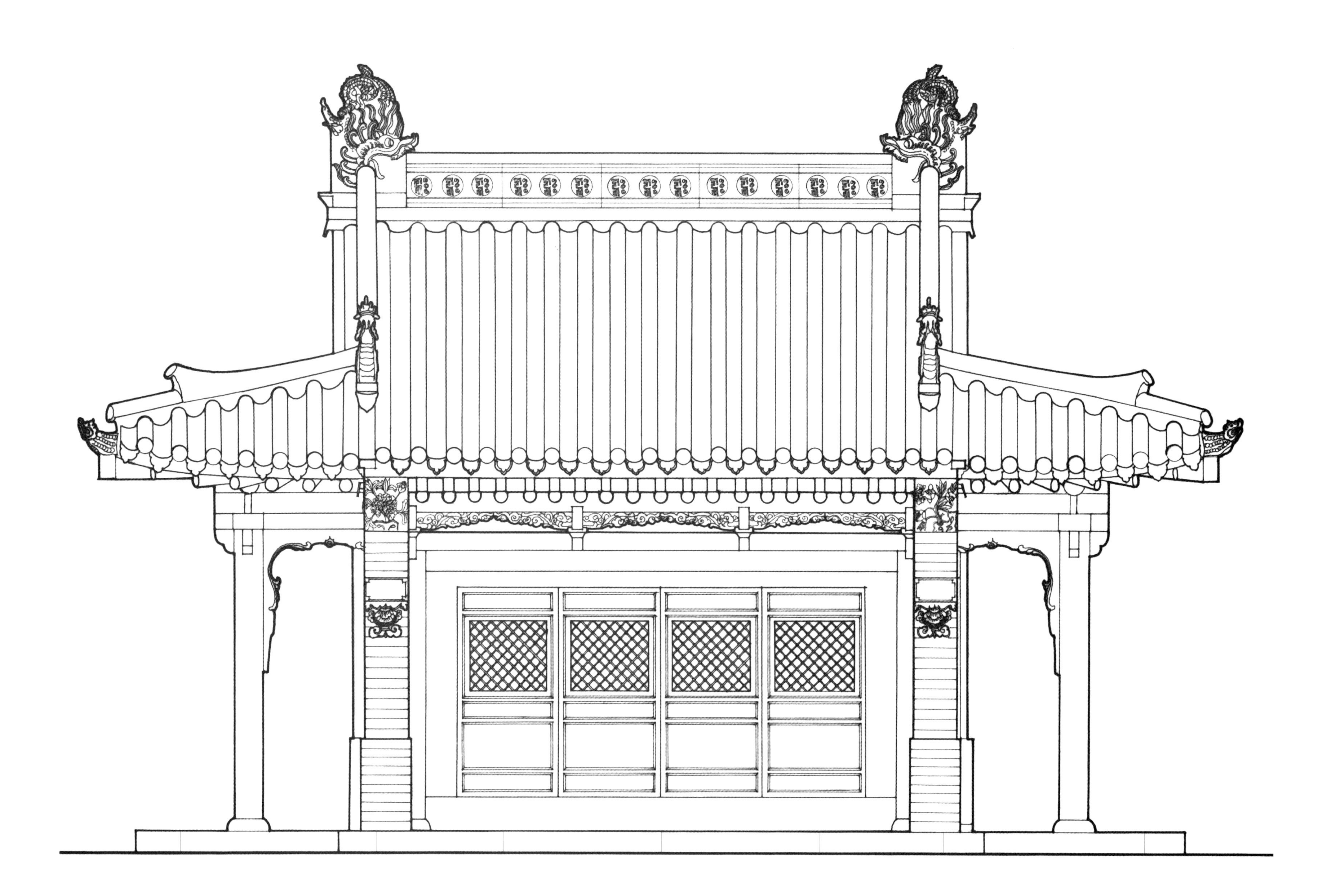

瞿昙殿东配殿西立面图
West Elevation of East Side Hall of Qutan Hall

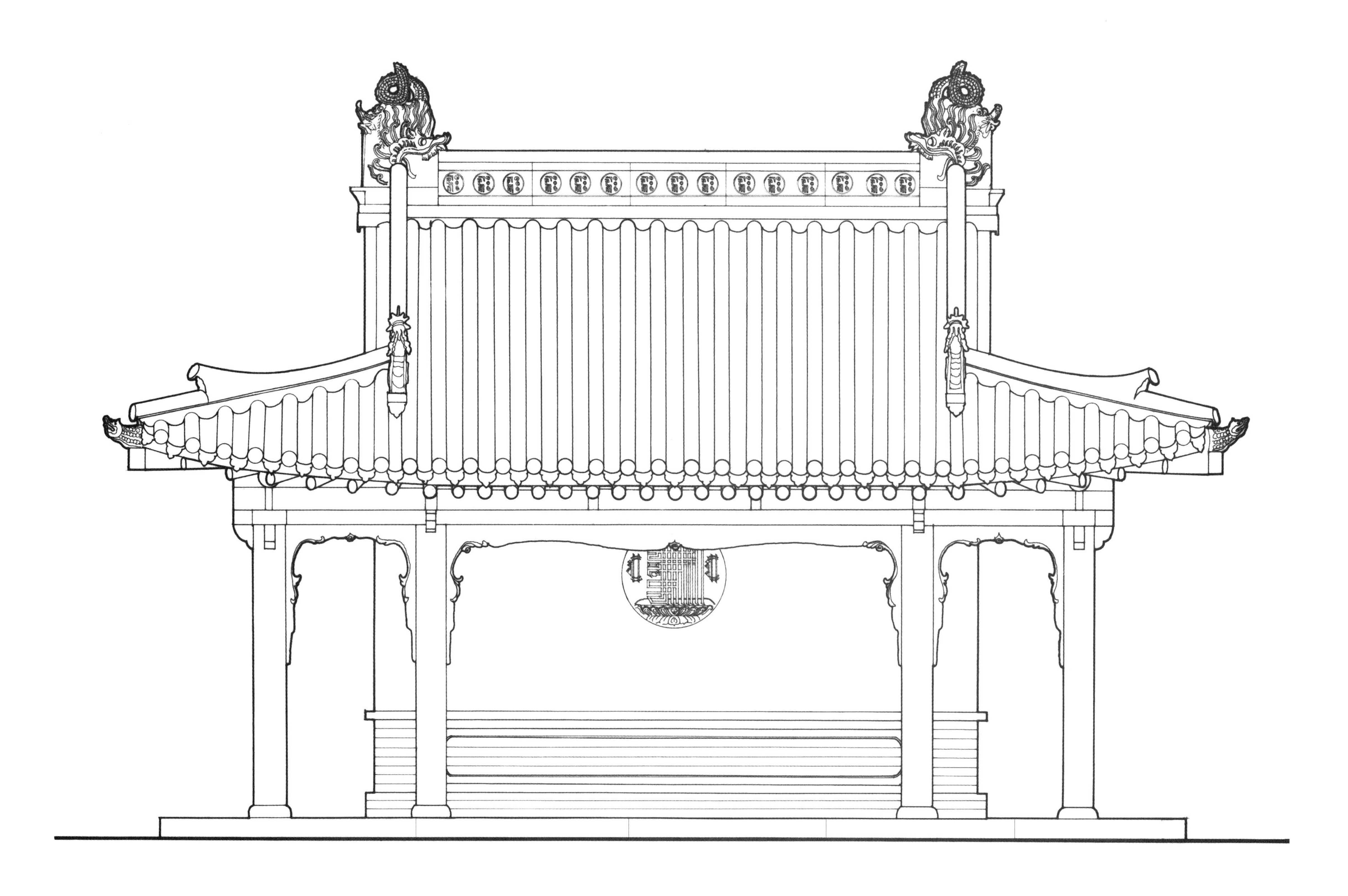

瞿昙殿东配殿东立面图
East Elevation of East Side Hall of Qutan Hall

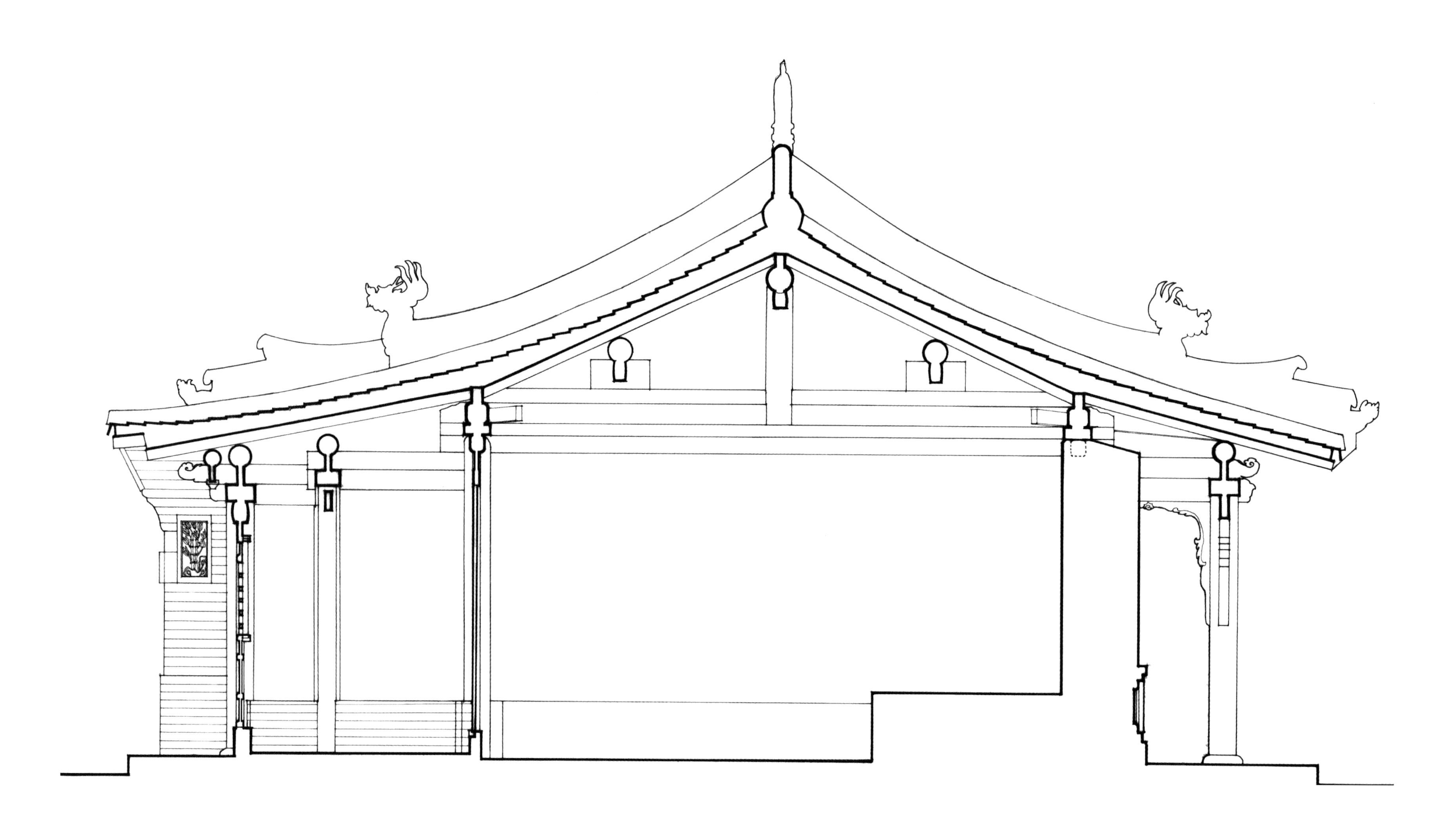

瞿昙殿东配殿横剖面图
Cross-section of East Side Hall of Qutan Hall

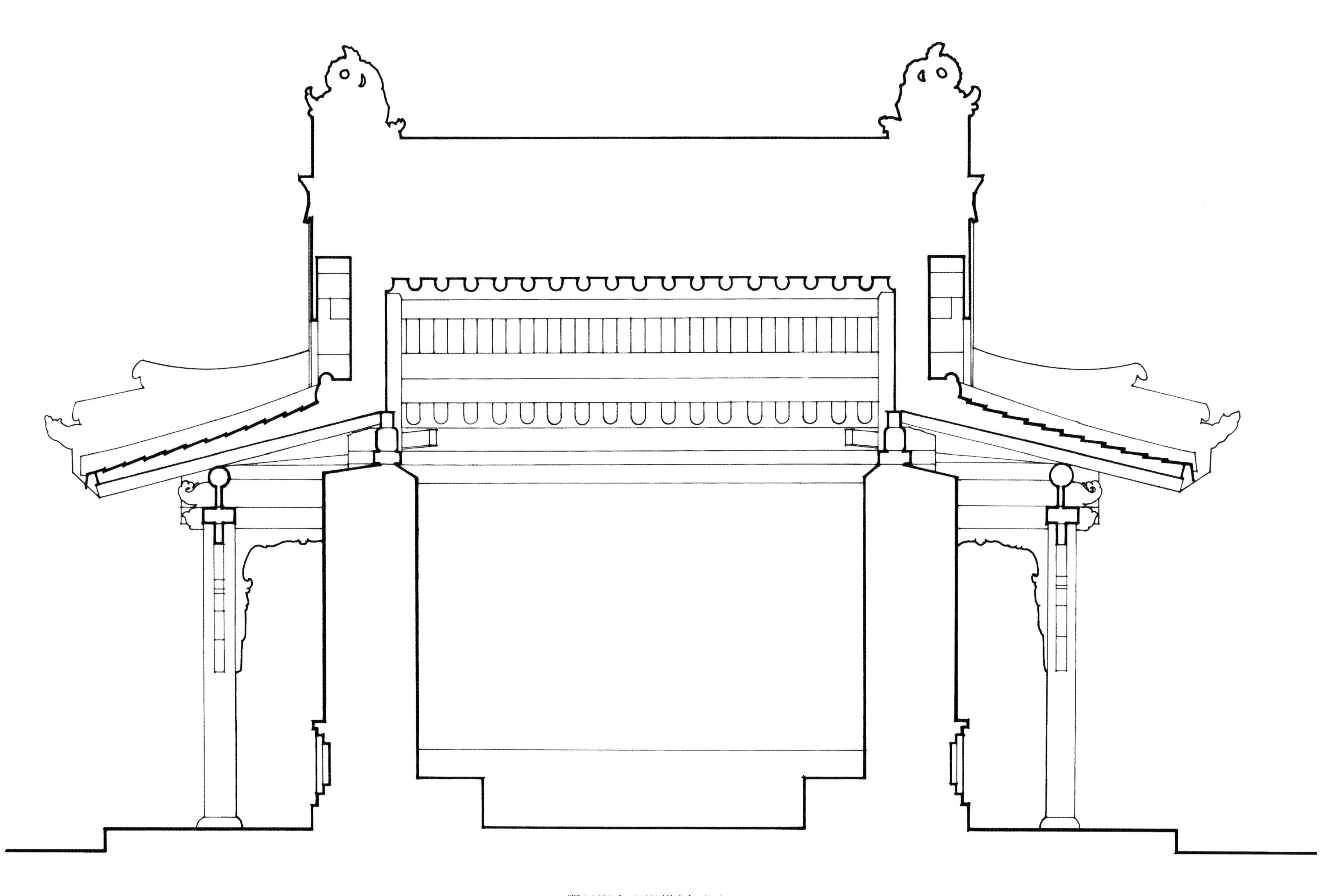

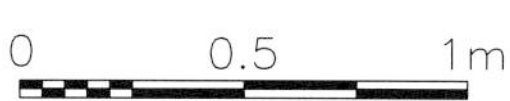

瞿昙殿东配殿纵剖面图
Longitudinal Section of East Side Hall of Qutan Hall

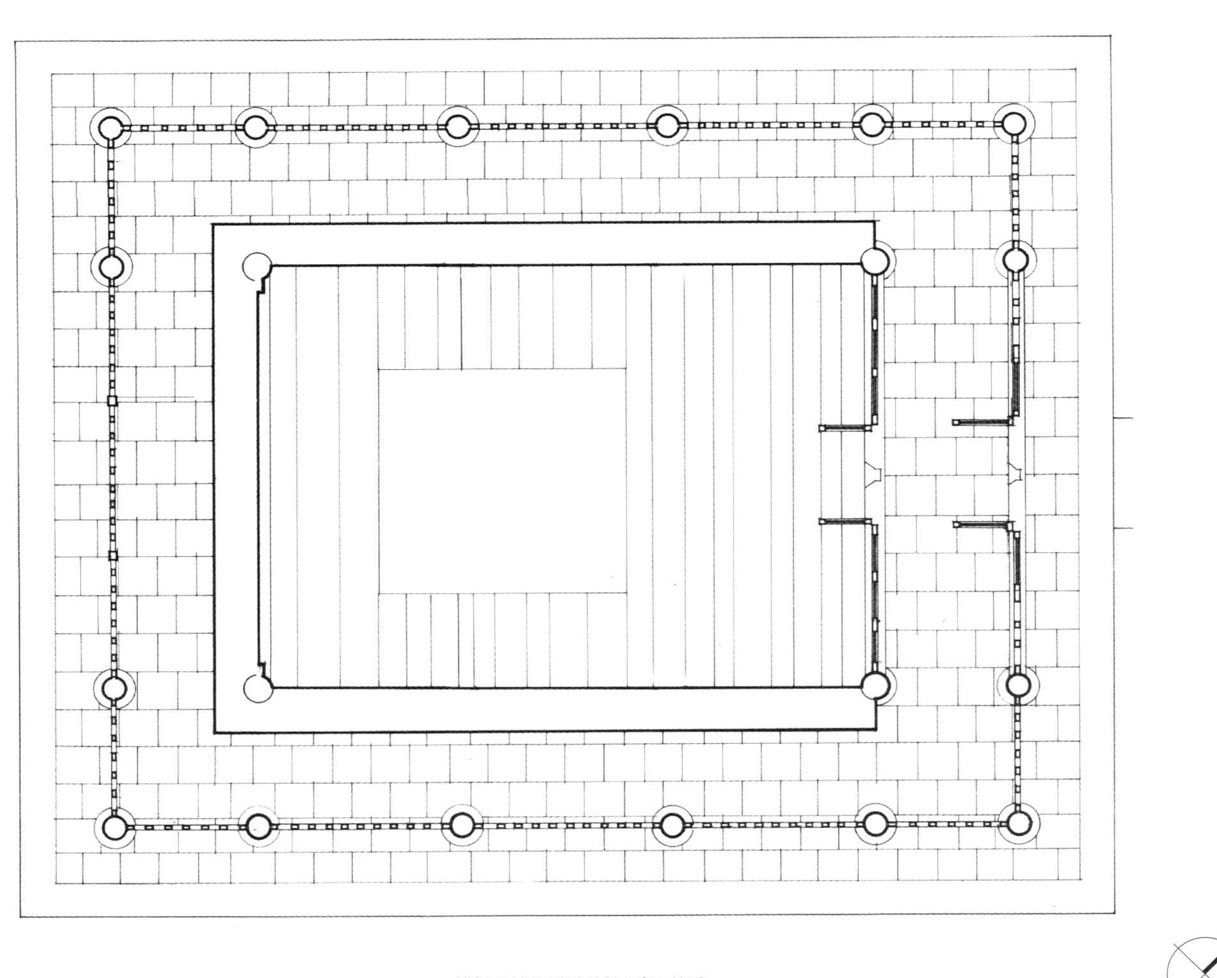

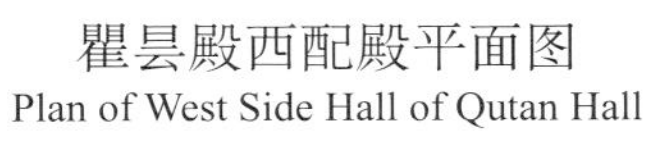
瞿昙殿西配殿平面图
Plan of West Side Hall of Qutan Hall

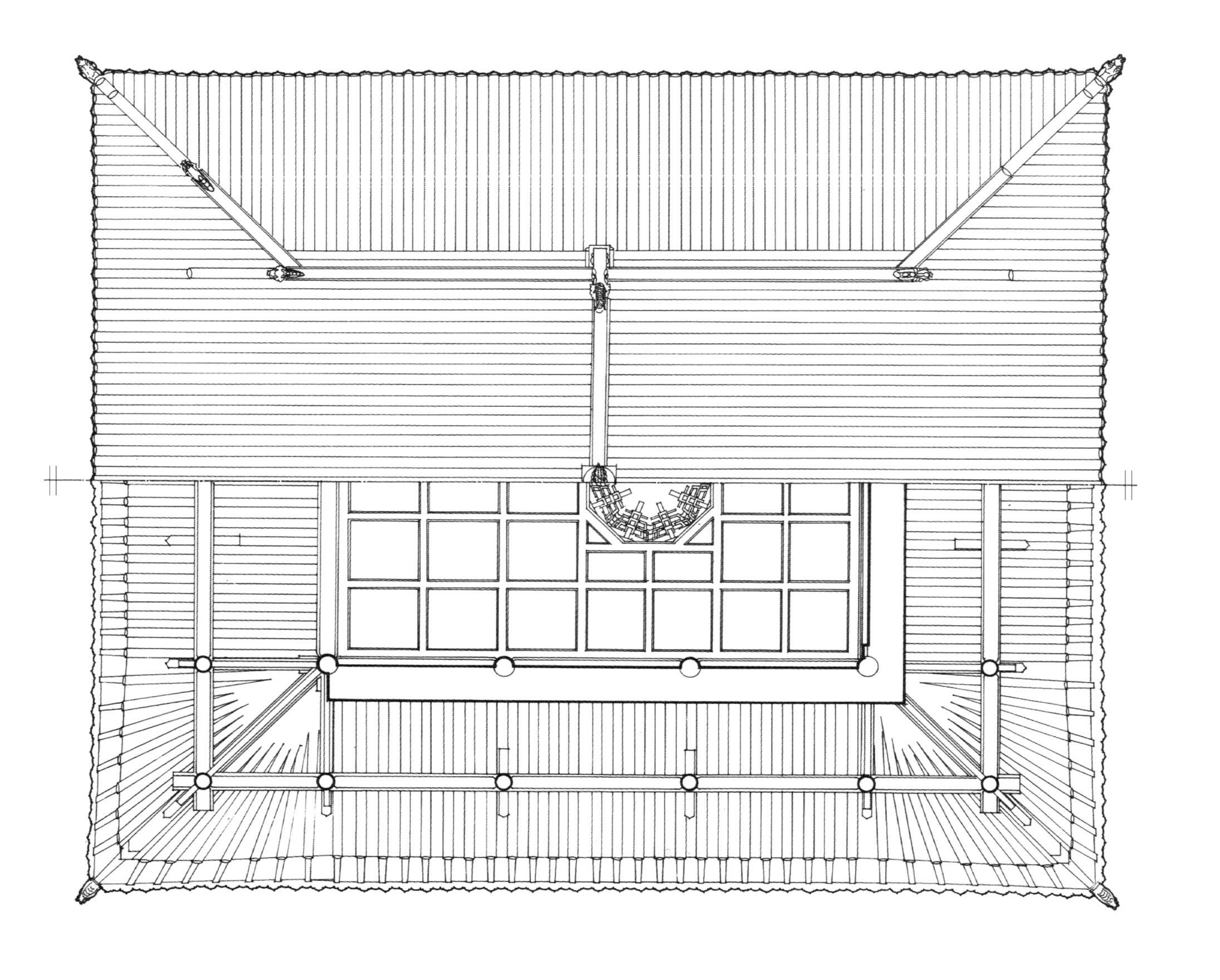

瞿昙殿西配殿屋顶平面及梁架仰视图
Roof Plan and Framework Plan of West Side Hall of Qutan Hall

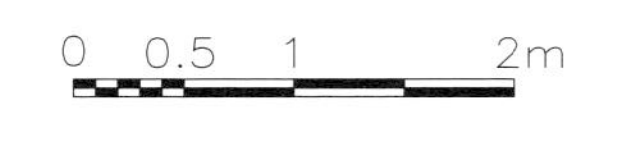

瞿昙殿西配殿东立面图
East Elevation of West Side Hall of Qutan Hall

瞿昙殿西配殿西立面图
West Elevation of West Side Hall of Qutan Hall

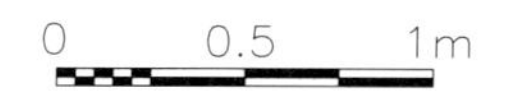

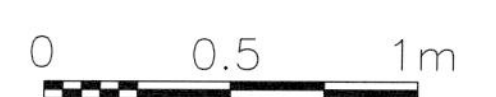

瞿昙殿西配殿南立面图
South Elevation of West Side Hall of Qutan Hall

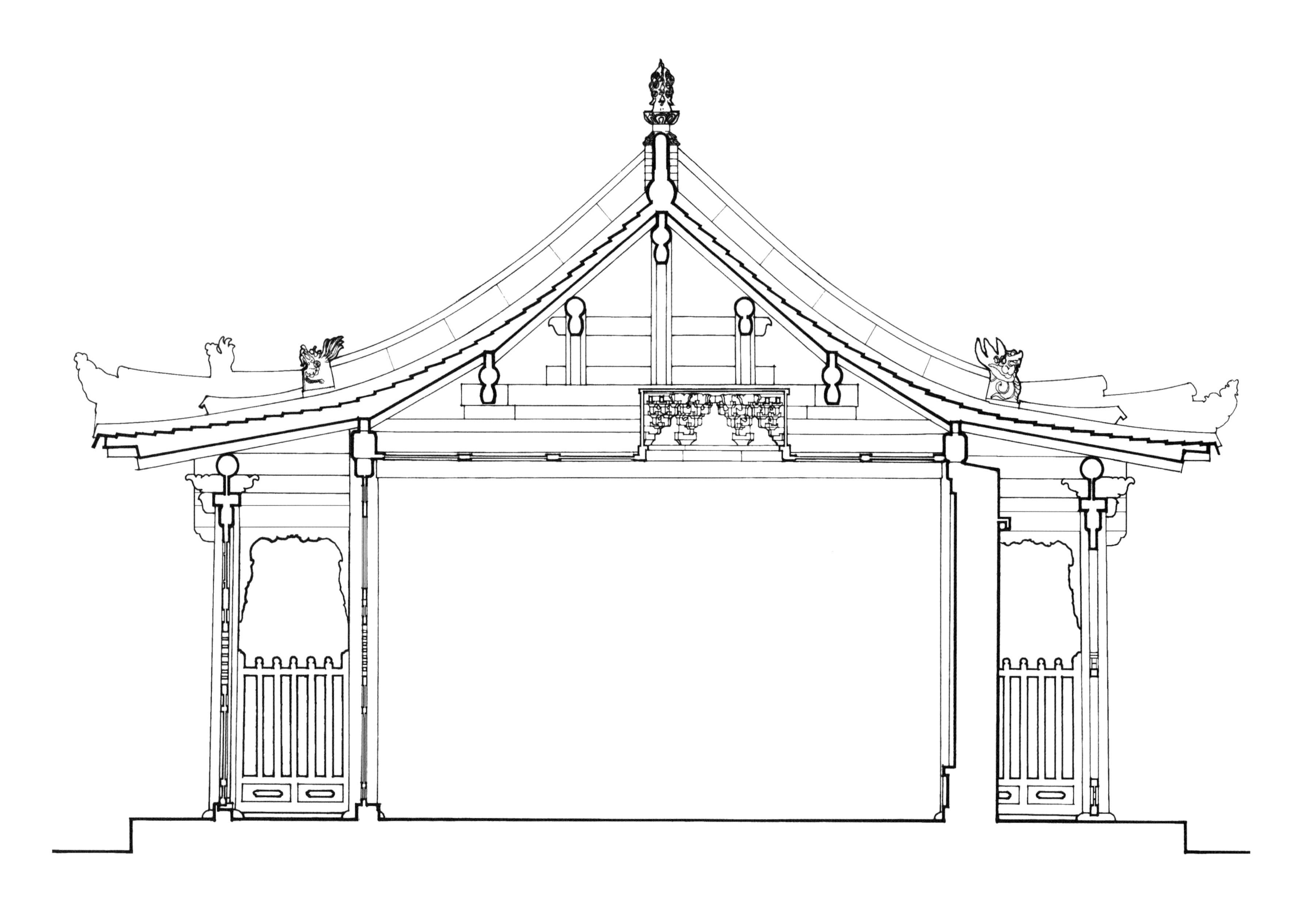

瞿昙殿西配殿横剖面图

Cross-section of West Side Hall of Qutan Hall

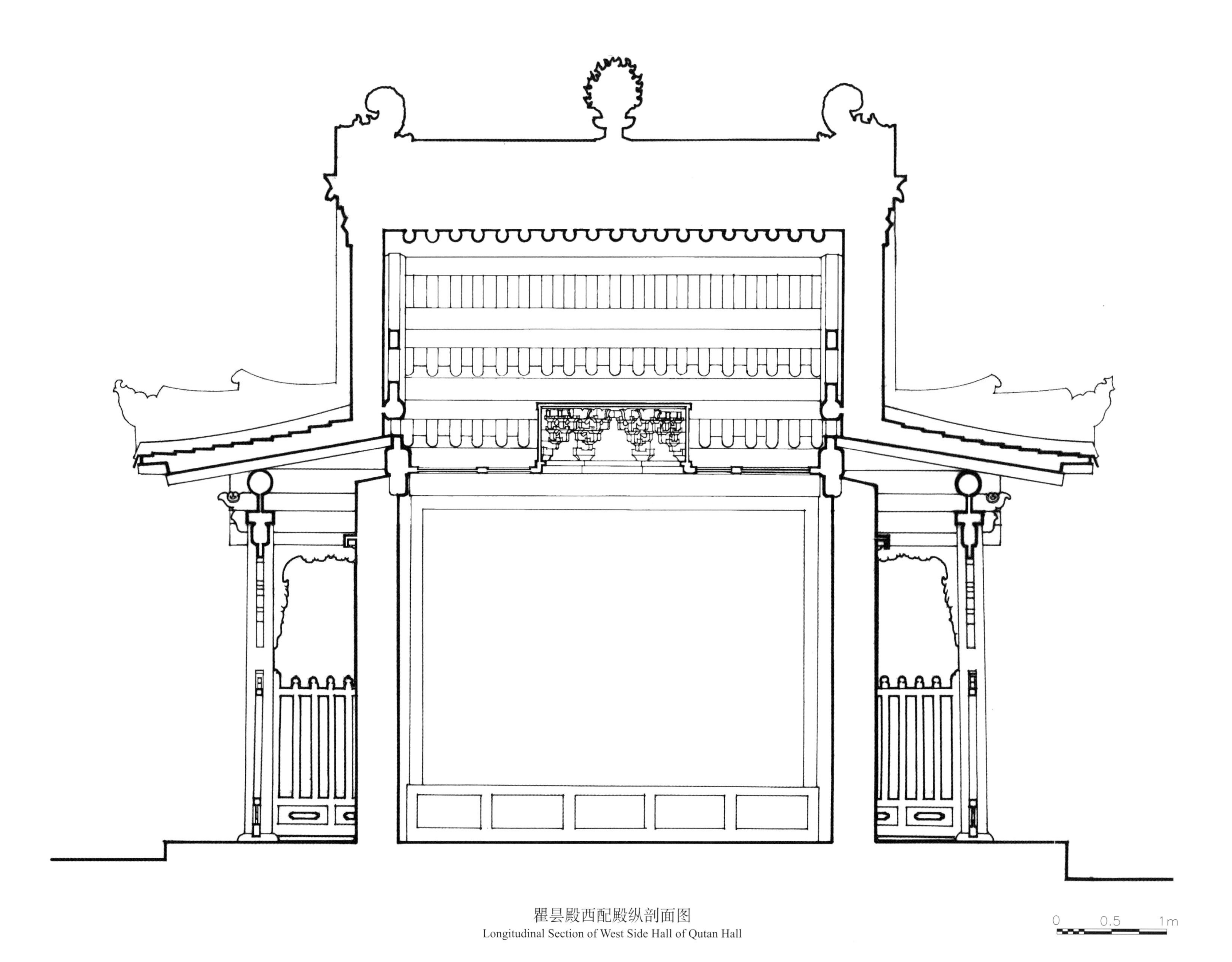

瞿昙殿西配殿纵剖面图
Longitudinal Section of West Side Hall of Qutan Hall

小鼓楼
Small Drum Tower

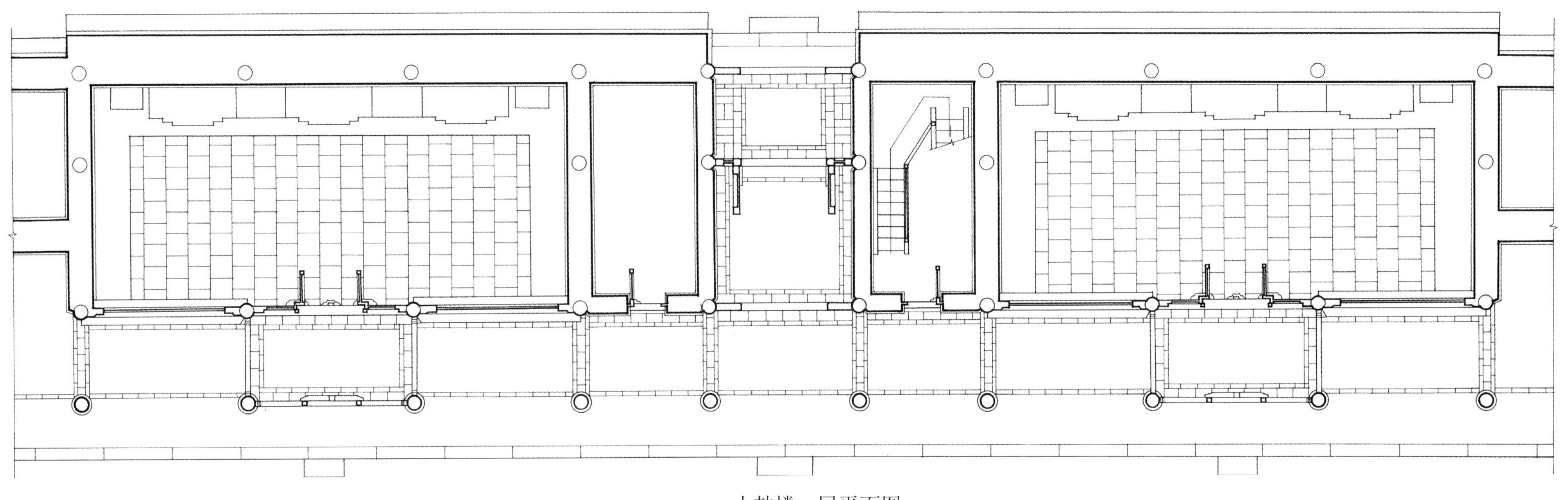

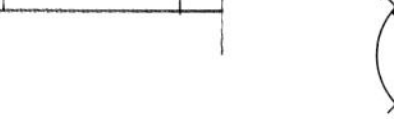

小鼓楼一层平面图
First Floor Plan of Small Drum Tower

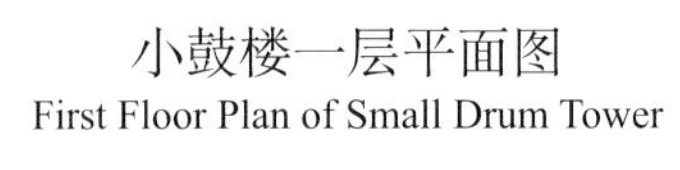

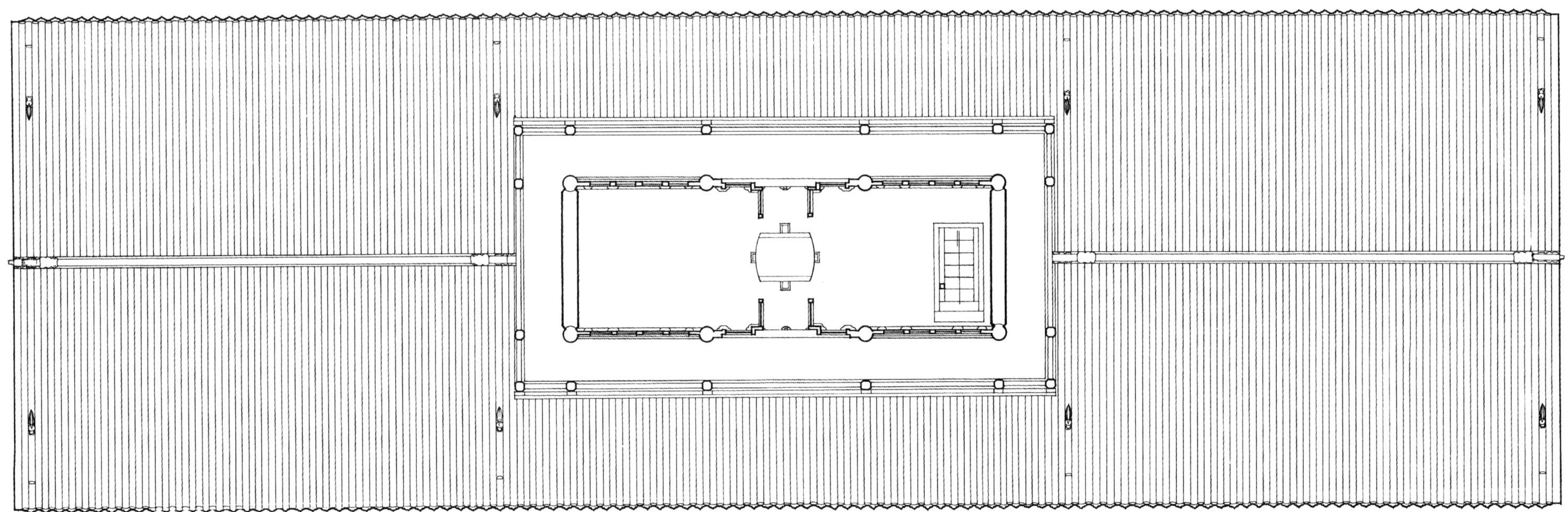

小鼓楼二层平面图
Second Floor Plan of Small Drum Tower

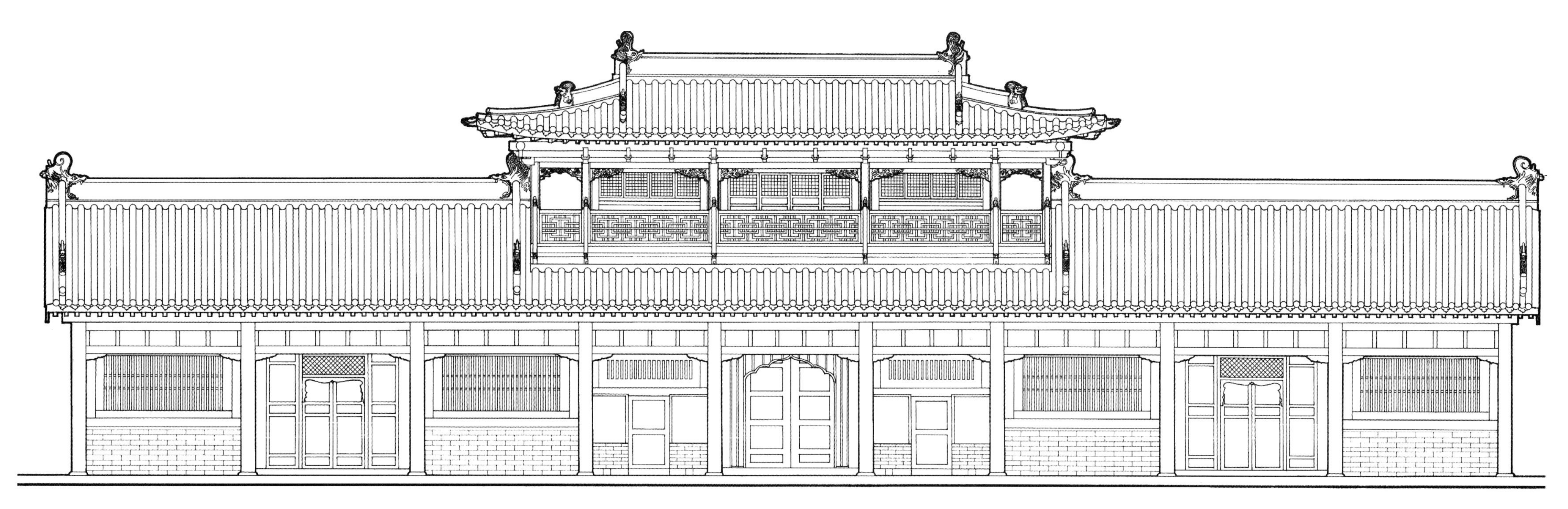

小鼓楼西立面图
West Elevation of Small Drum Tower

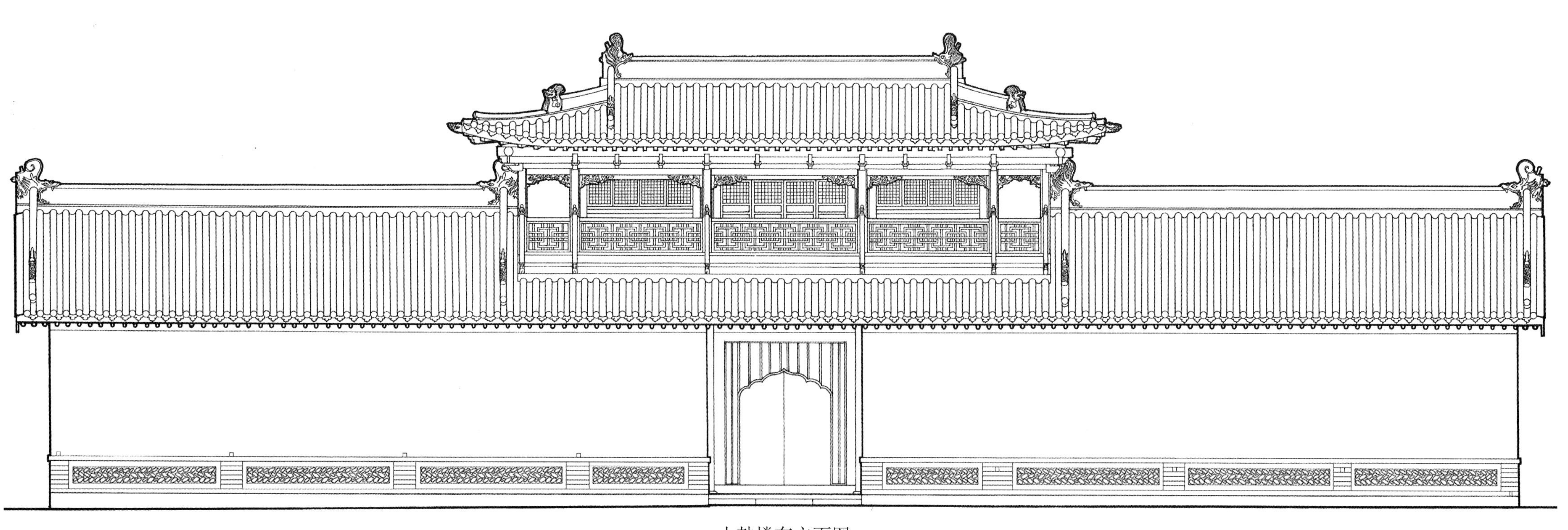

小鼓楼东立面图
East Elevation of Small Drum Tower

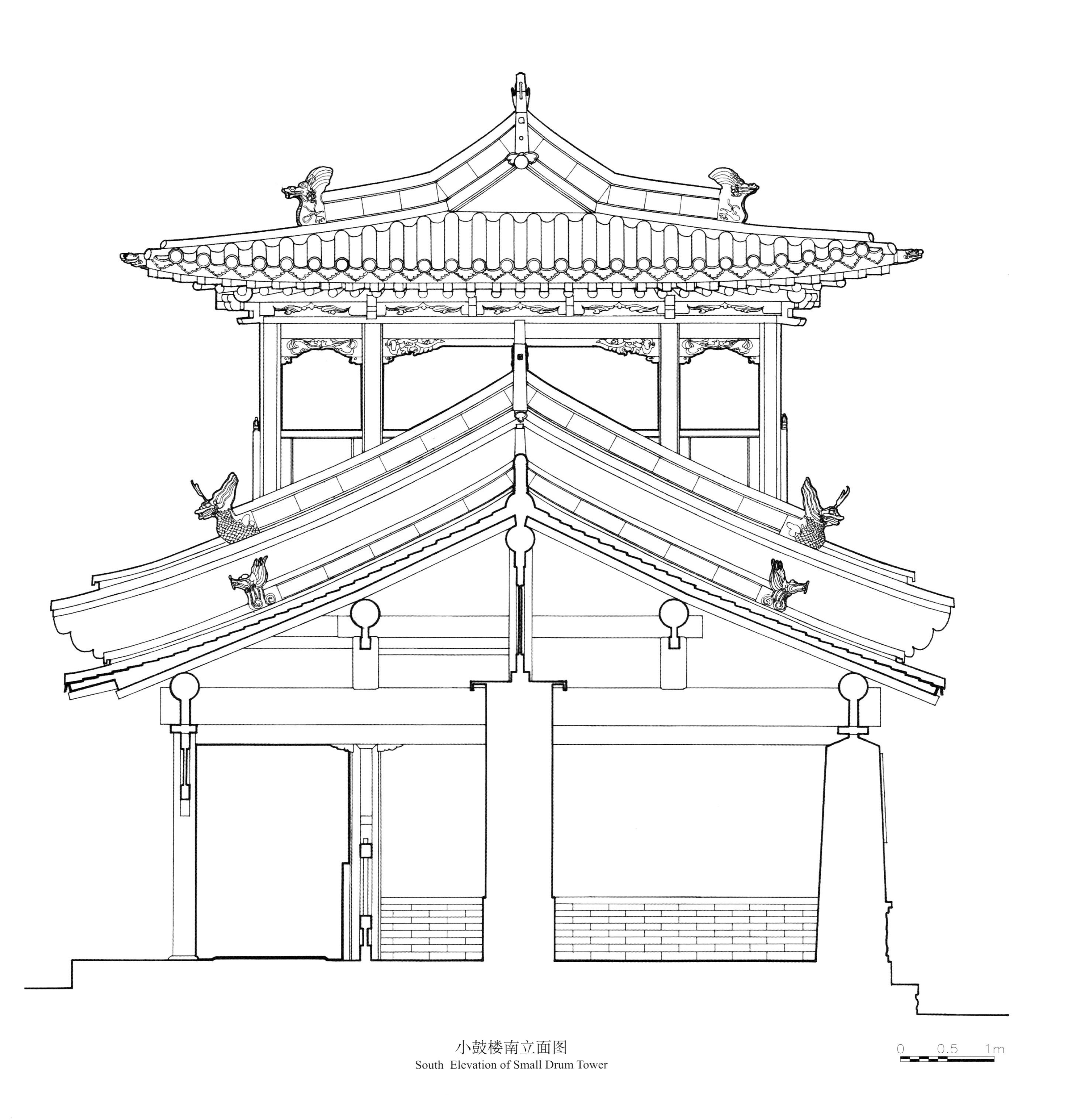

小鼓楼南立面图
South Elevation of Small Drum Tower

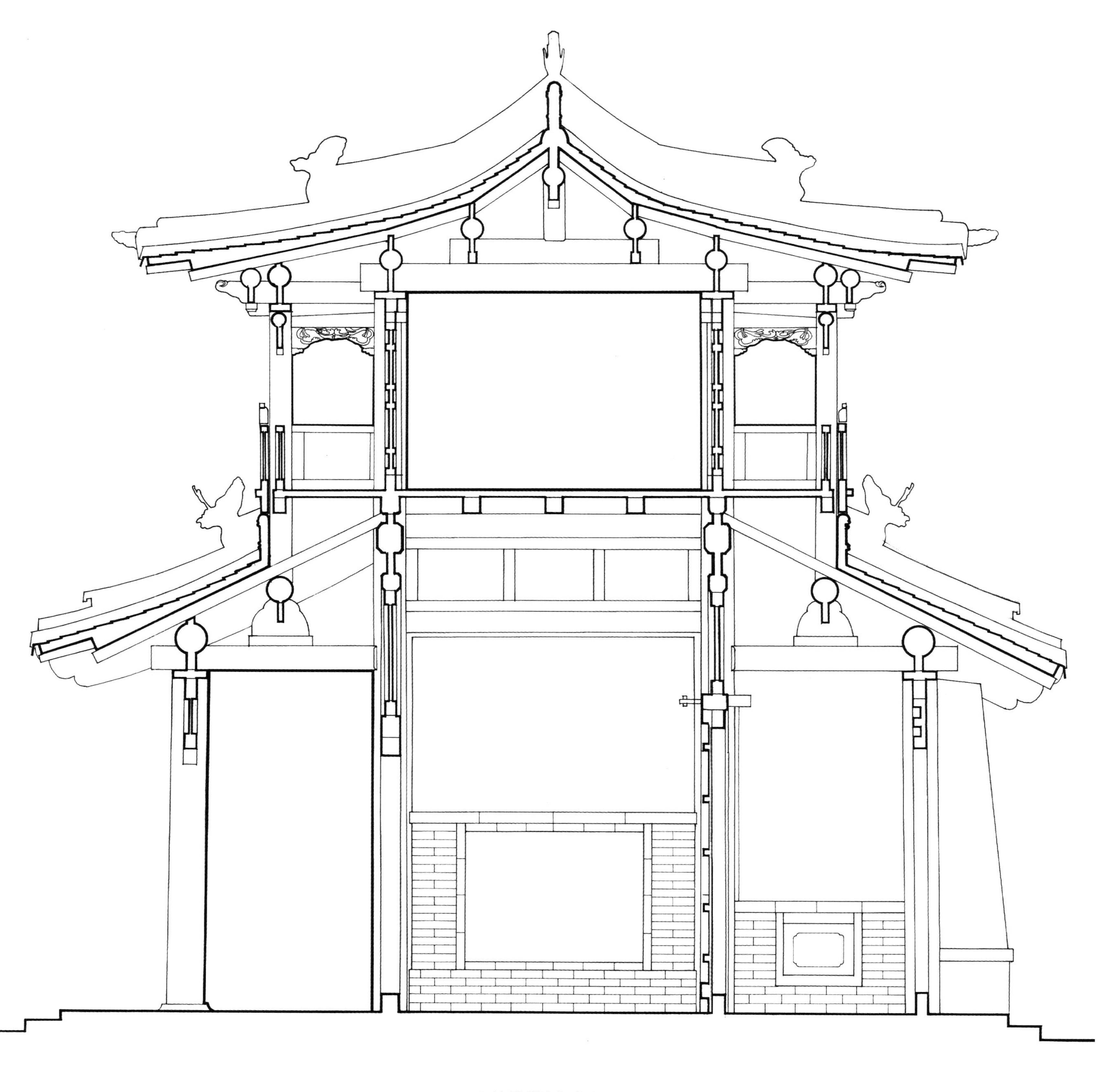

小鼓楼横剖面图
Cross-section of Small Drum Tower

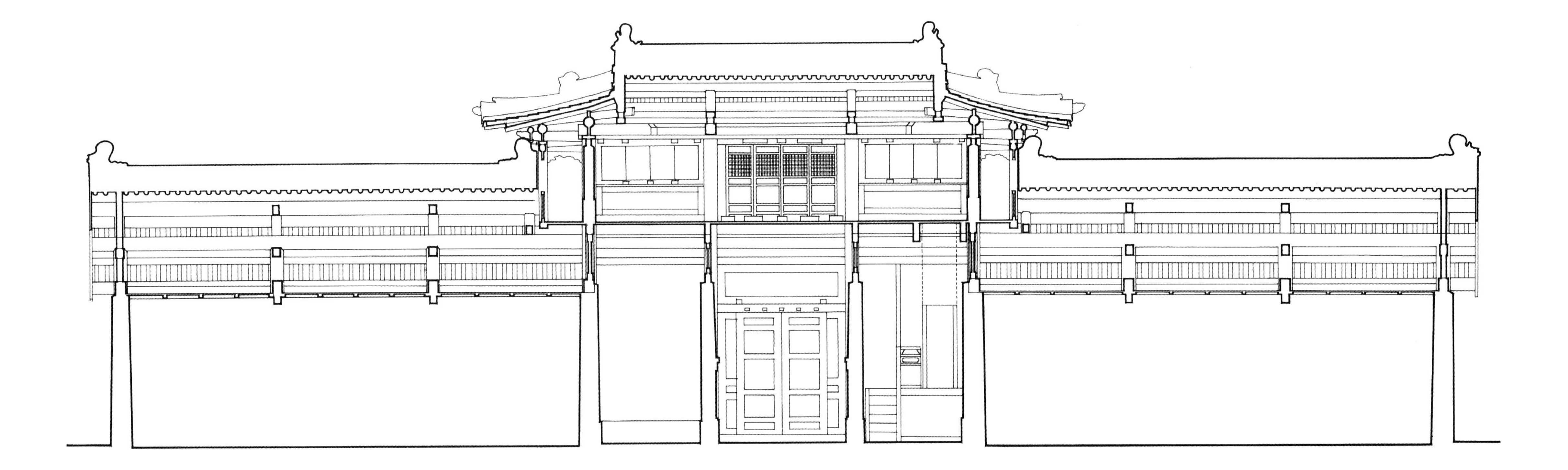

小鼓楼纵剖面图
Longitudinal Section of Small Drum Tower

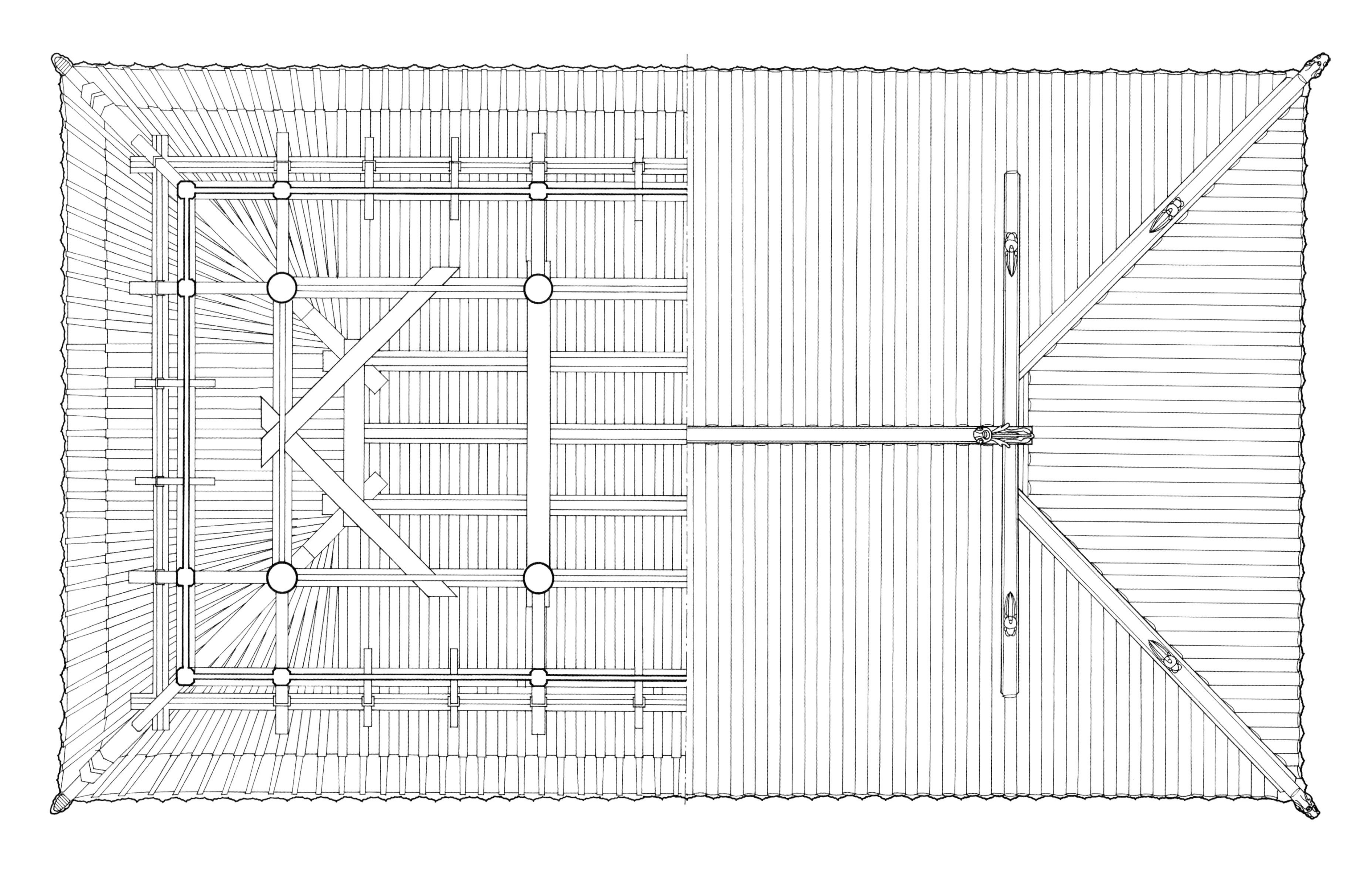

小鼓楼梁架仰视及屋顶平面图
Roof Plan and Framework Plan of Small Drum Tower

0 0.5 1m

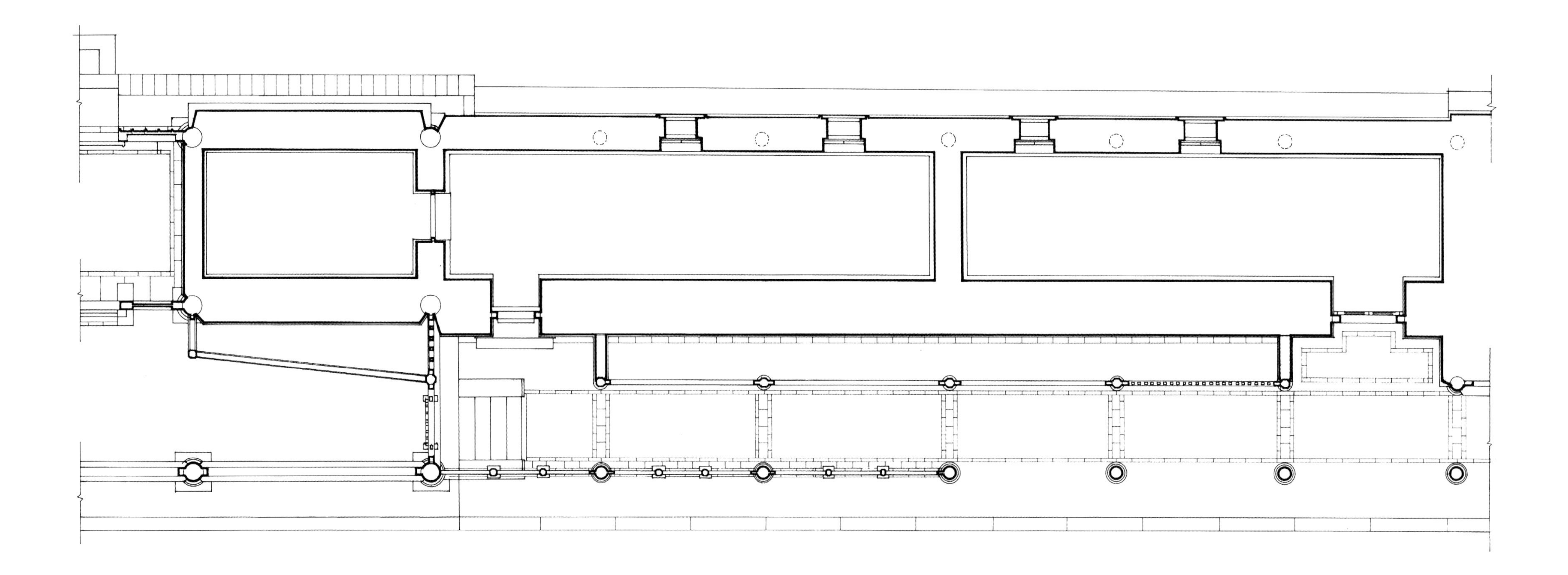

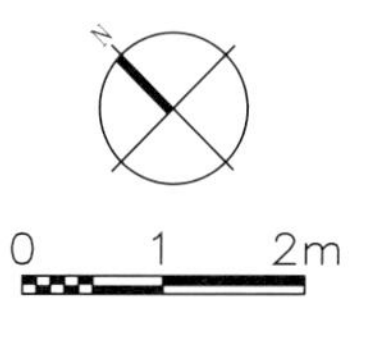

小鼓楼北廊平面图
Plan of North Corridor of Small Drum Tower

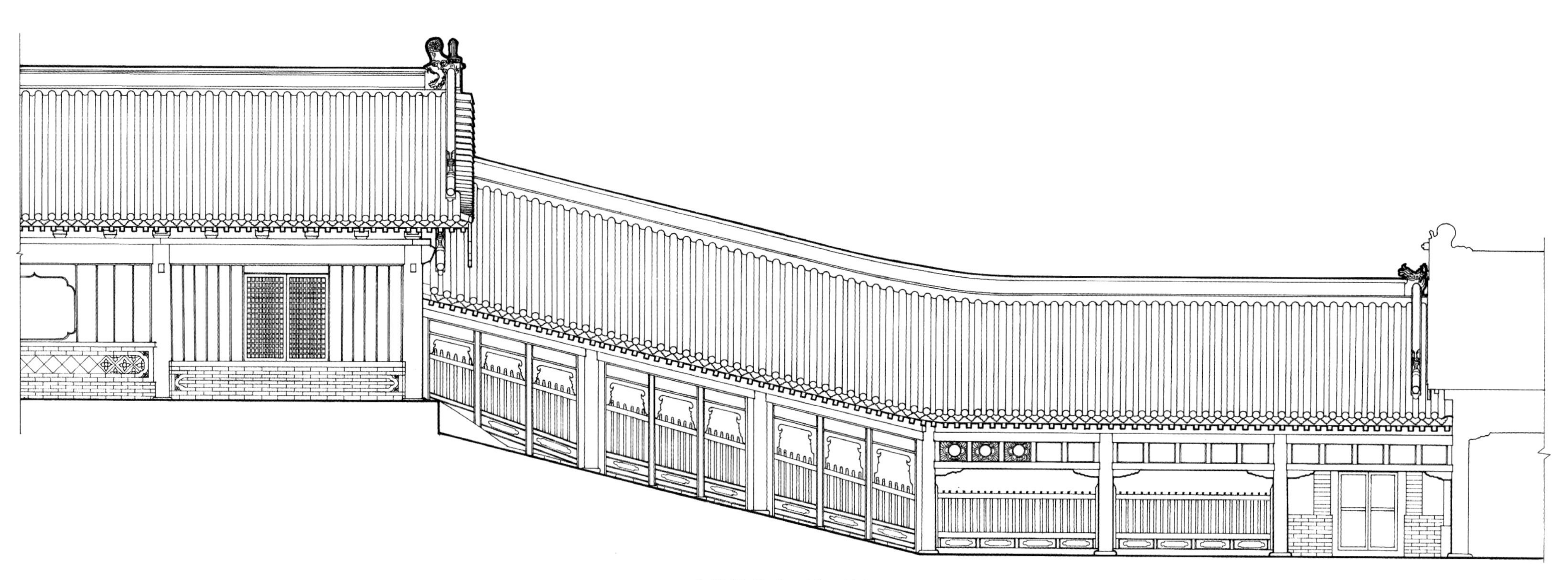

小鼓楼北廊西立面图
West Elevation of North Corridor of Small Drum Tower

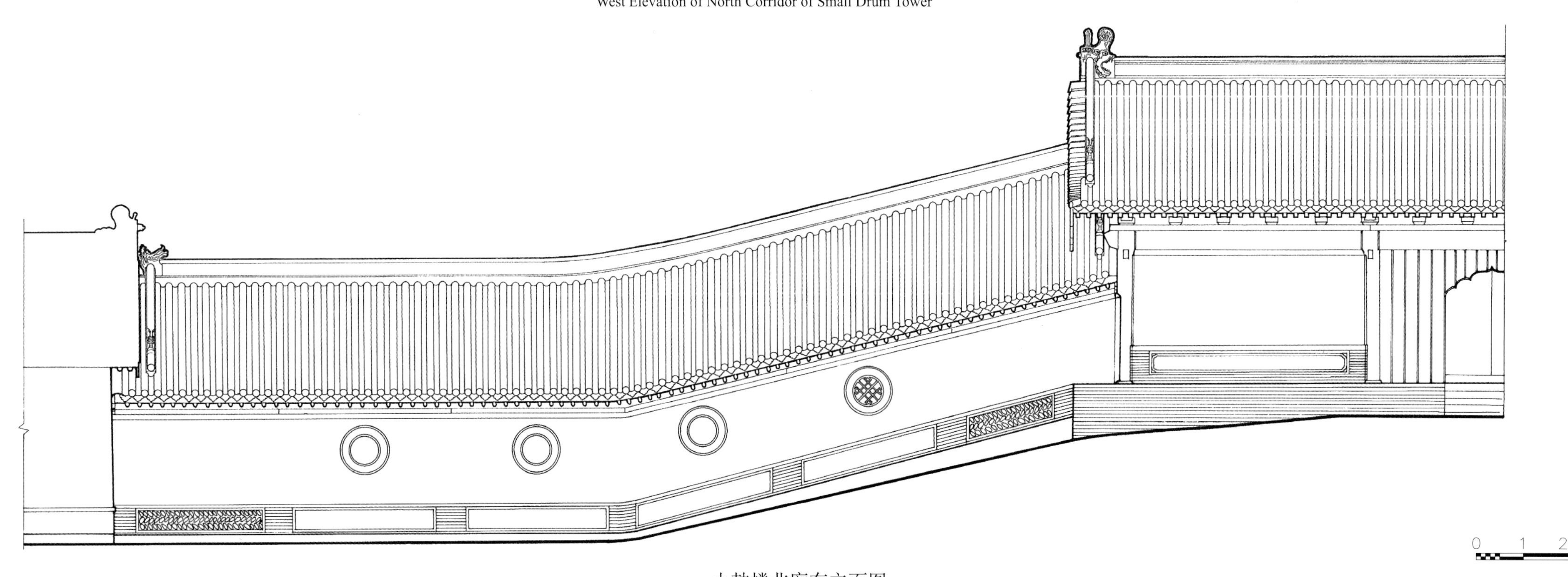

小鼓楼北廊东立面图
East Elevation of North Corridor of Small Drum Tower

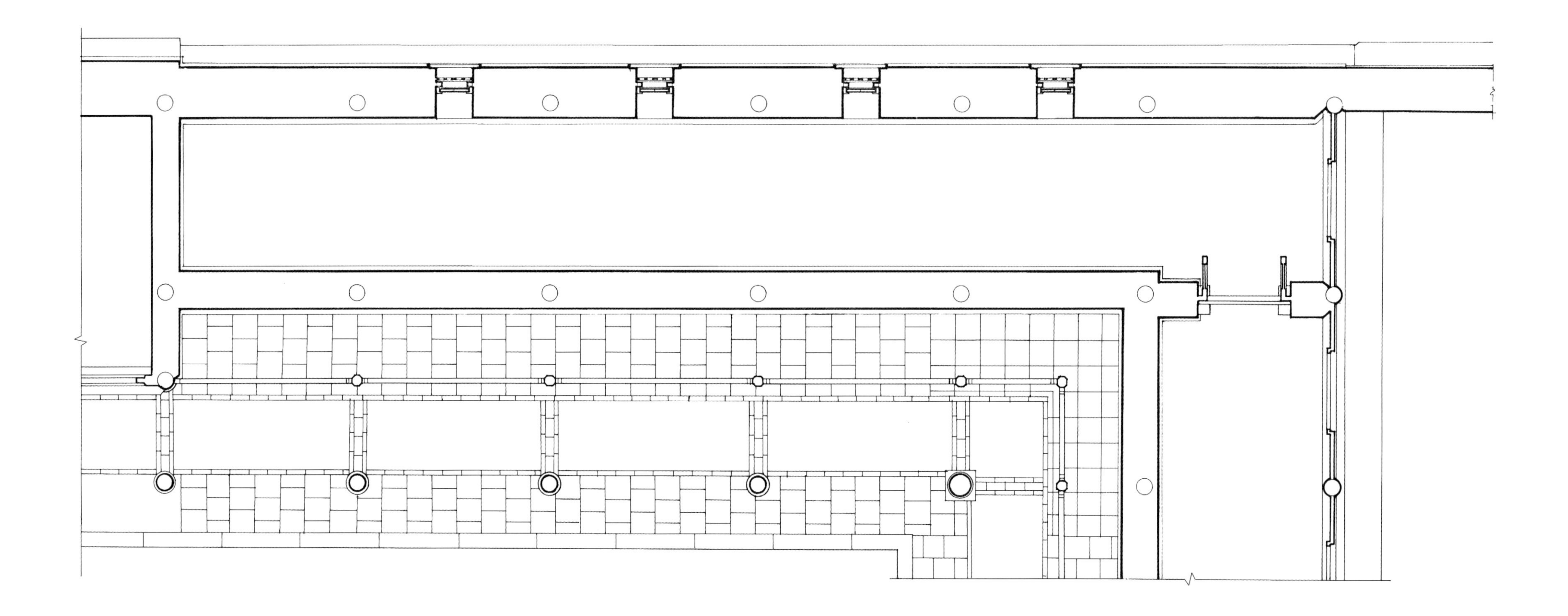

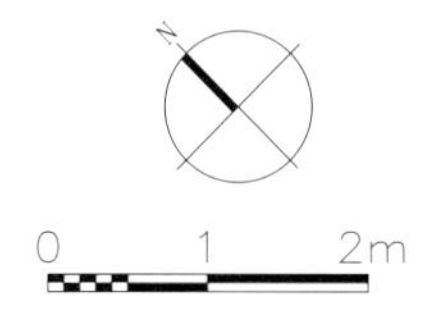

三世殿南廊平面图
Plan of South Corridor of Sanshi Hall

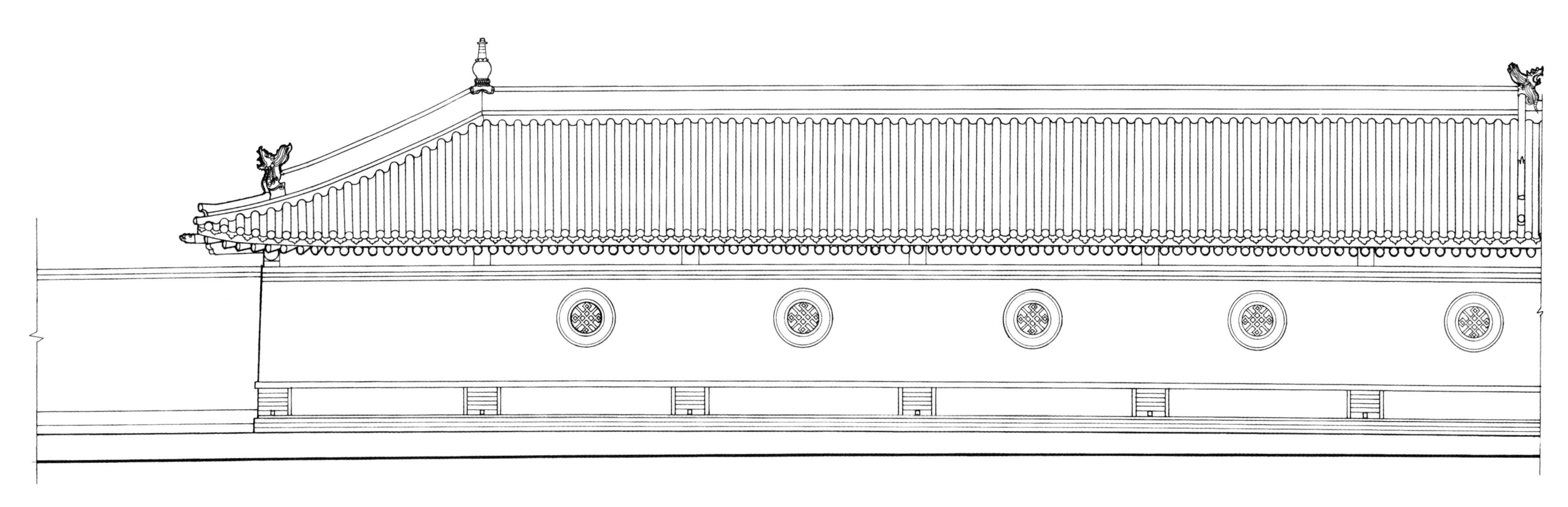

三世殿南廊东立面图
East Elevation of South Corridor of Sanshi Hall

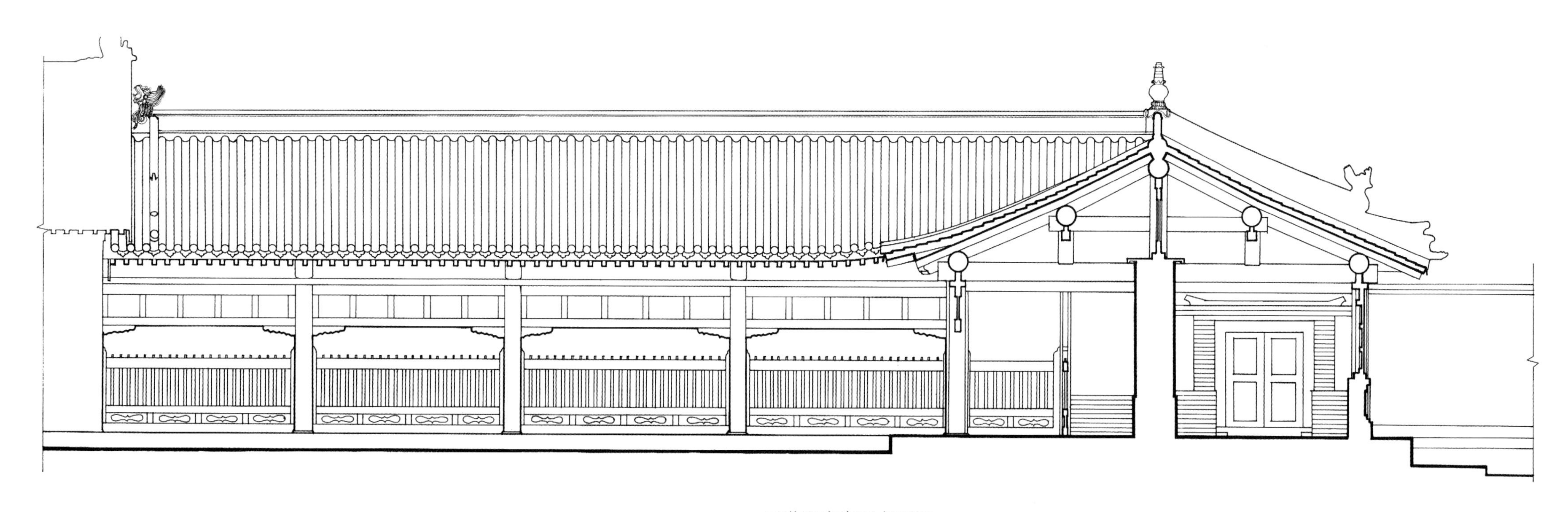

三世殿南廊西立面图
West Elevation of South Corridor of Sanshi Hall

小钟楼
Small Bell Tower

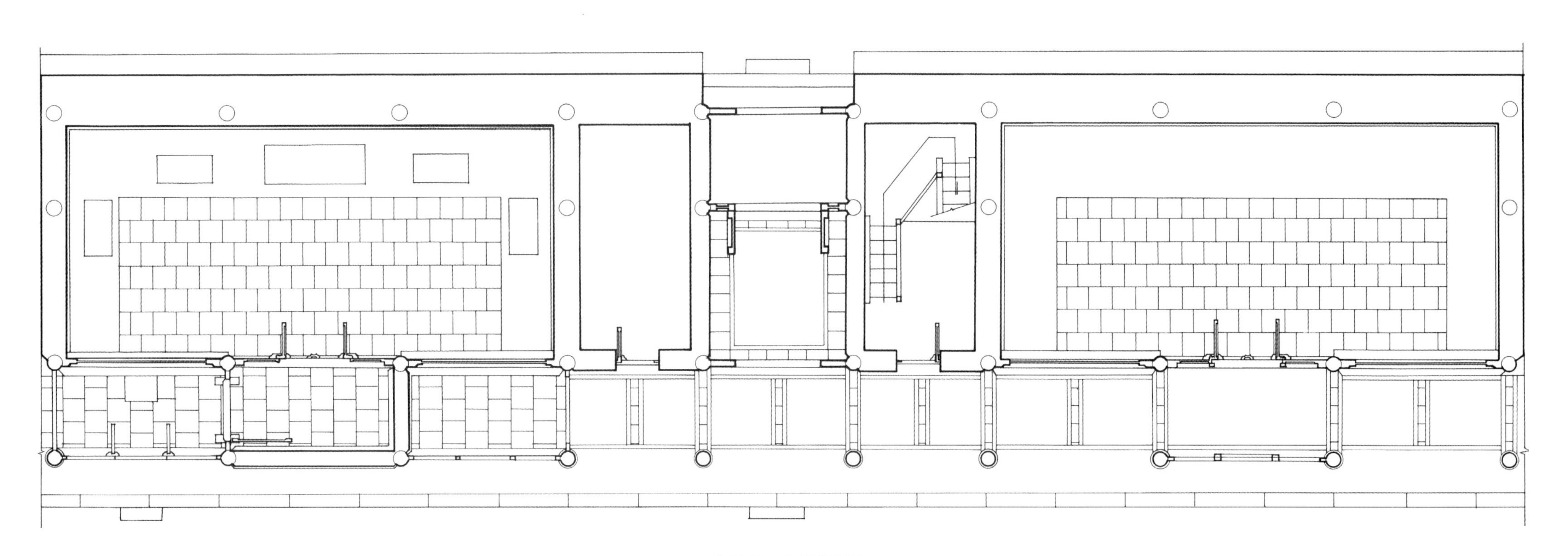

小钟楼一层平面图
First Floor Plan of Small Bell Tower

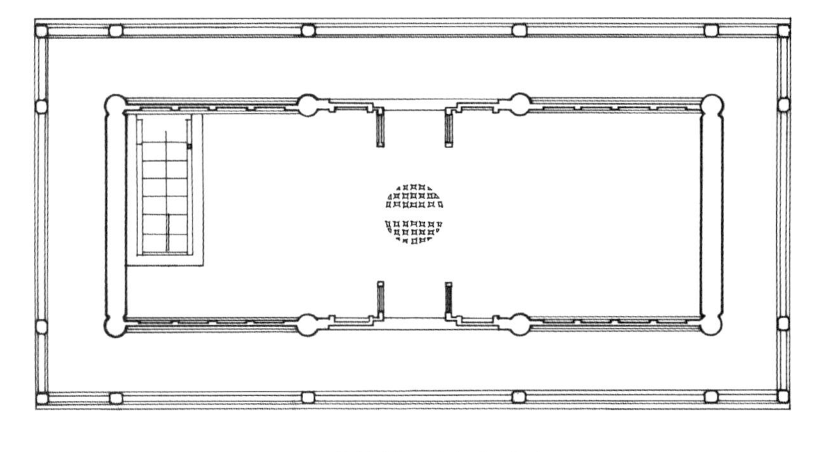

小钟楼二层平面图
Second Floor Plan of Small Bell Tower

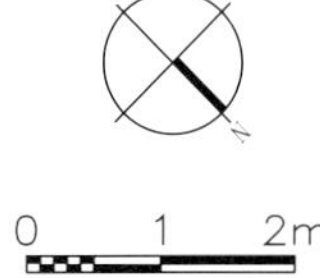

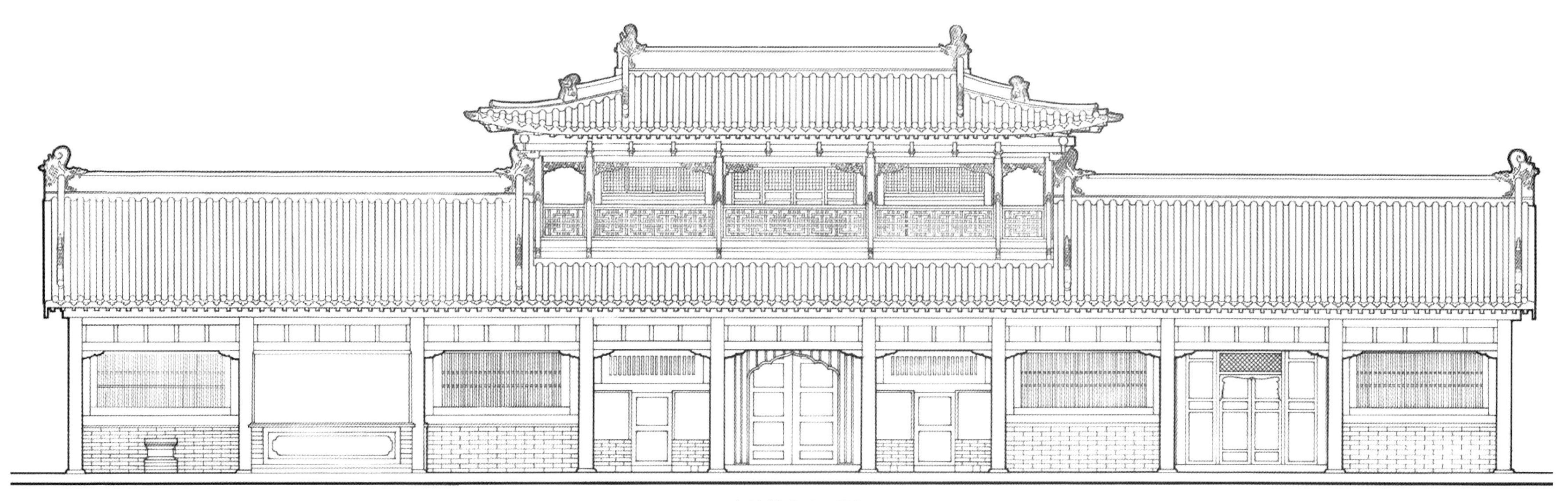

小钟楼东立面图
East Elevation of Small Bell Tower

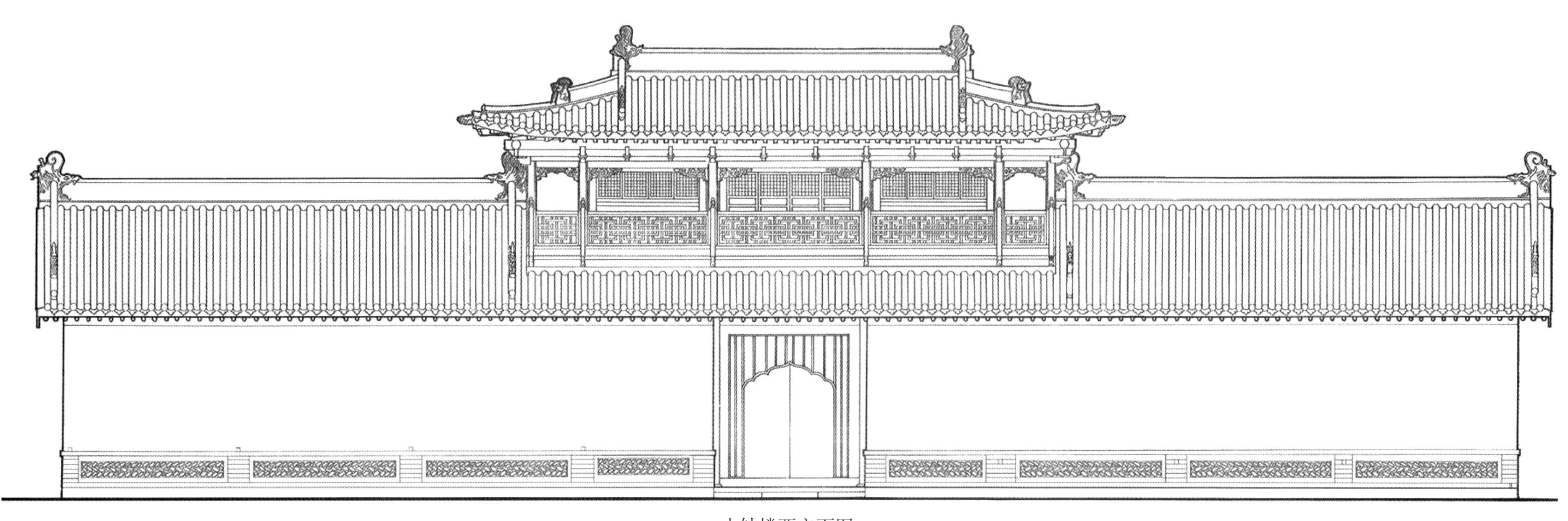

小钟楼西立面图
West Elevation of Small Bell Tower

小钟楼南立面图
South Elevation of Small Bell Tower

0 0.5 1m

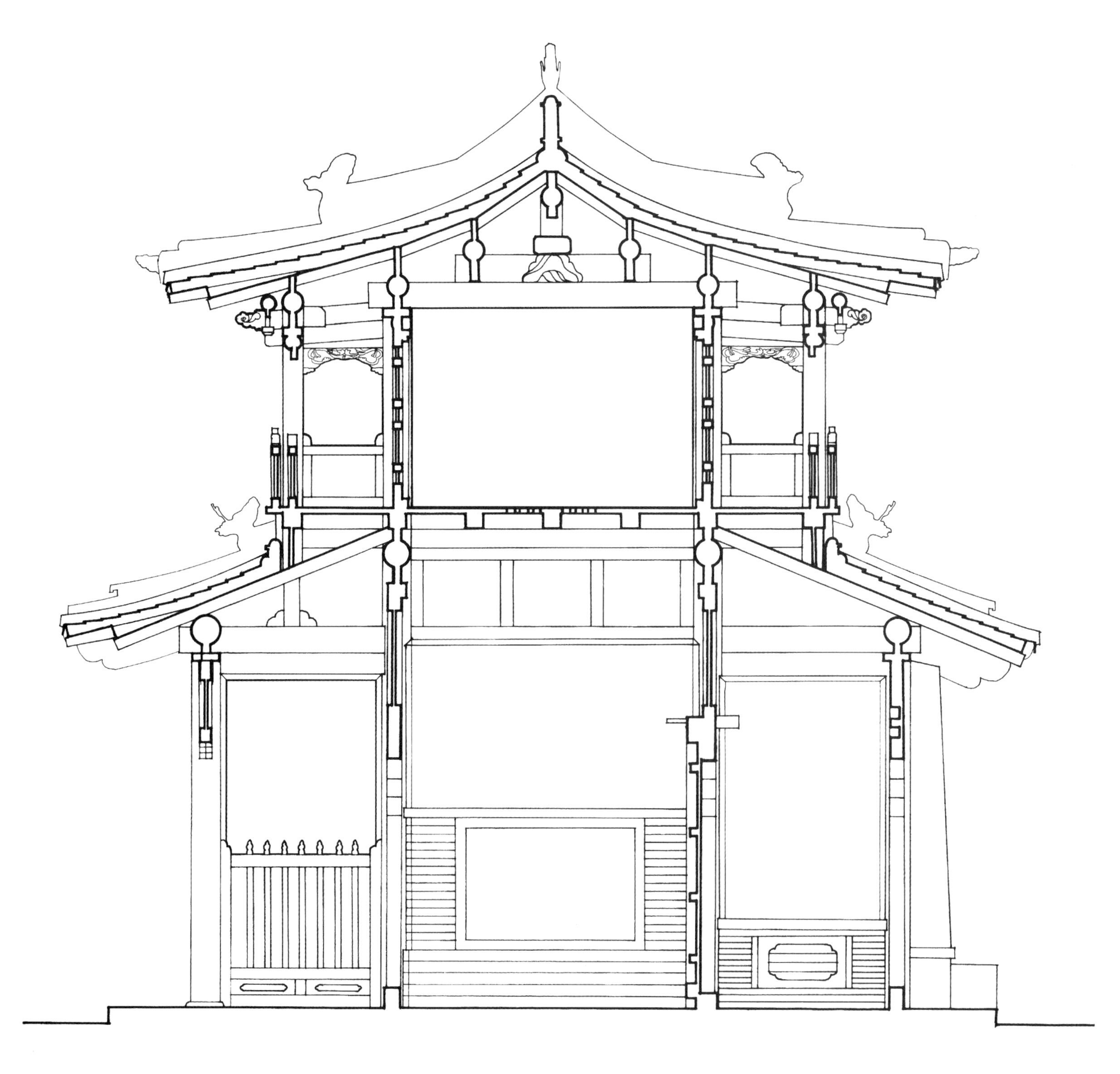

小钟楼横剖面图

Cross-section of Small Bell Tower

小钟楼纵剖面图
Longitudinal Section of Small Bell Tower

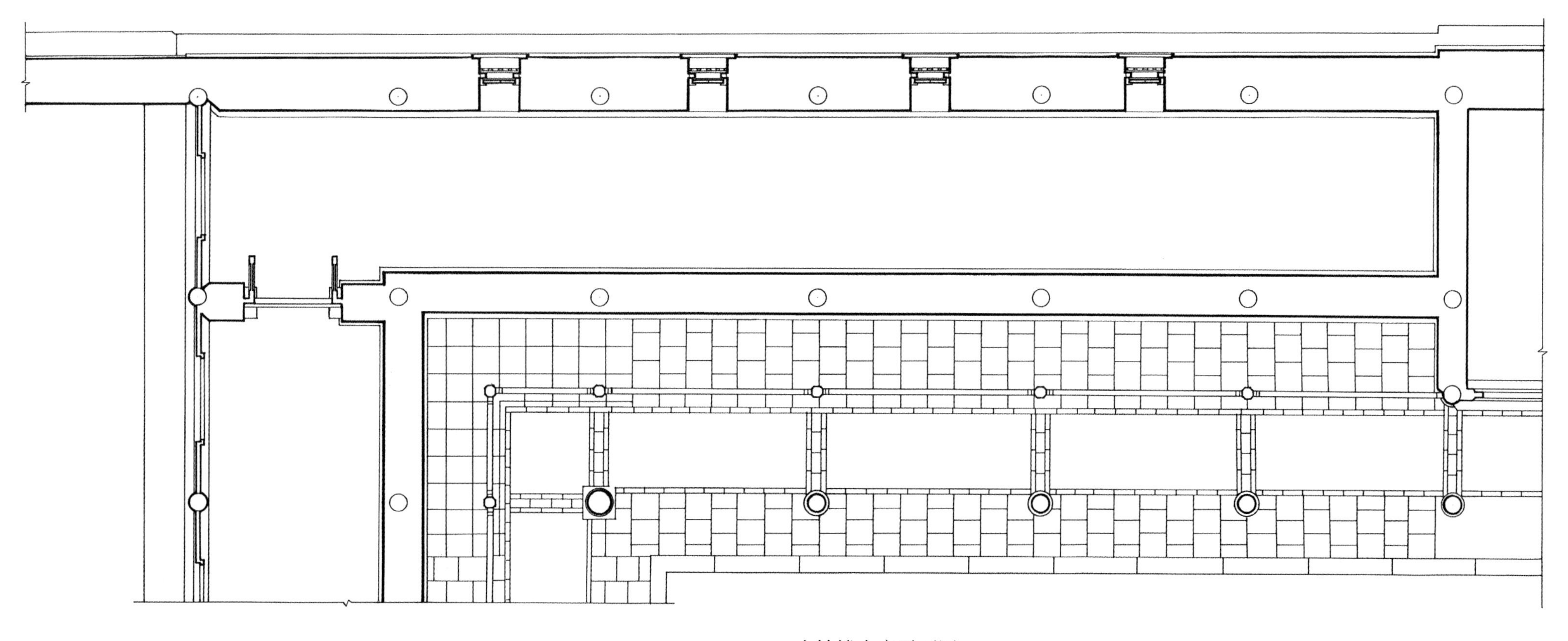

小钟楼南廊平面图
Plan of South Corridor of Small Bell Tower

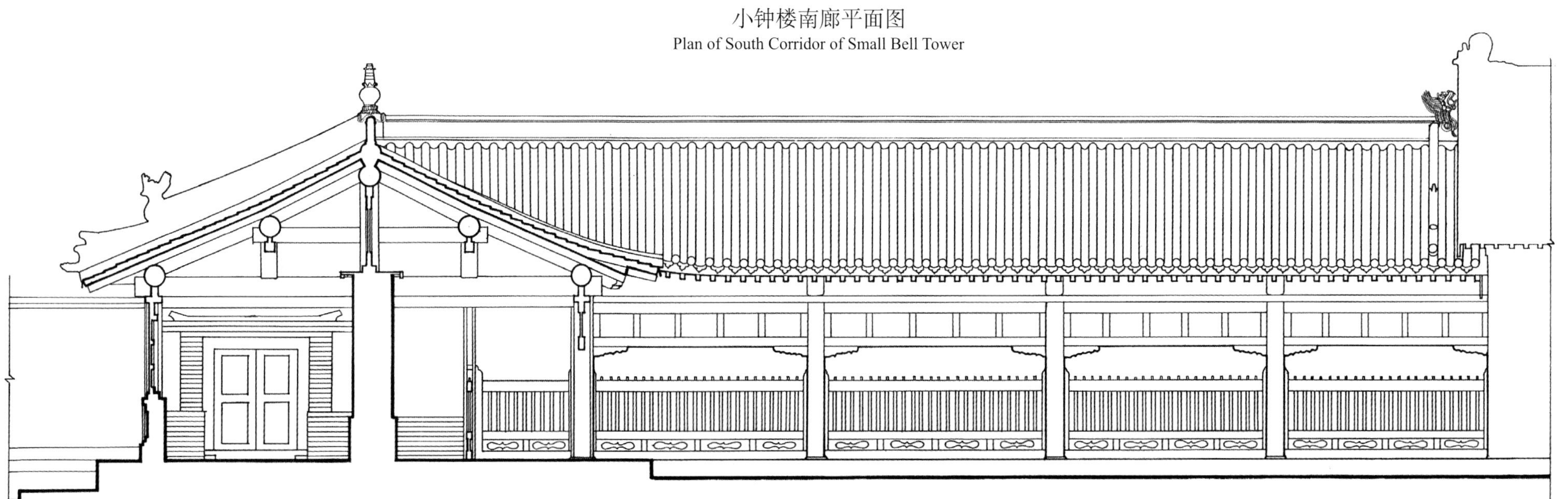

小钟楼南廊东立面图
East Elevation of South Corridor of Small Bell Tower

0 1 2m

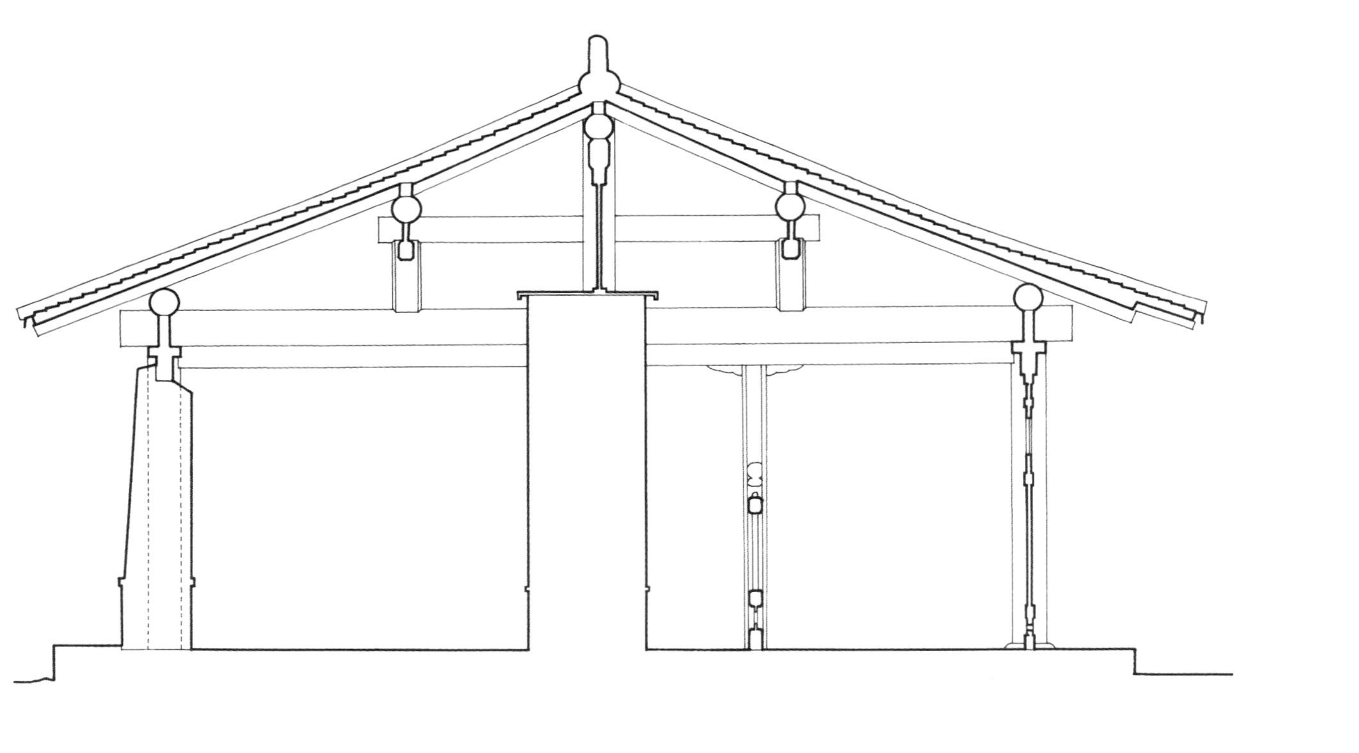

小钟楼北廊横剖面图
Cross-section of North Corridor of Small Bell Tower

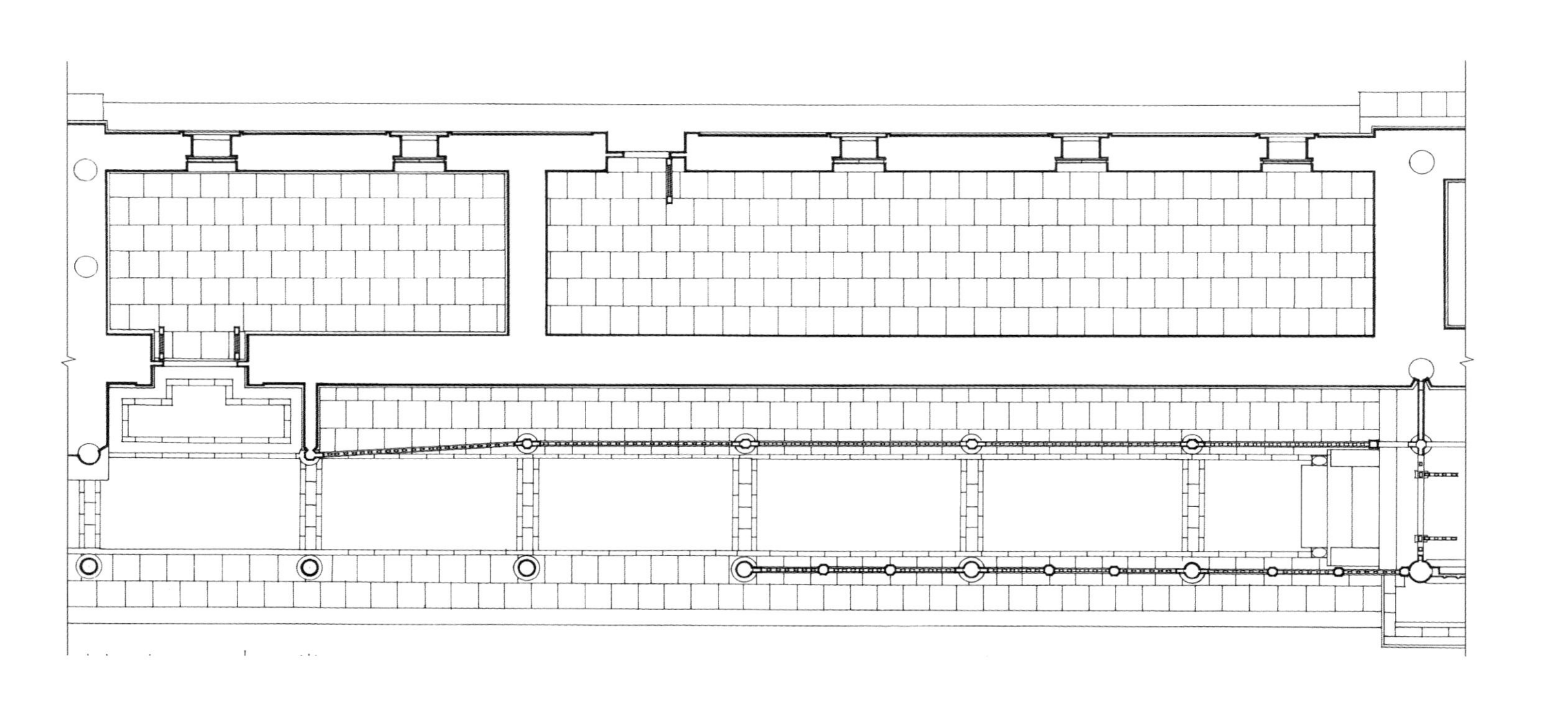

小钟楼北廊平面图
Plan of North Corridor of Small Bell Tower

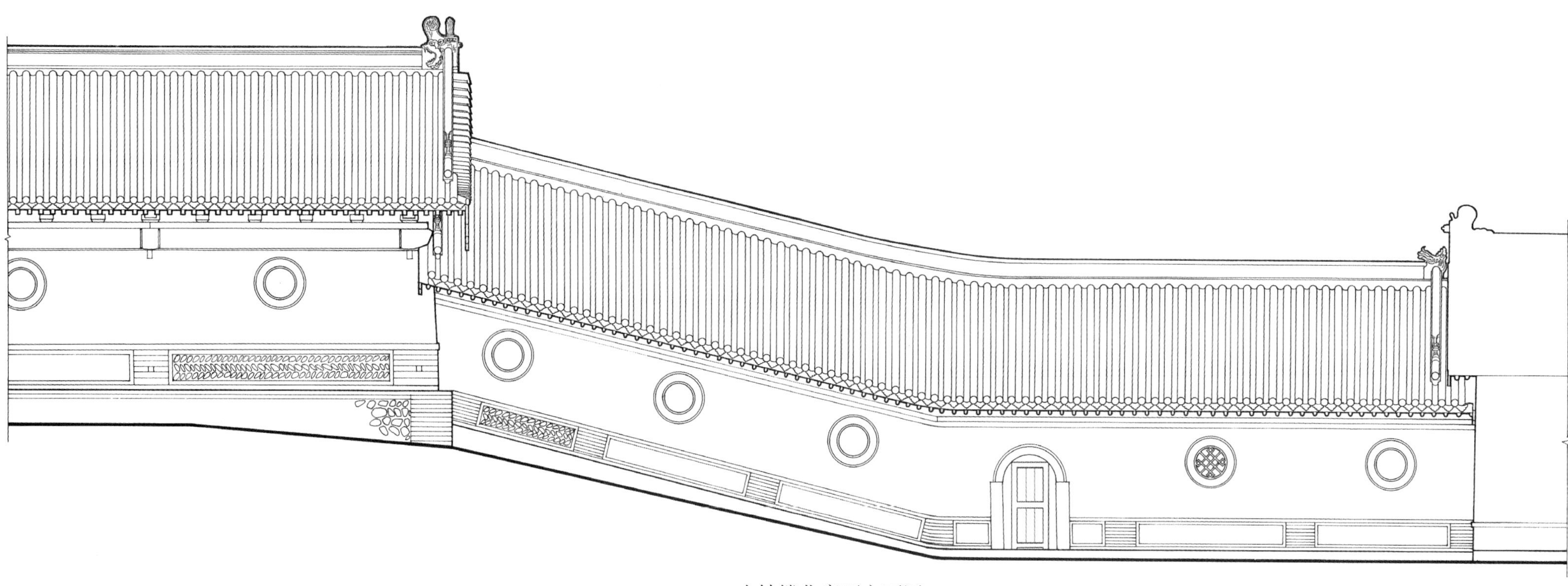

小钟楼北廊西立面图
West Elevation of North Corridor of Small Bell Tower

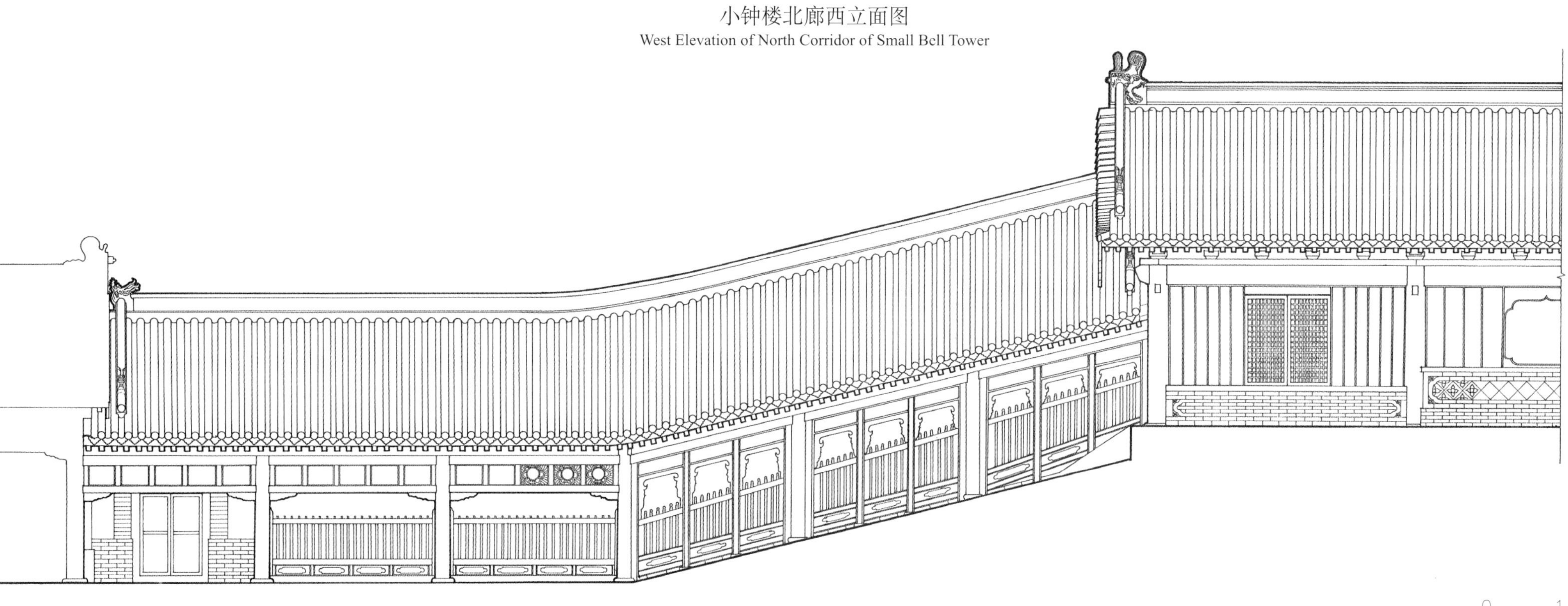

小钟楼北廊东立面图
East Elevation of North Corridor of Small Bell Tower

宝光殿
Baoguang Hall

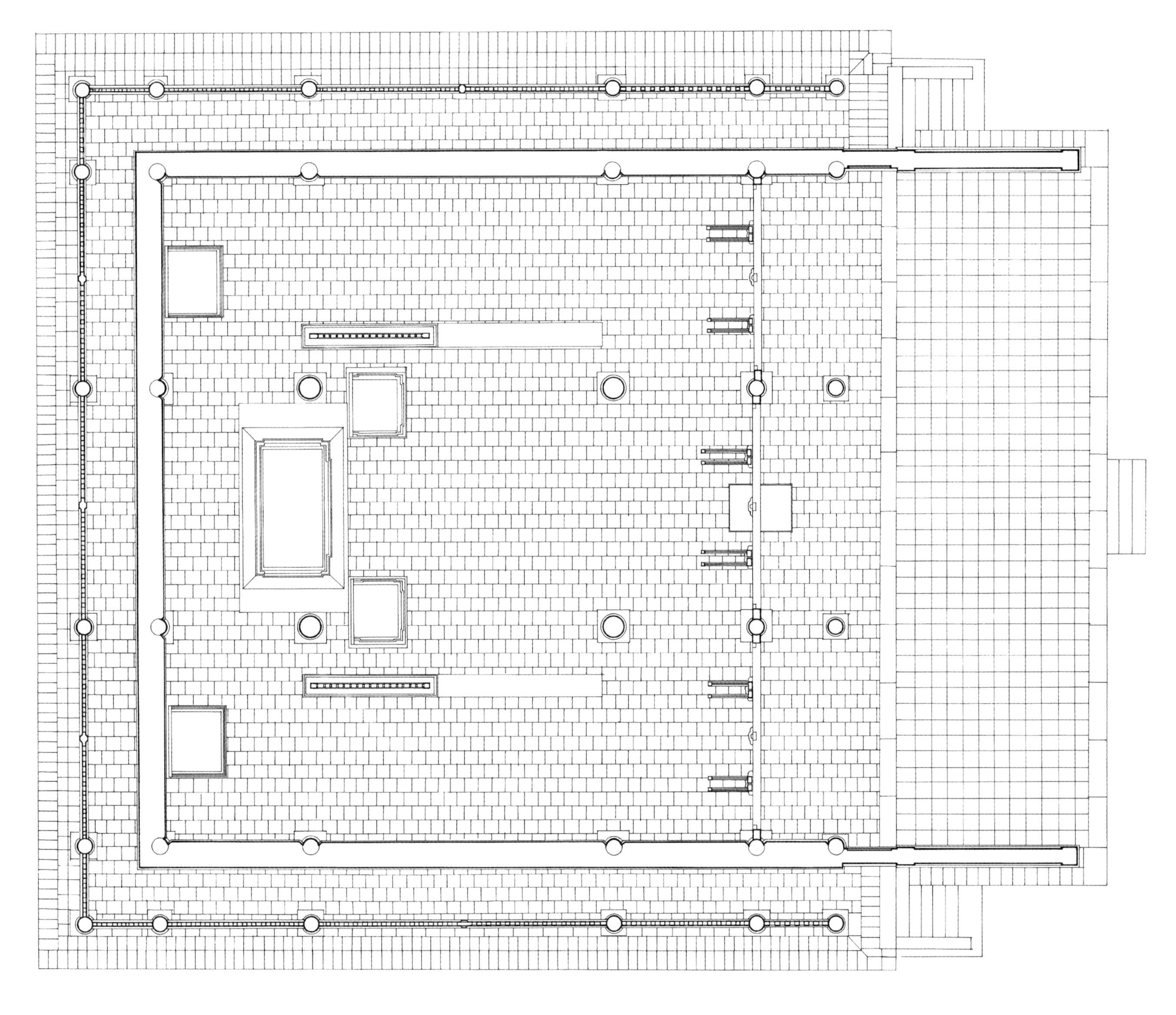

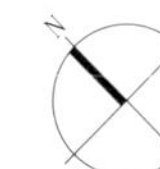

宝光殿平面图
Plan of Baoguang Hall

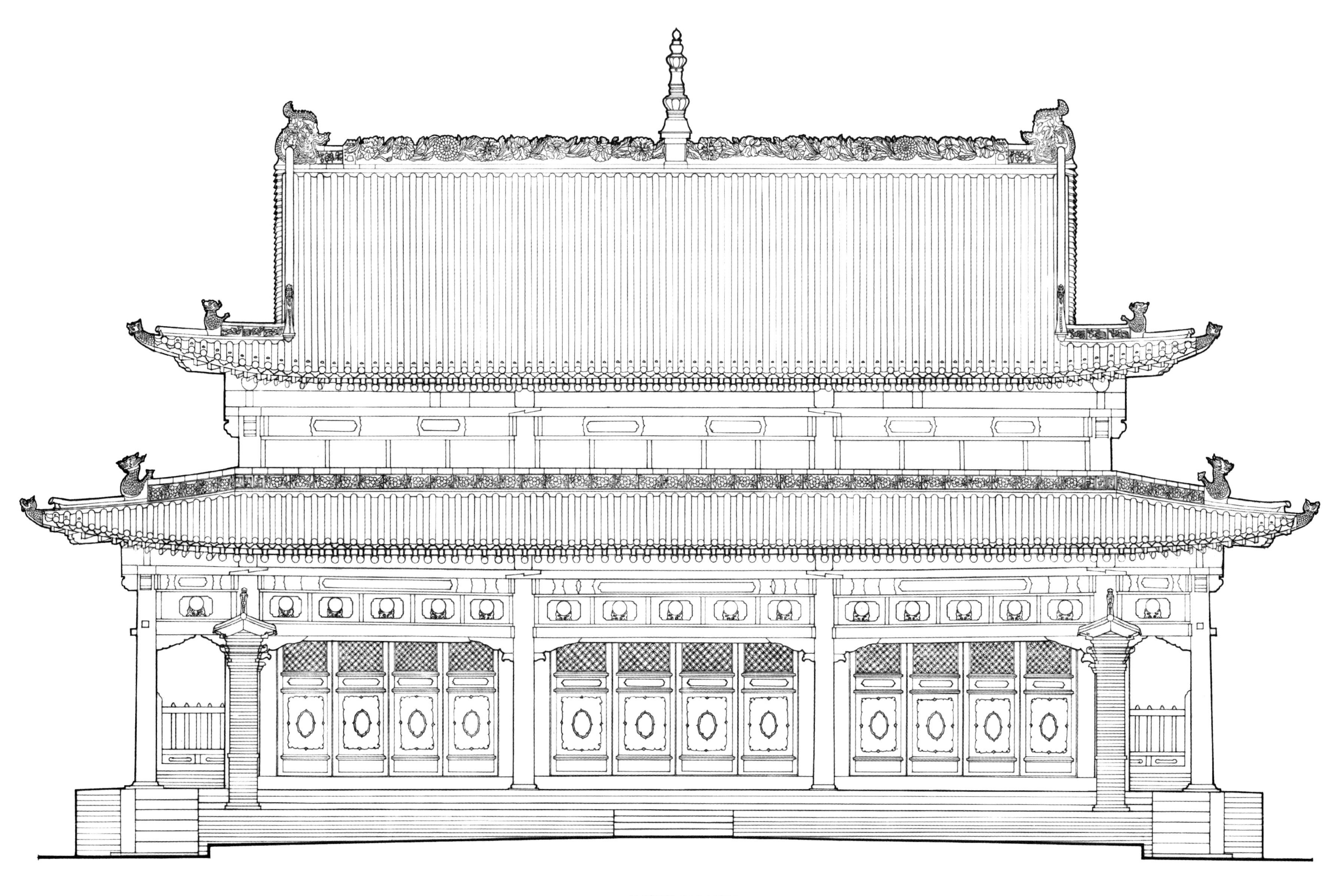

宝光殿南立面图
South Elevation of Baoguang Hall

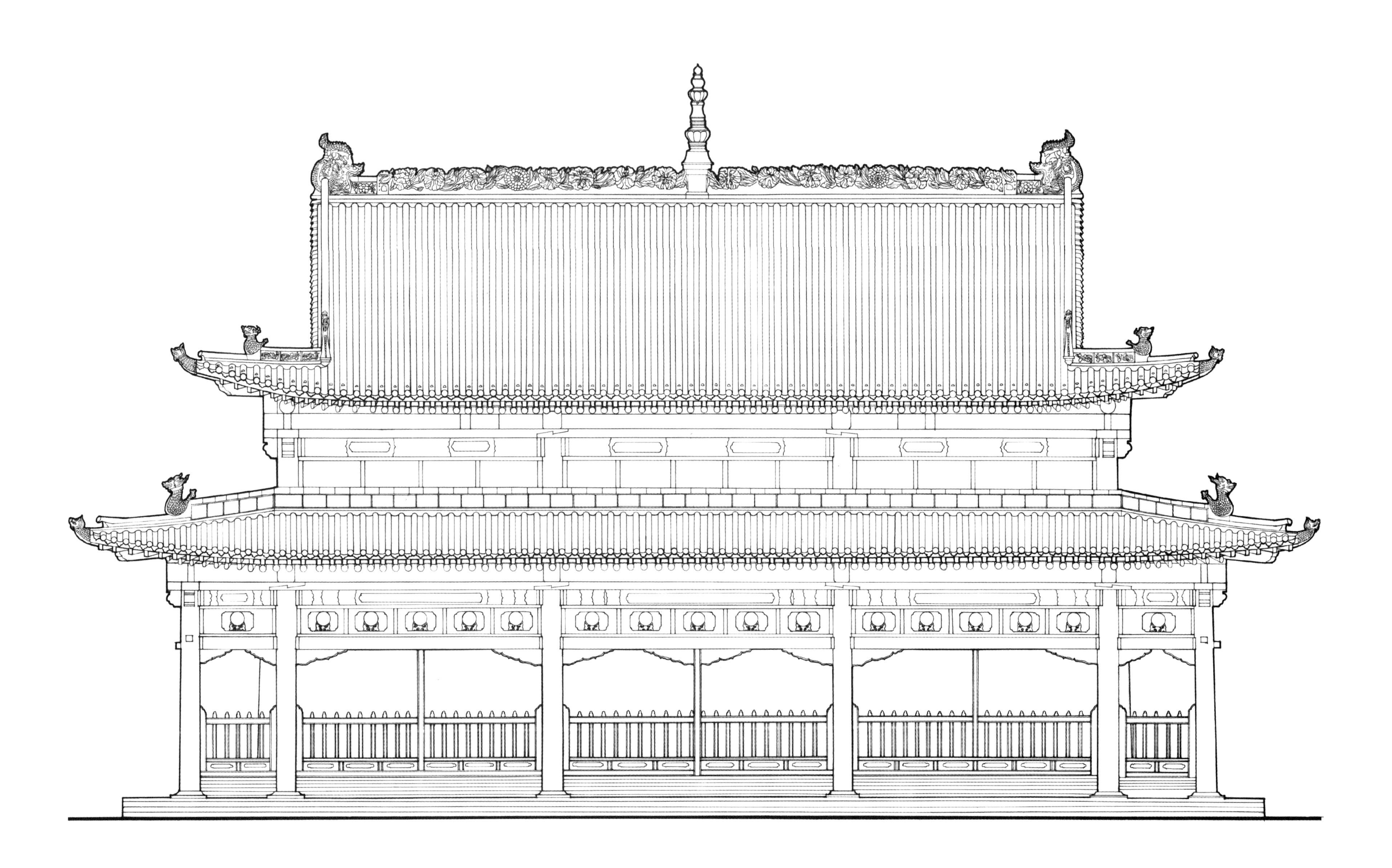

宝光殿北立面图
North Elevation of Baoguang Hall

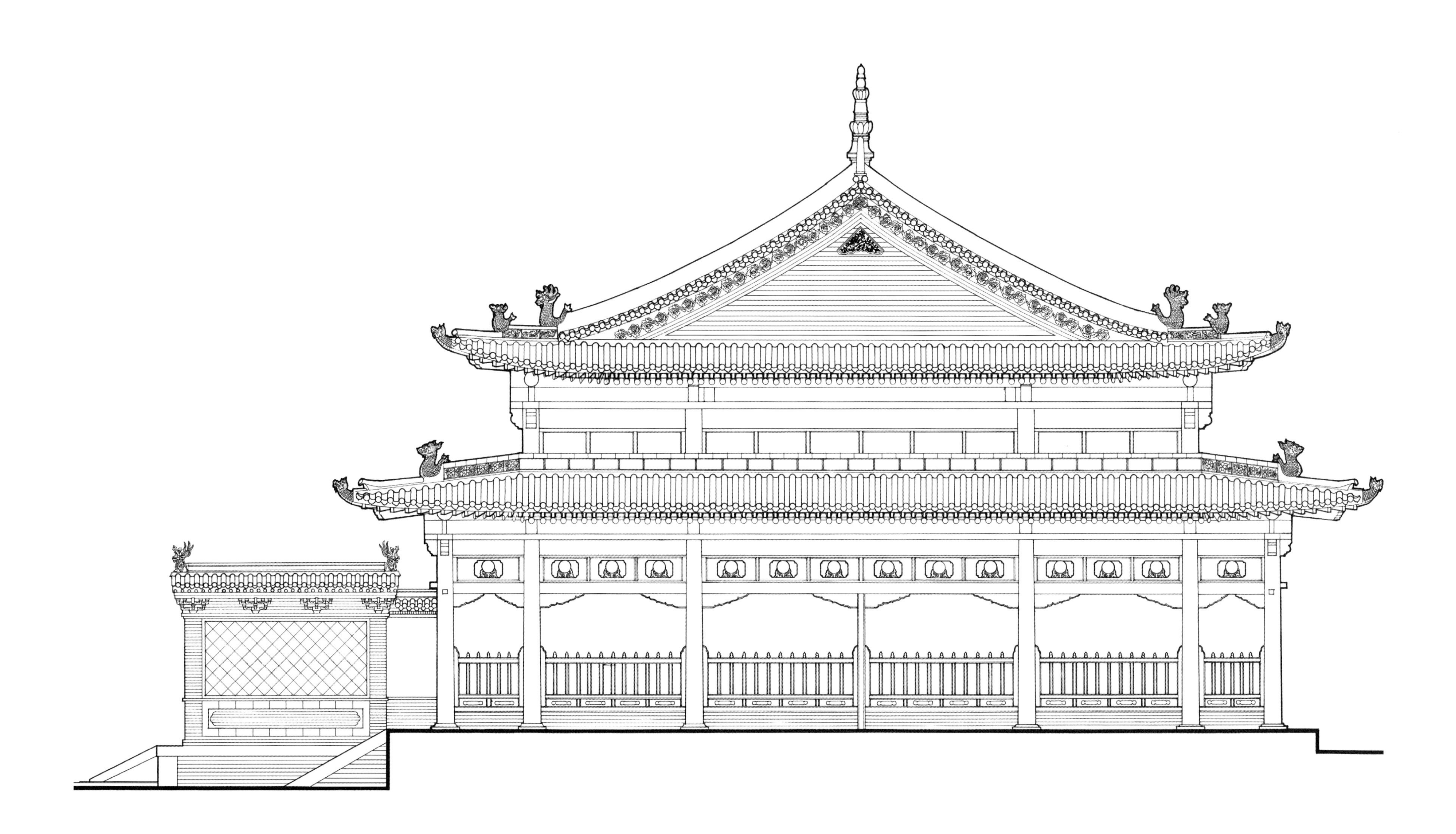

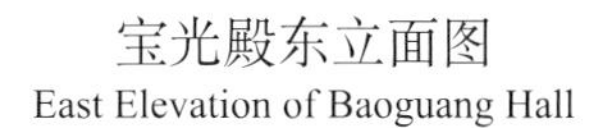

宝光殿东立面图

East Elevation of Baoguang Hall

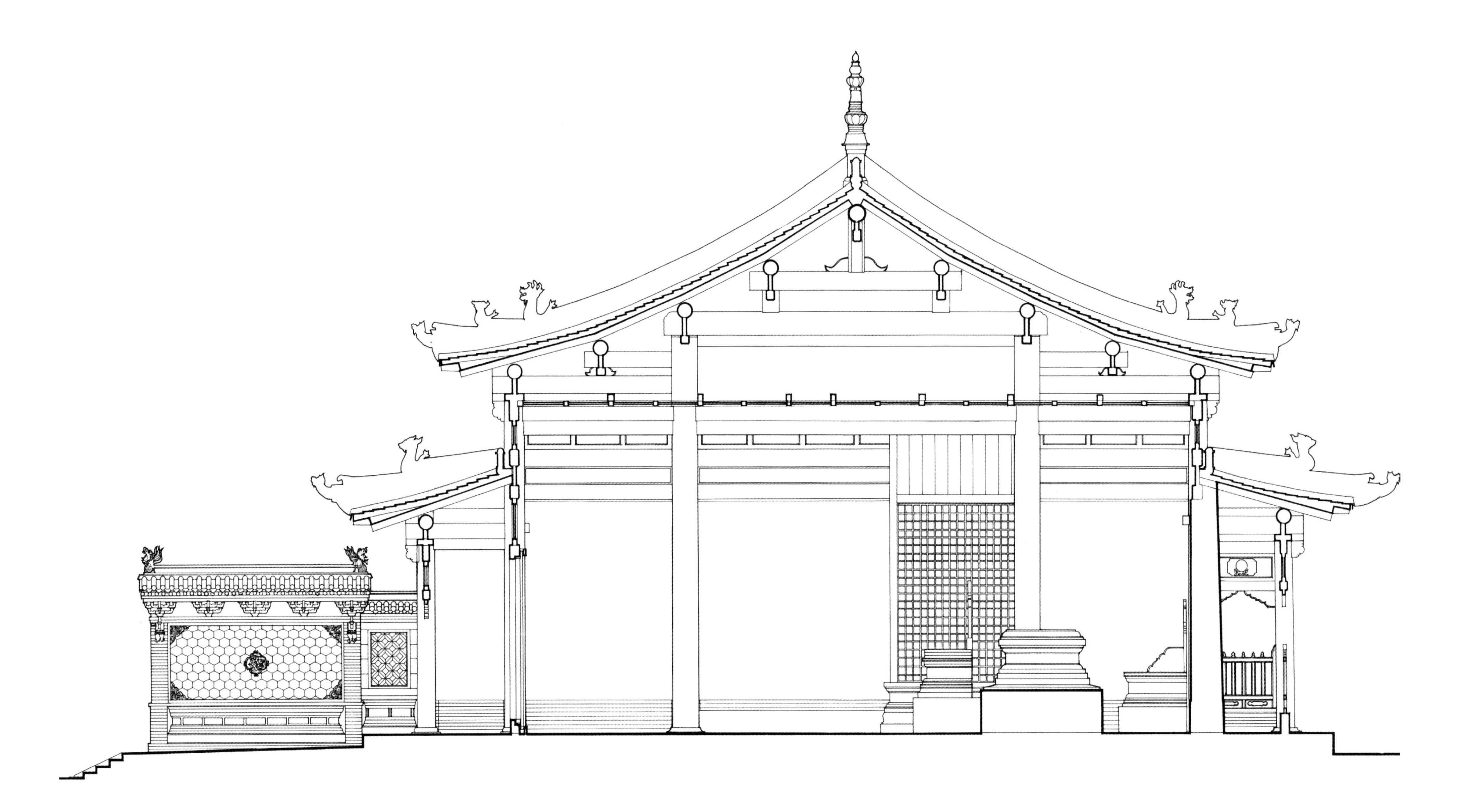

宝光殿明间剖面图
Mingjian Cross-section of Baoguang Hall

0 1 2m

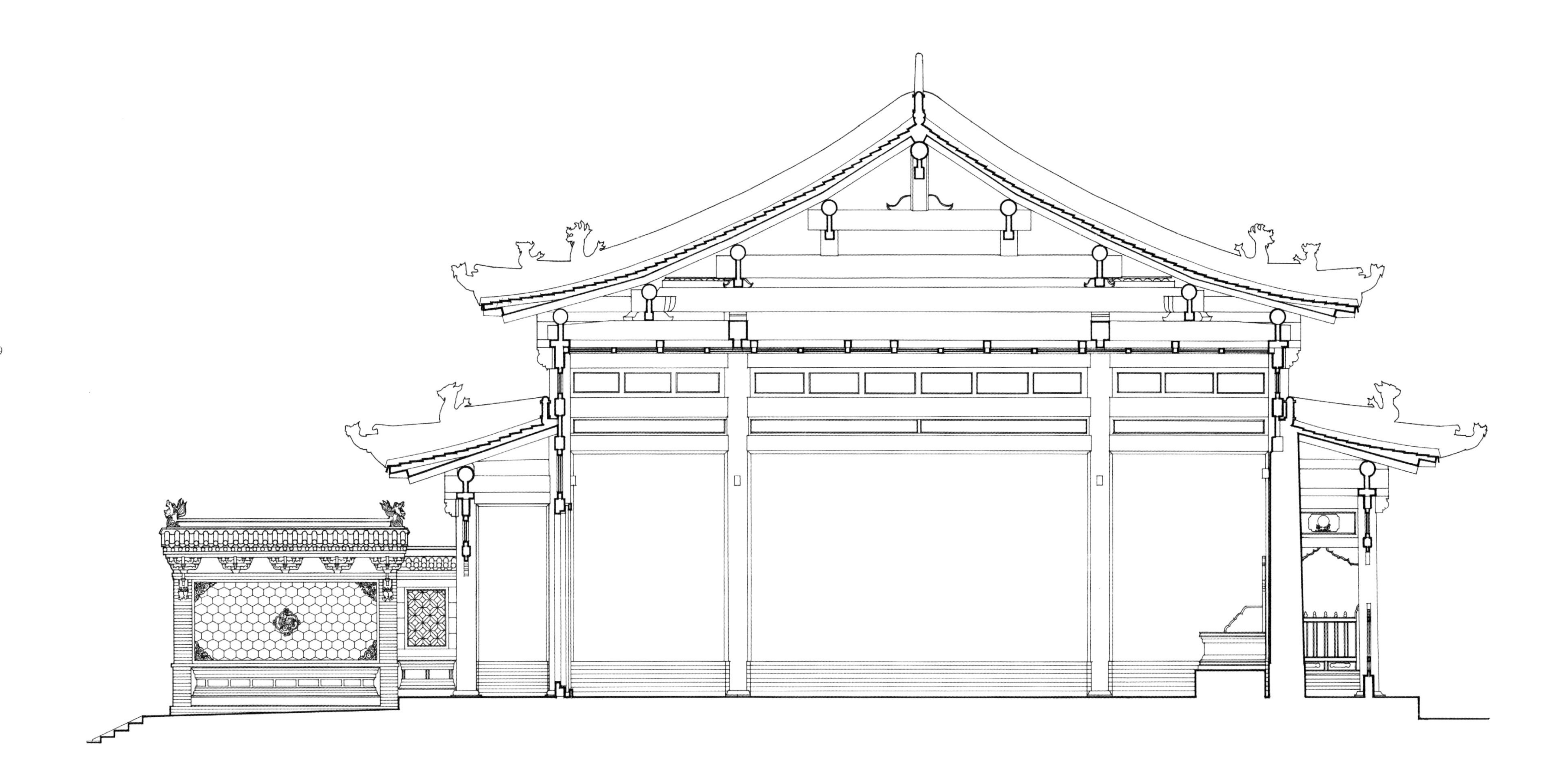

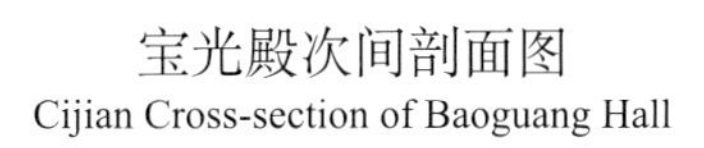

宝光殿次间剖面图
Cijian Cross-section of Baoguang Hall

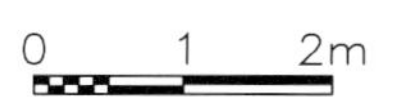

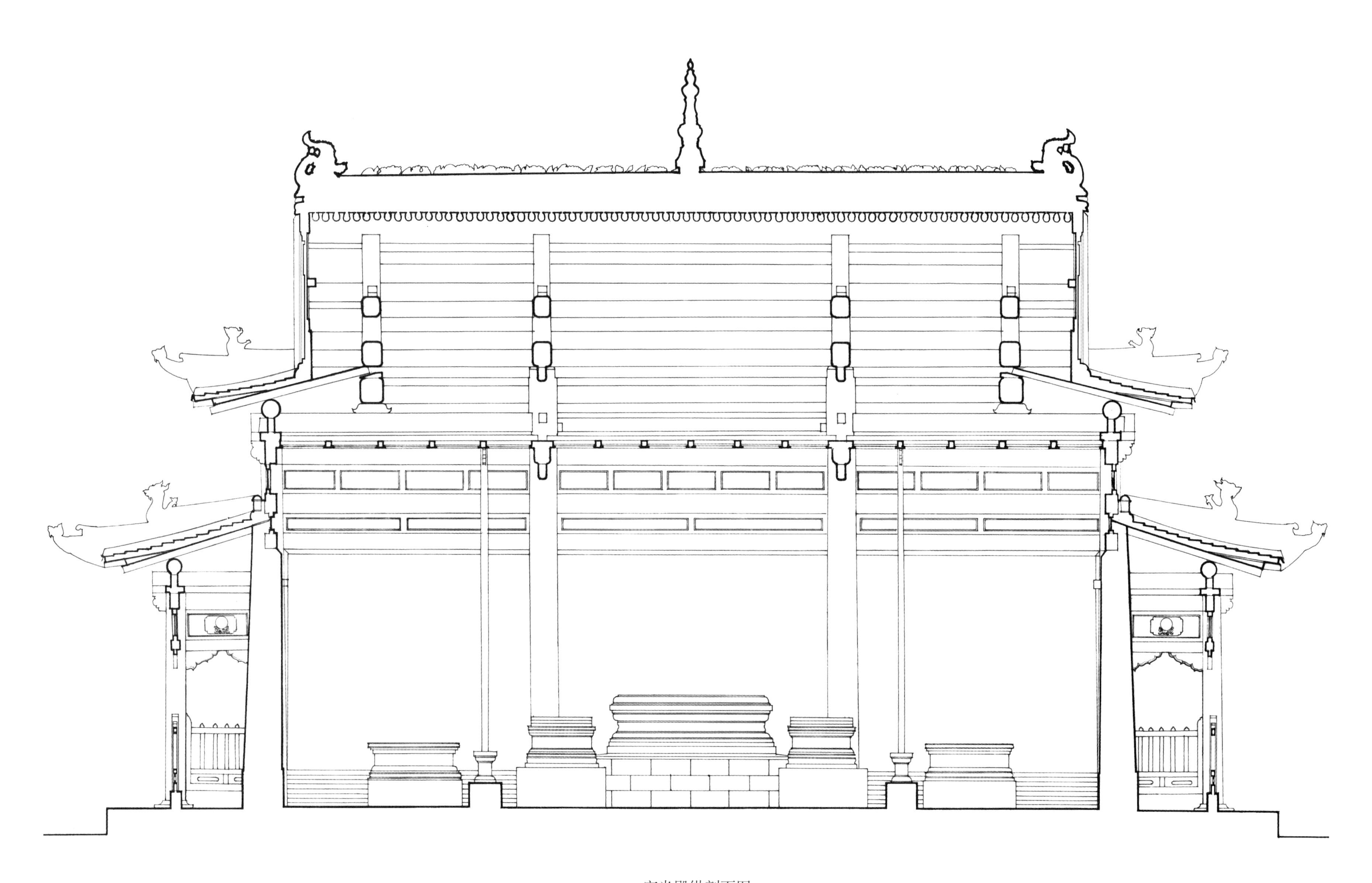

宝光殿纵剖面图
Longitudinal Section of Baoguang Hall

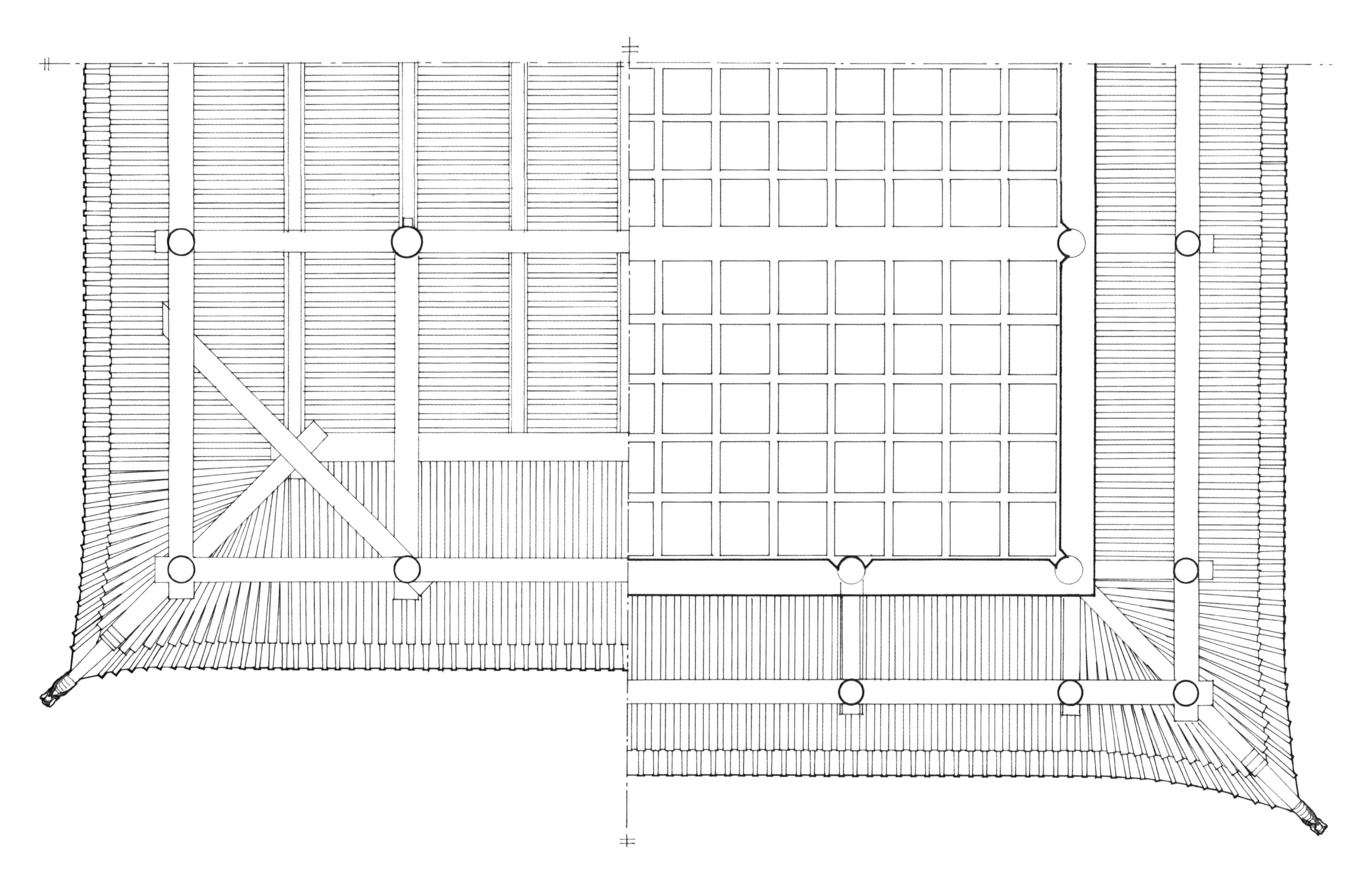

宝光殿梁架仰视图
Framework Plan of Baoguang Hall

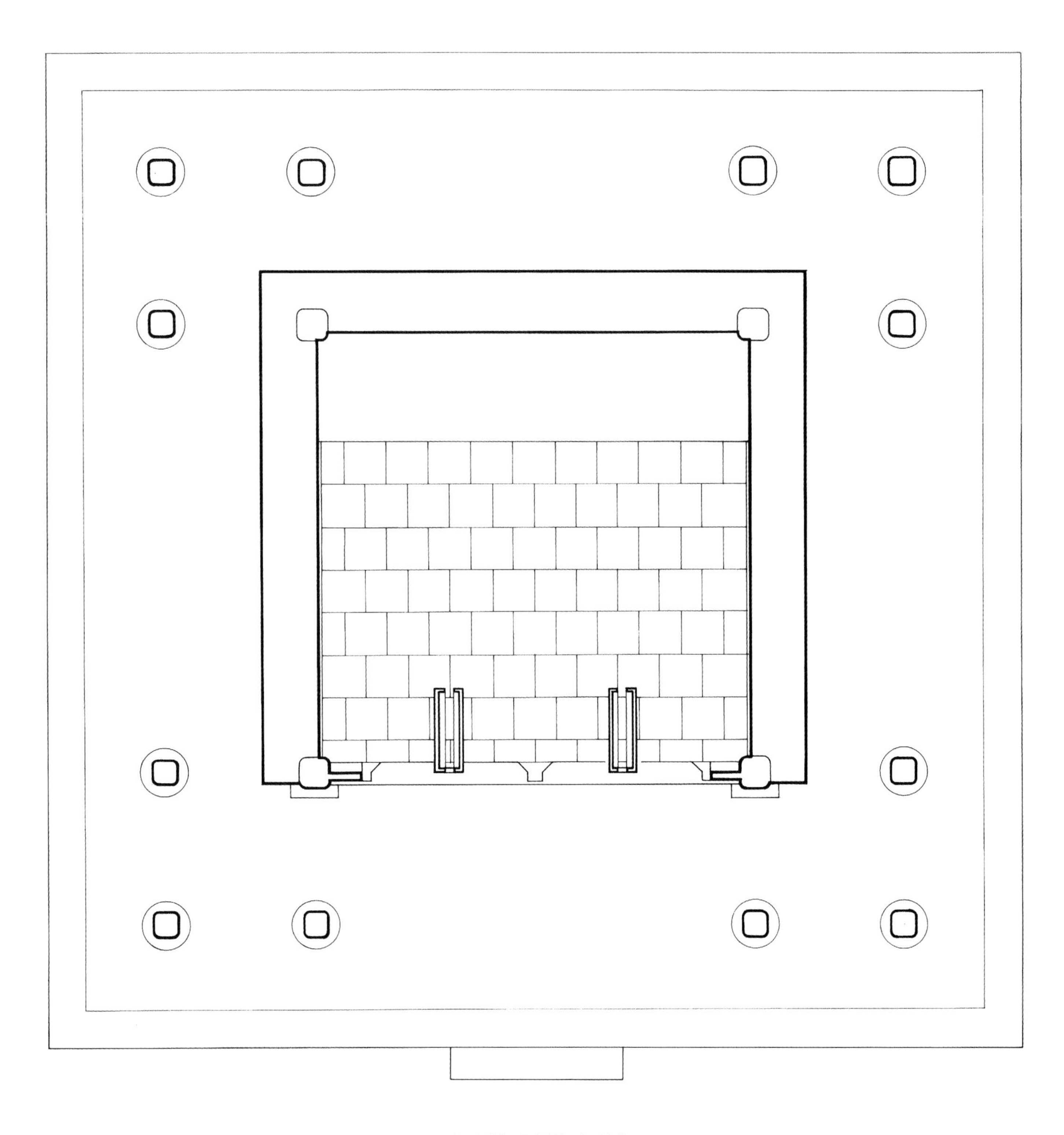

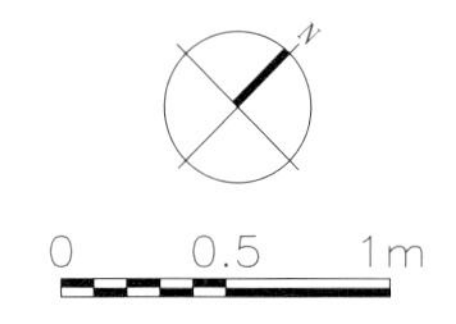

宝光殿西配殿平面图
Plan of West Side Hall of Baoguang Hall

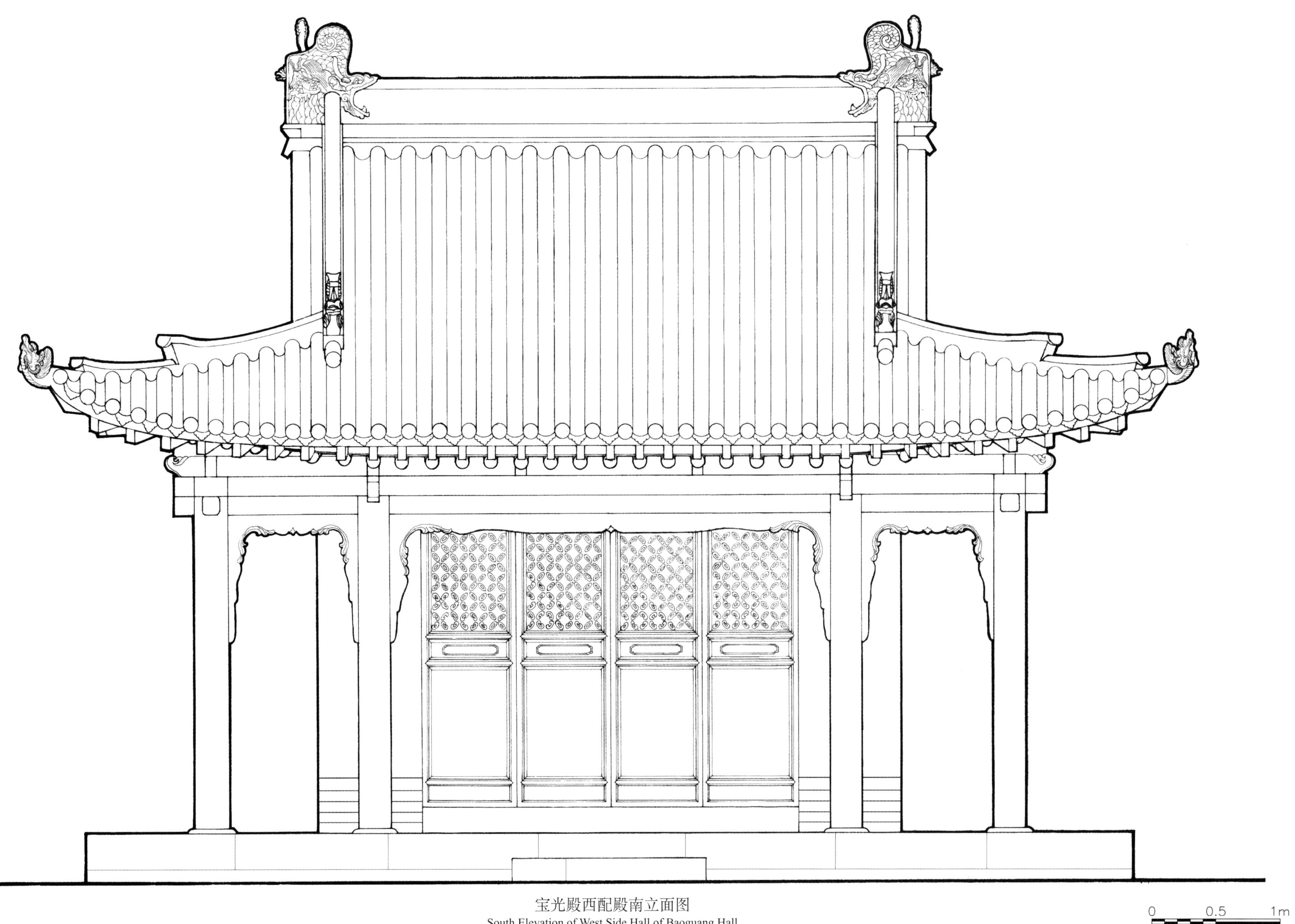

宝光殿西配殿南立面图
South Elevation of West Side Hall of Baoguang Hall

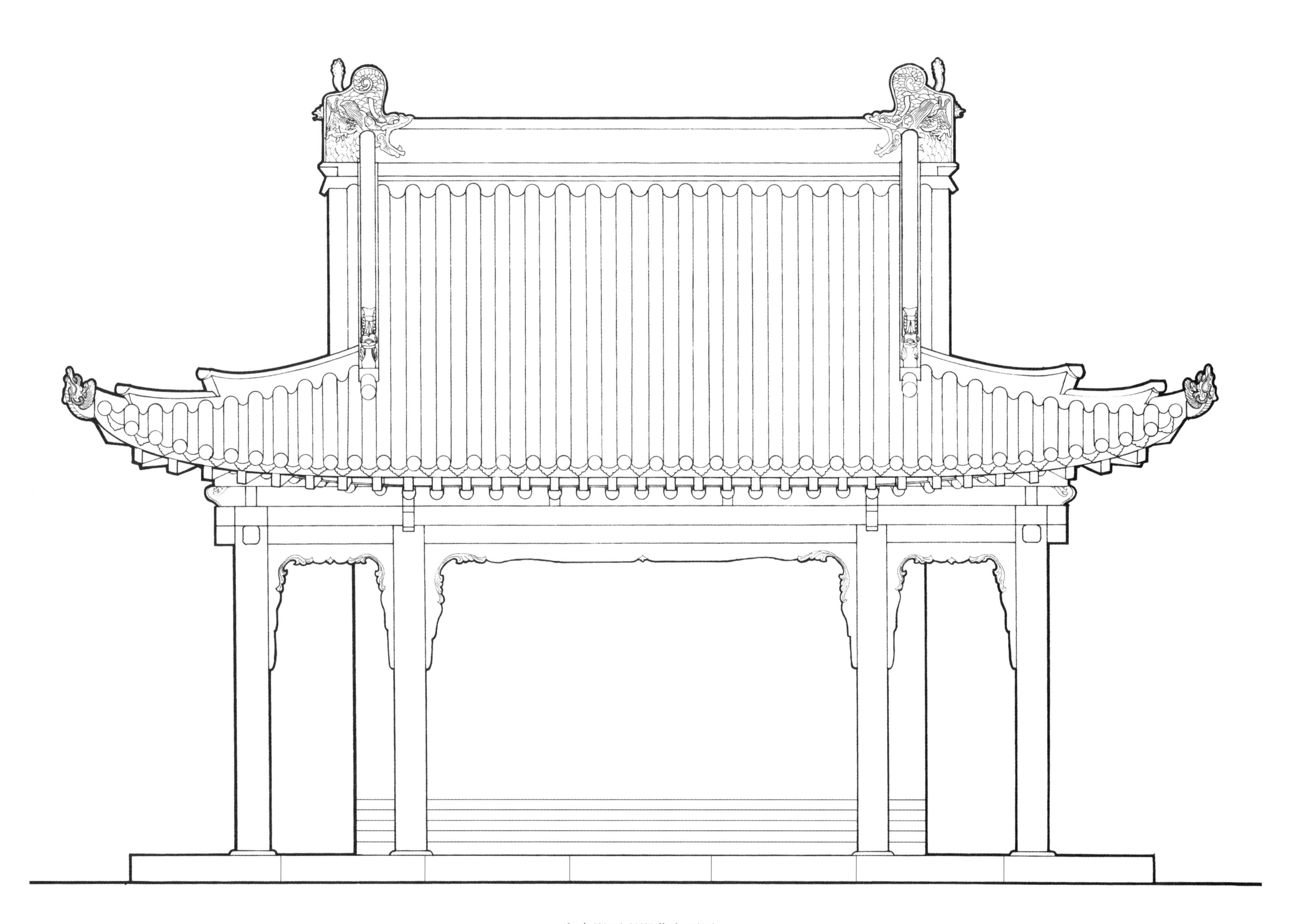

宝光殿西配殿北立面图
North Elevation of West Side Hall of Baoguang Hall

宝光殿西配殿东立面图
East Elevation of West Side Hall of Baoguang Hall

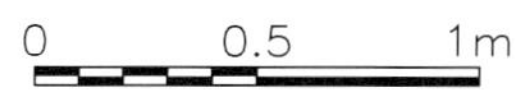

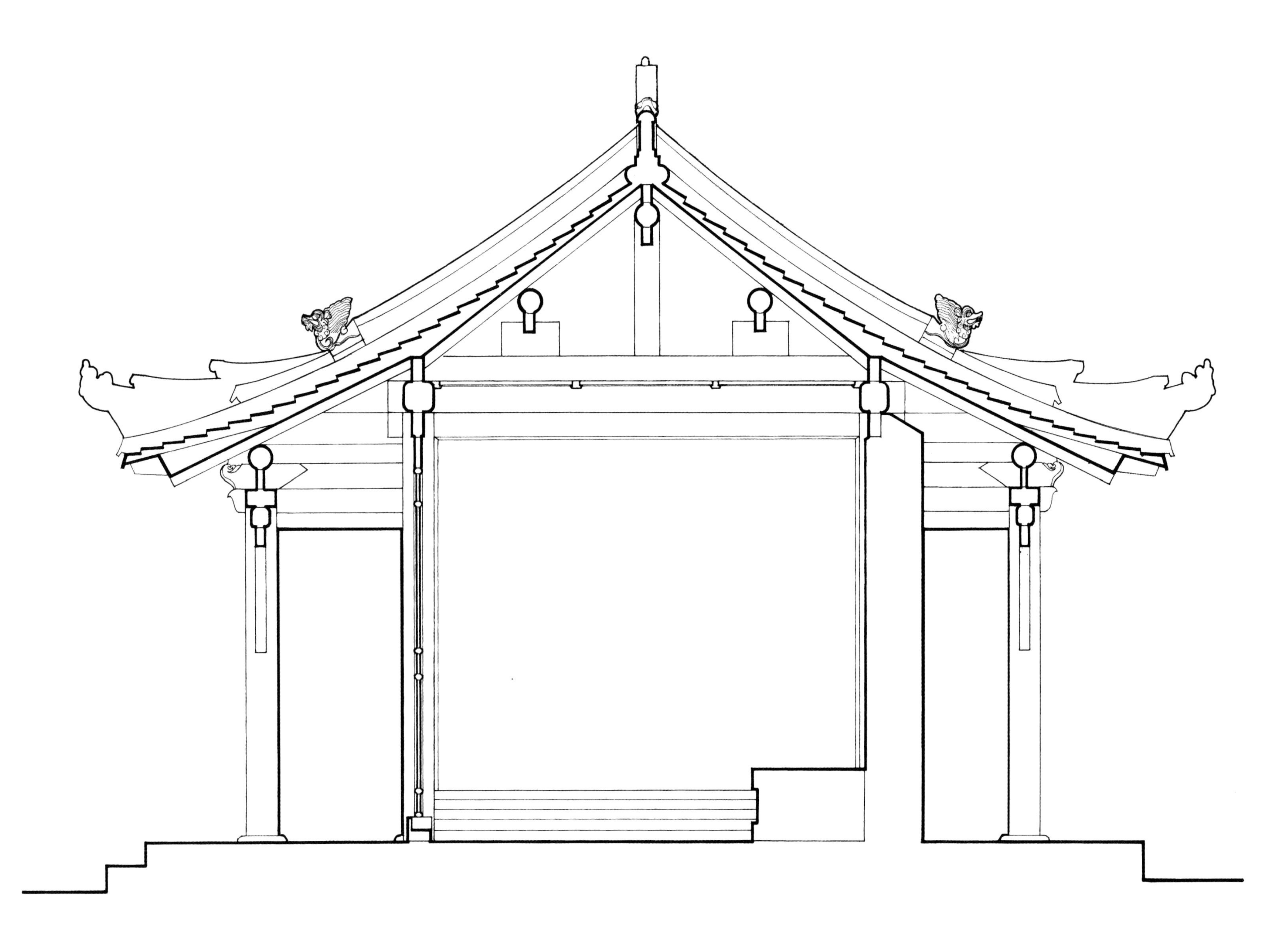

宝光殿西配殿横剖面图
Cross-section of West Side Hall of Baoguang Hall

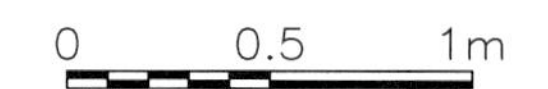

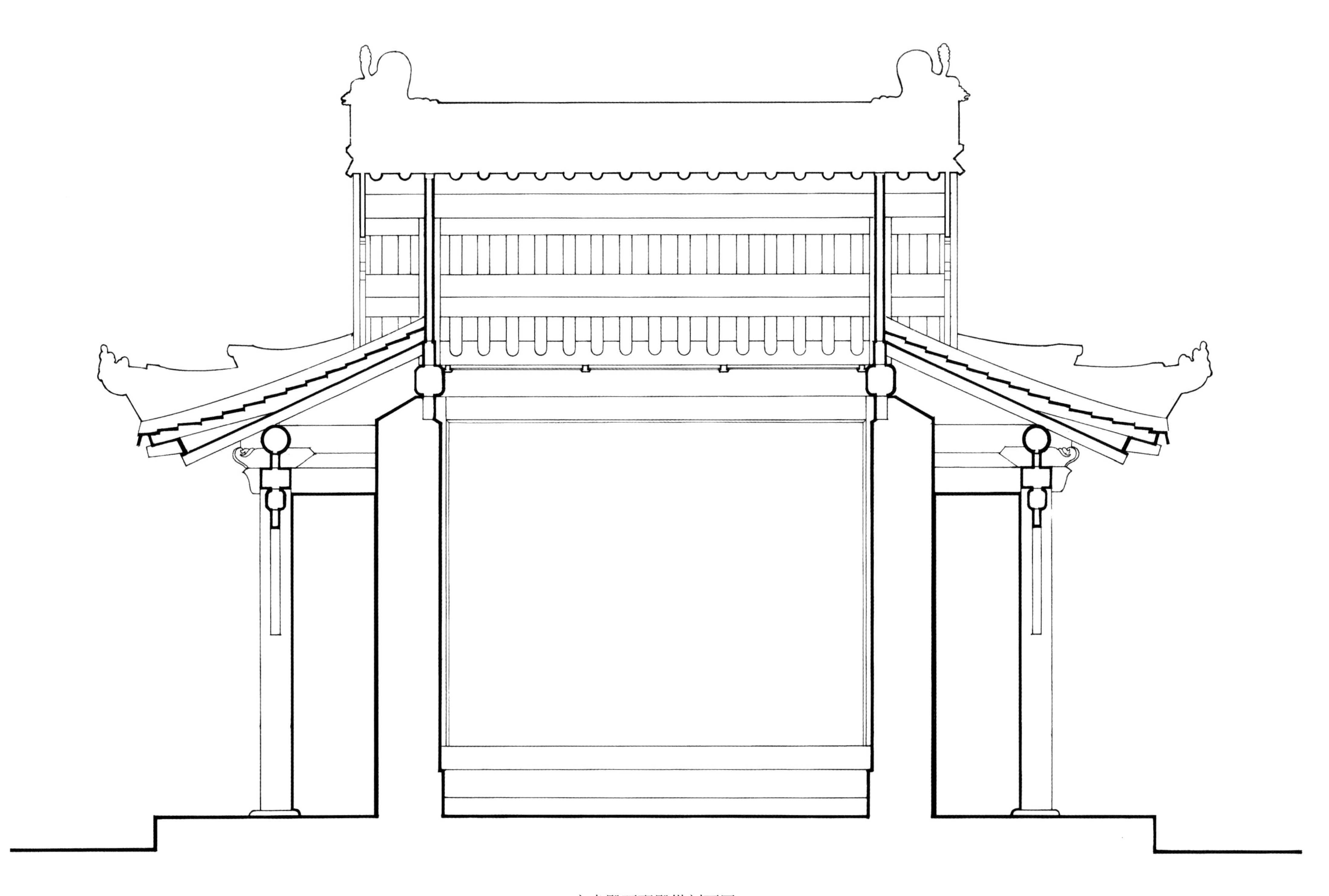

宝光殿西配殿纵剖面图
Longitudinal Section of West Side Hall of Baoguang Hall

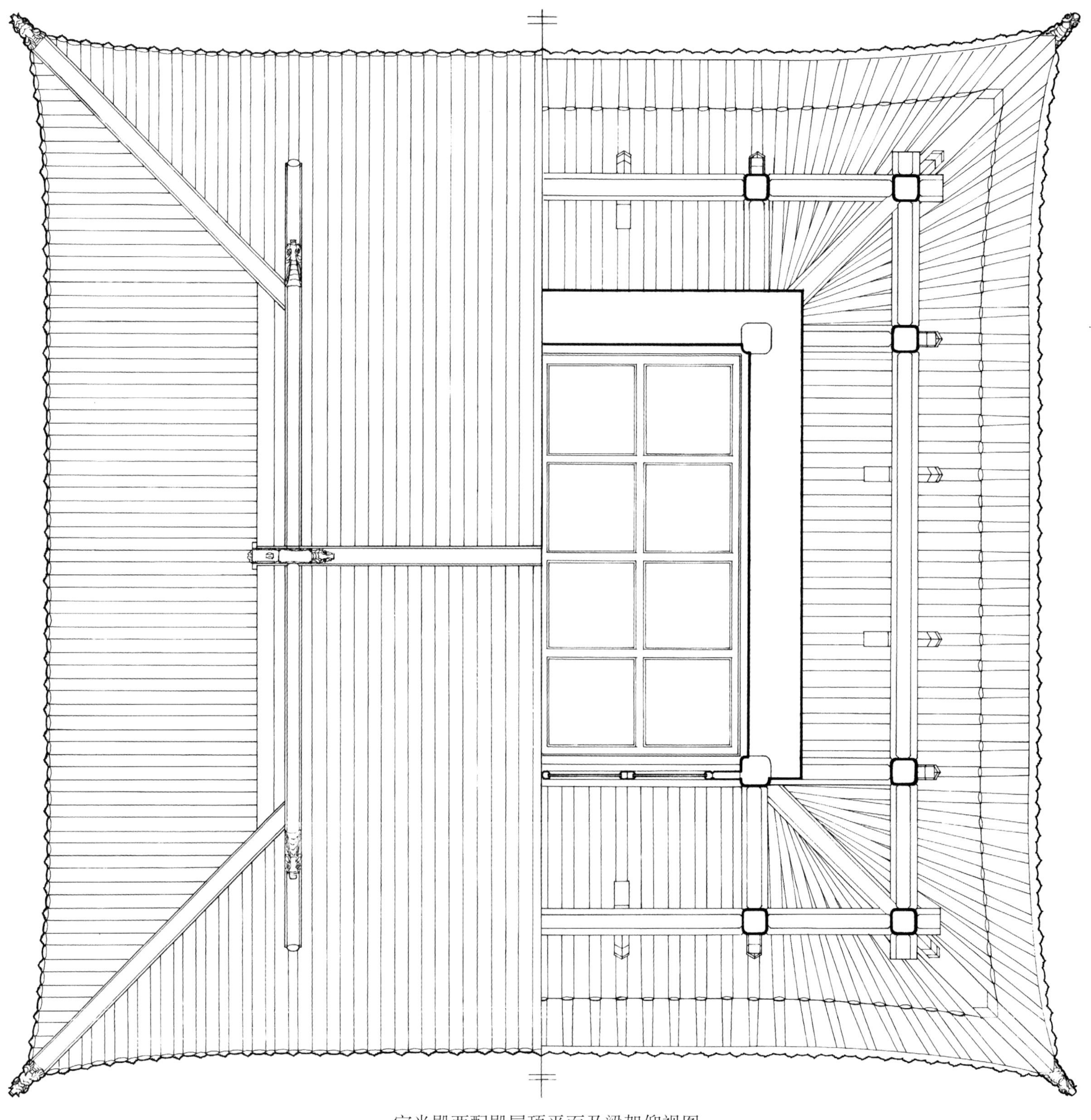

宝光殿西配殿屋顶平面及梁架仰视图

Roof Plan and Framework Plan of West Side Hall of Baoguang Hall

大鼓楼
Great Drum Tower

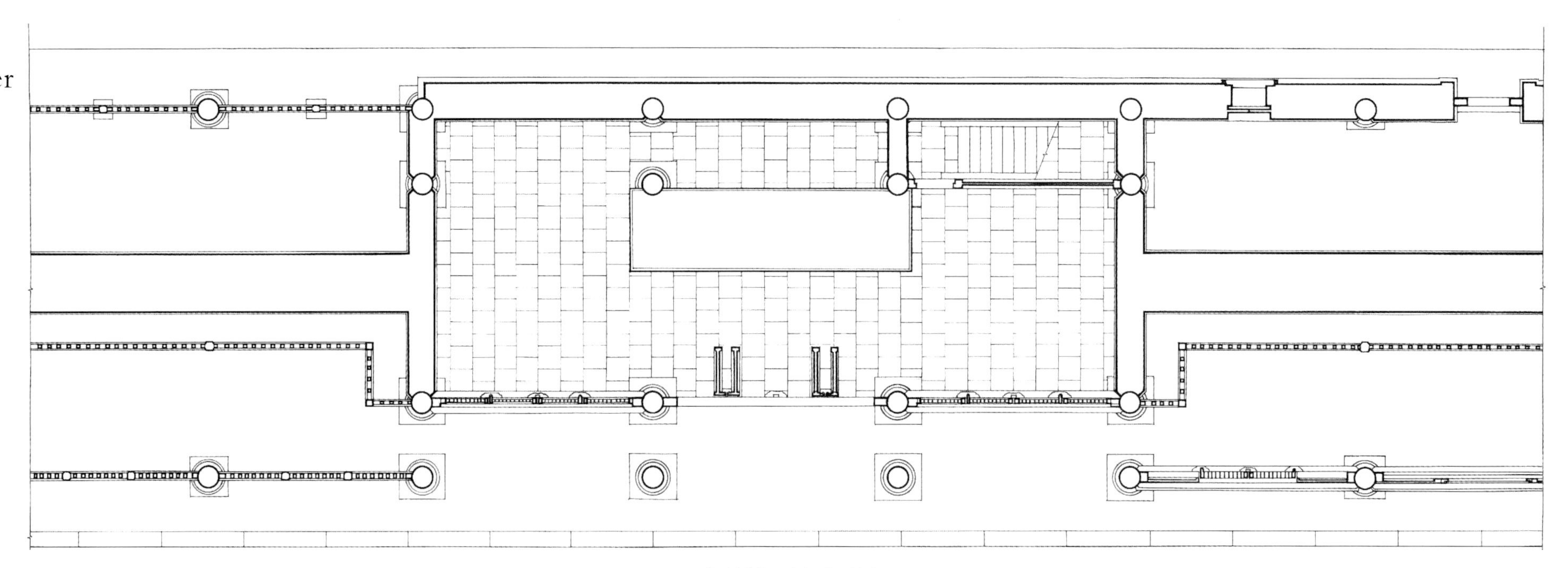

大鼓楼一层平面图
First Floor Plan of Great Drum Tower

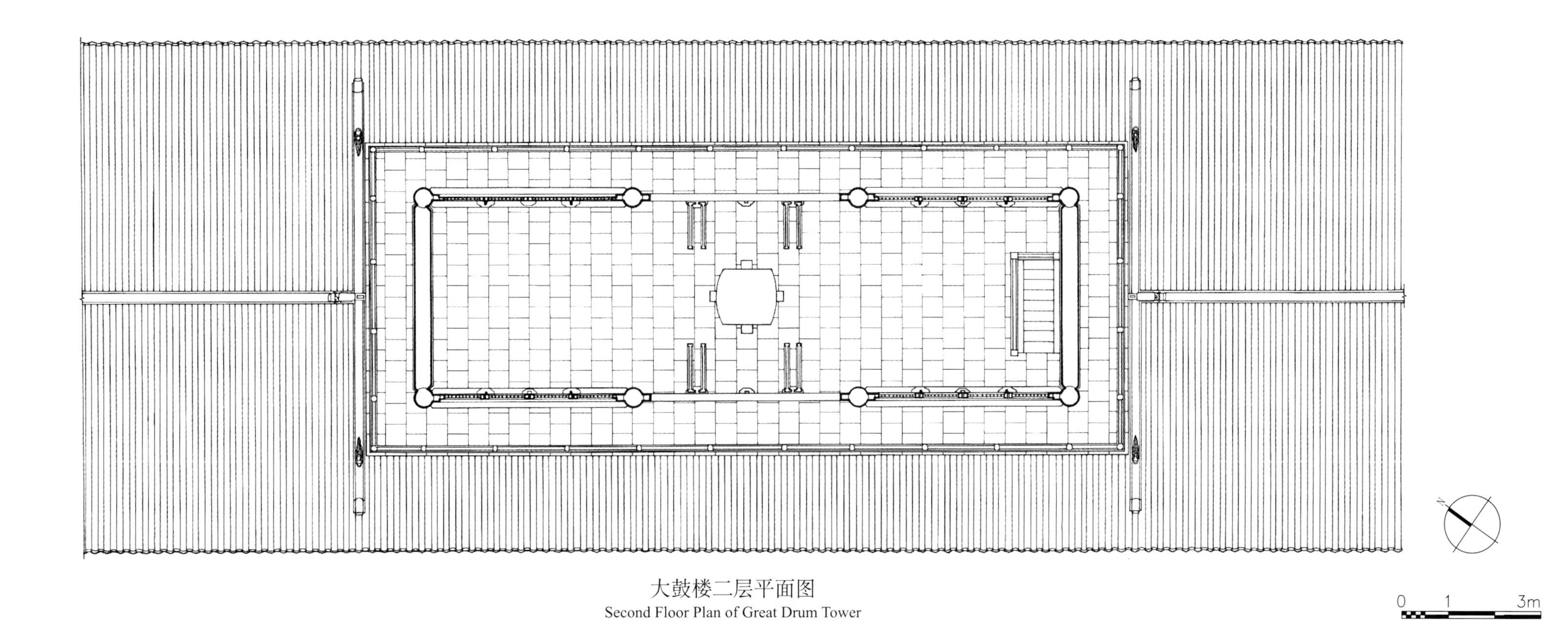

大鼓楼二层平面图
Second Floor Plan of Great Drum Tower

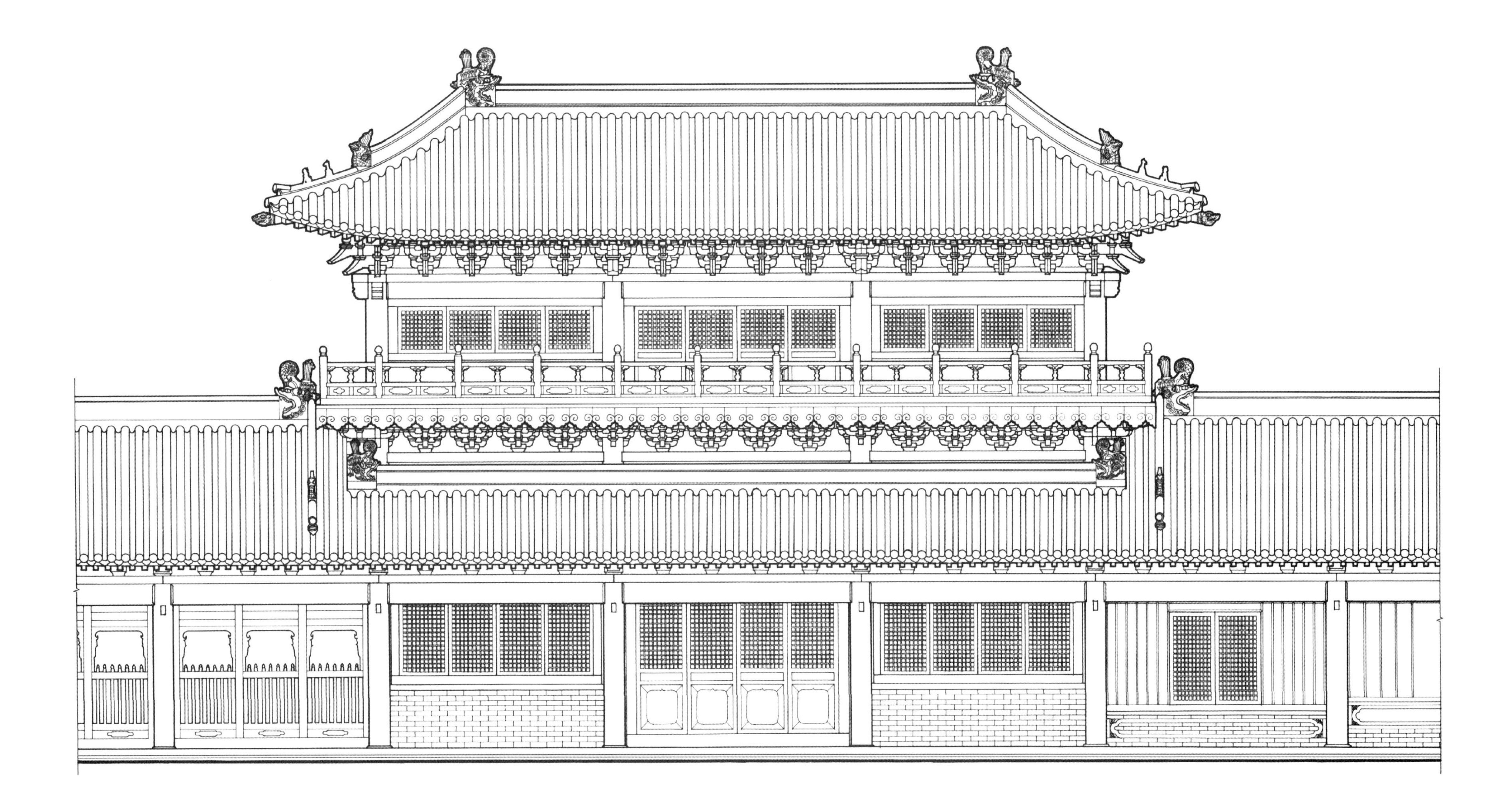

大鼓楼西立面图
West Elevation of Great Drum Tower

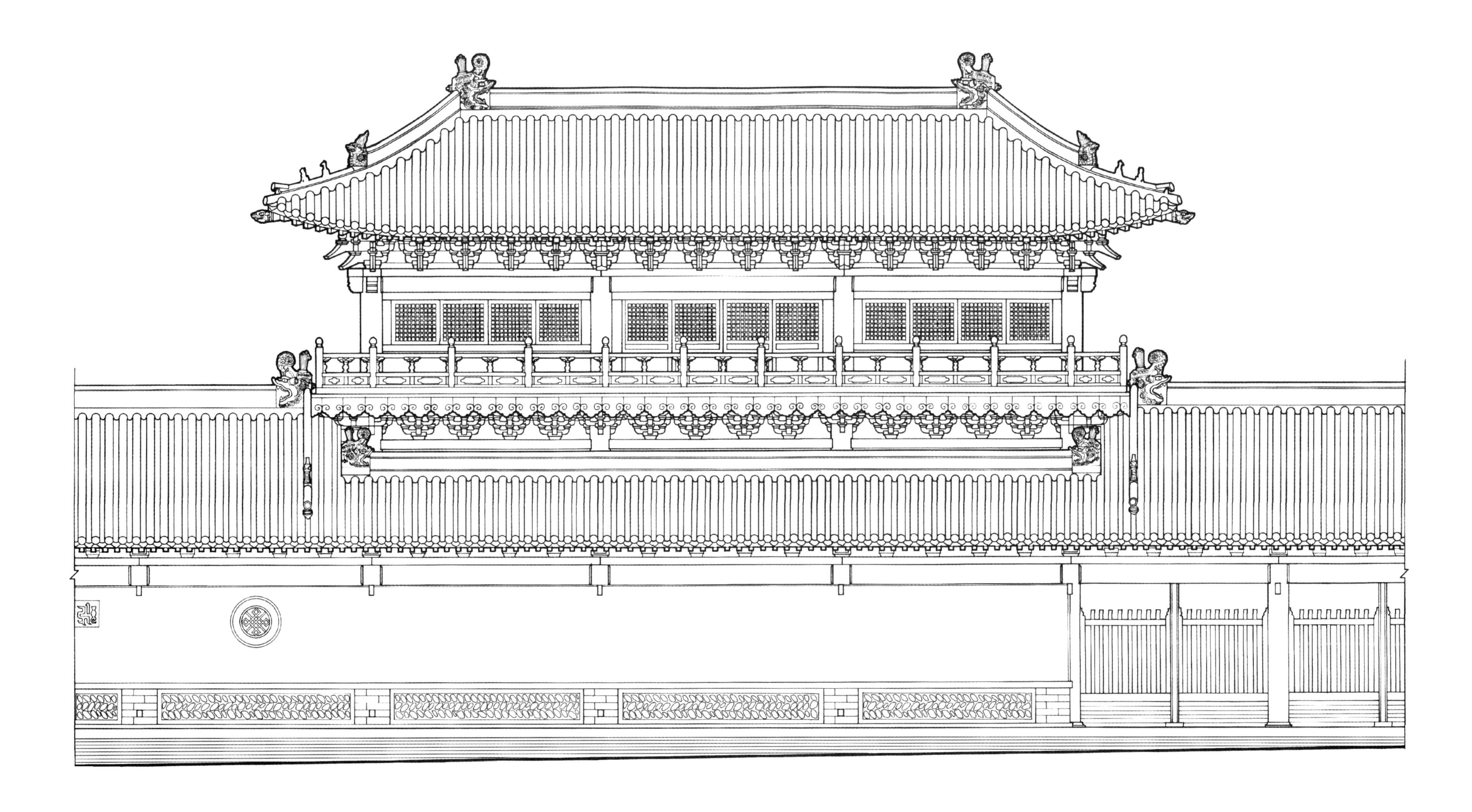

大鼓楼东立面图
East Elevation of Great Drum Tower

大鼓楼南立面图
South Elevation of Great Drum Tower

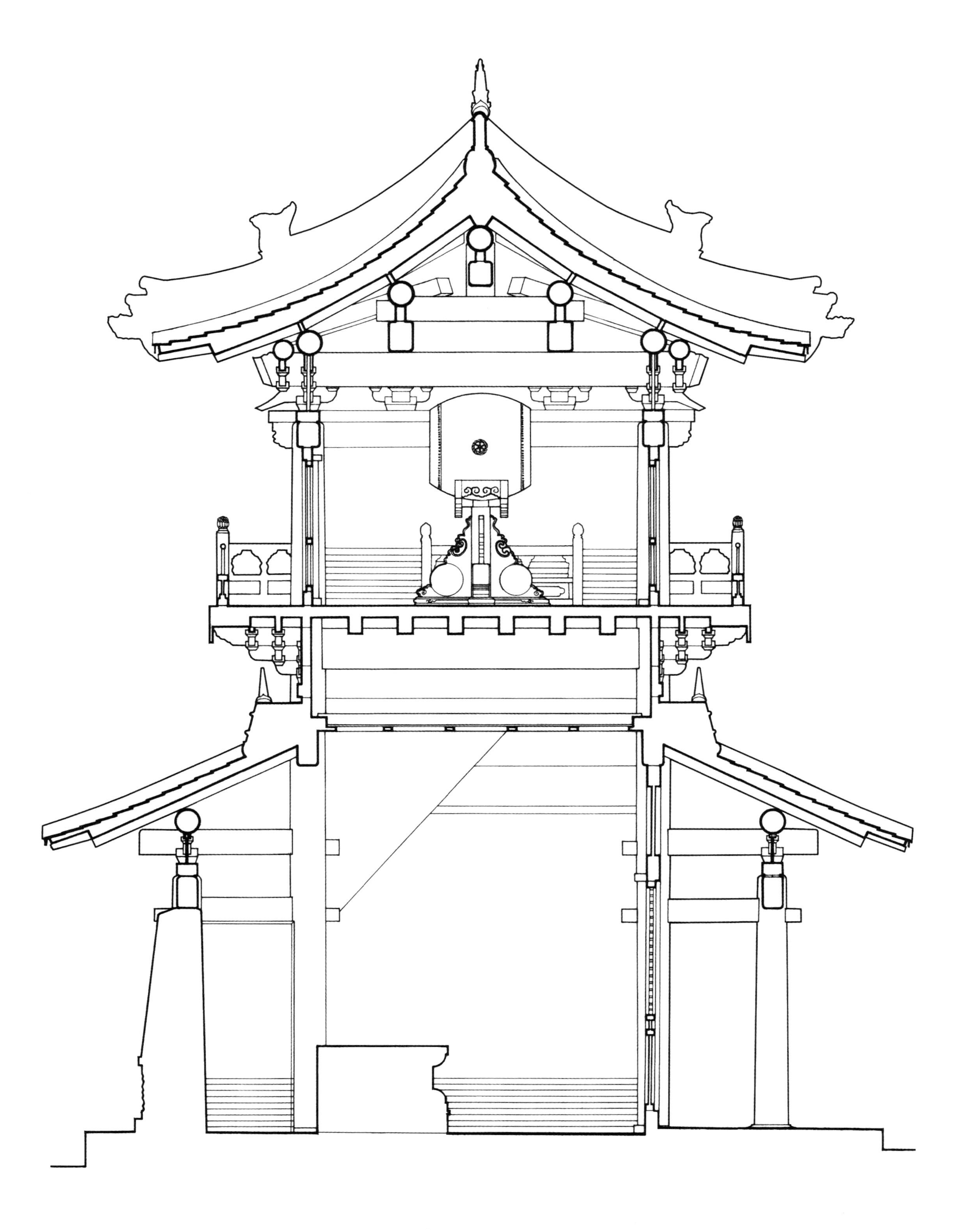

大鼓楼明间剖面图
Mingjian Cross-section of Great Drum Tower

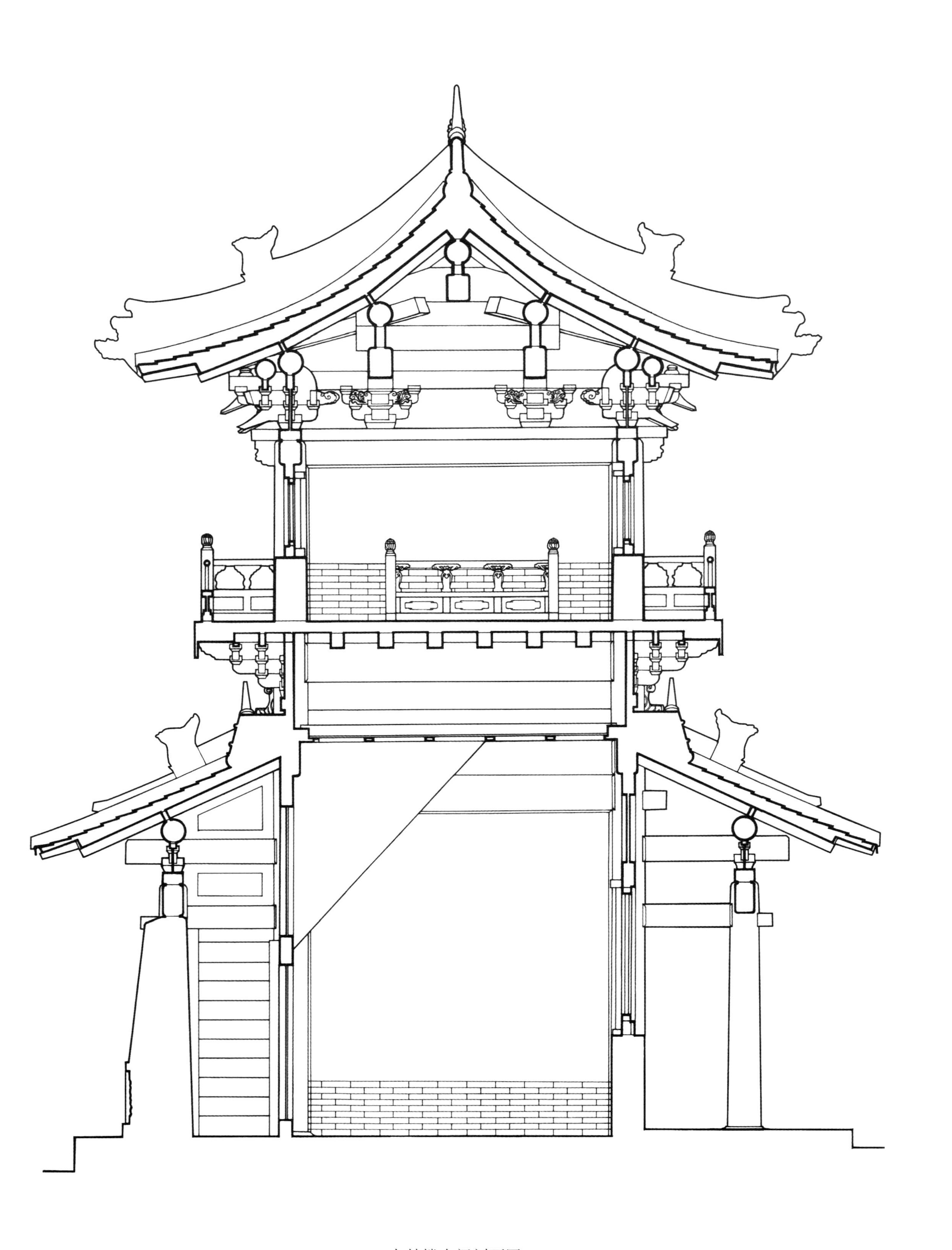

大鼓楼次间剖面图
Cijian Cross-section of Great Drum Tower

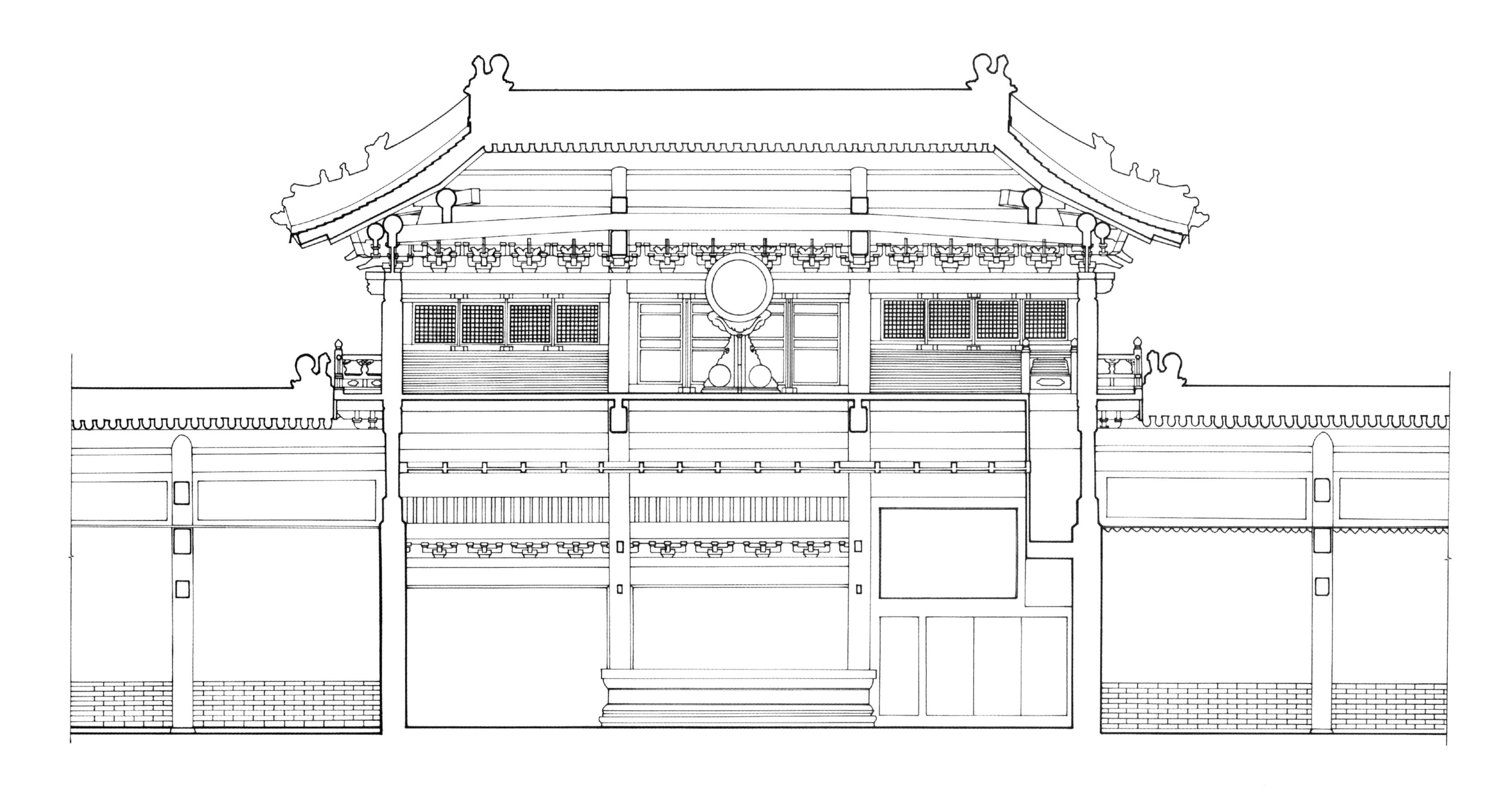

大鼓楼纵剖面图
Longitudinal Section of Great Drum Tower

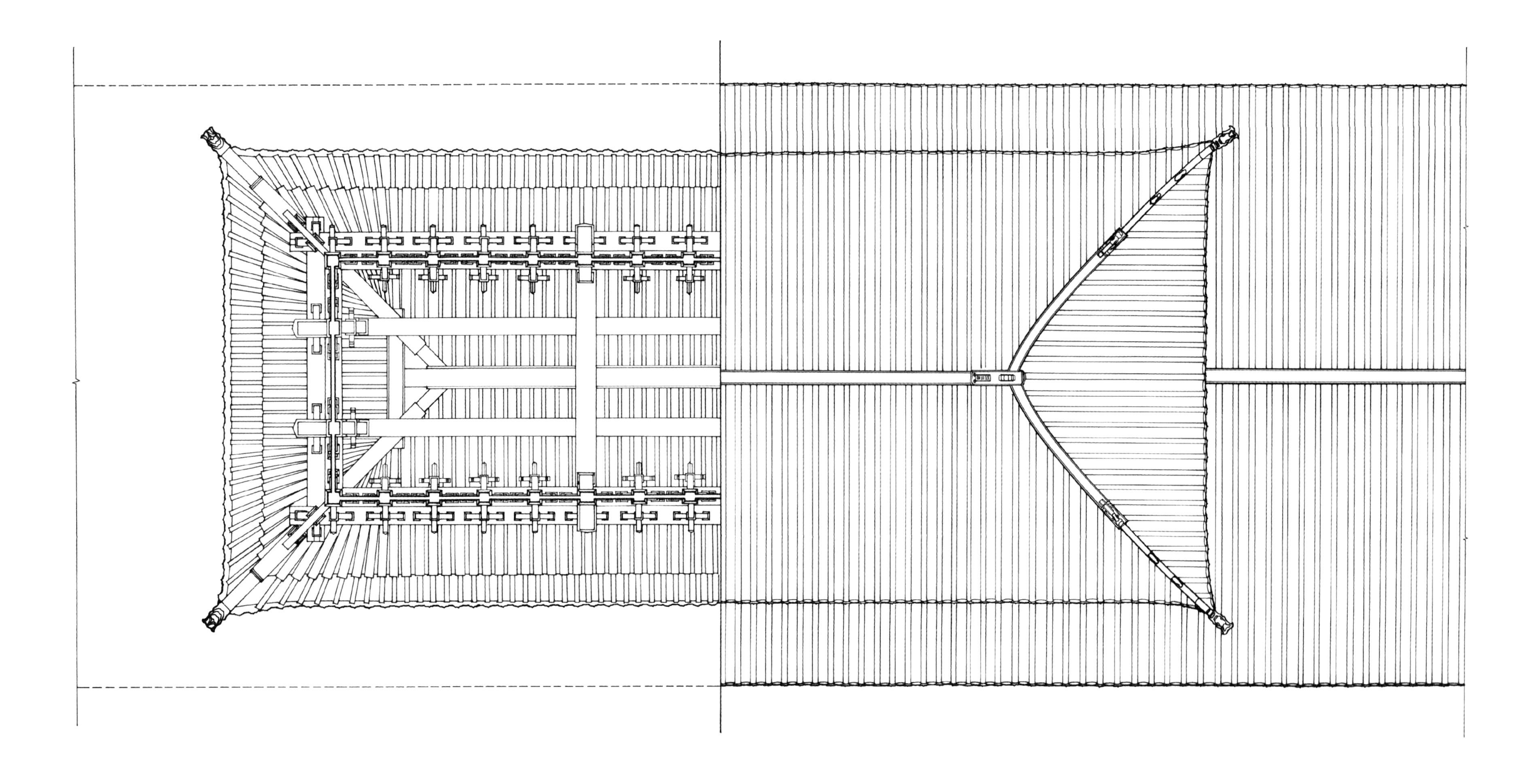

大鼓楼二层梁架仰视图及屋顶平面图
Framework Plan and Roof Plan of Second Floor of Great Drum Tower

0 0.5 1 2m

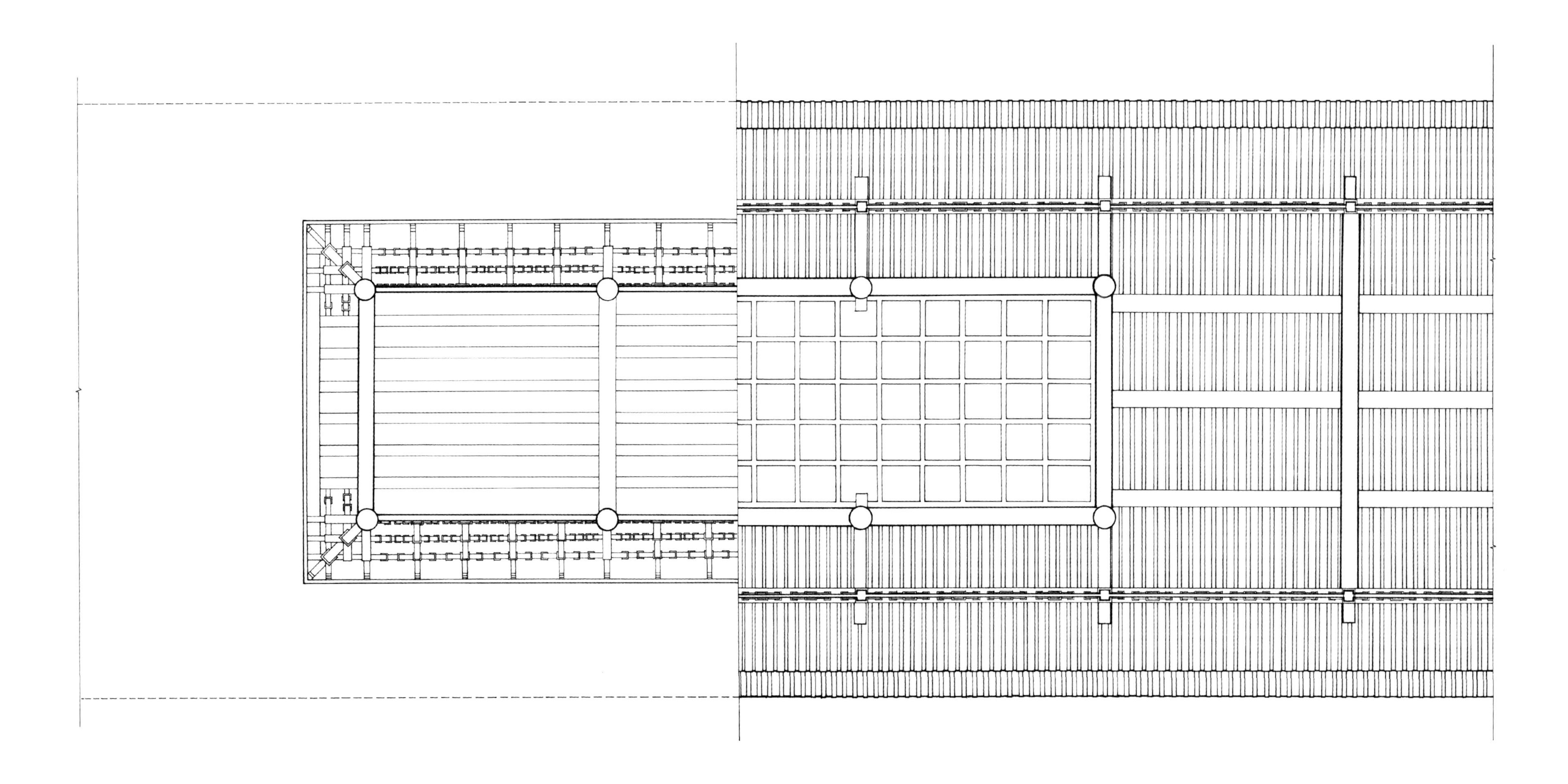

大鼓楼一层及平坐梁架仰视图

Framework Plan of First Floor and Pingzuo Layer of Great Drum Tower

大鼓楼栏杆大样图
Handrail of Great Drum Tower

0 0.05 0.25m

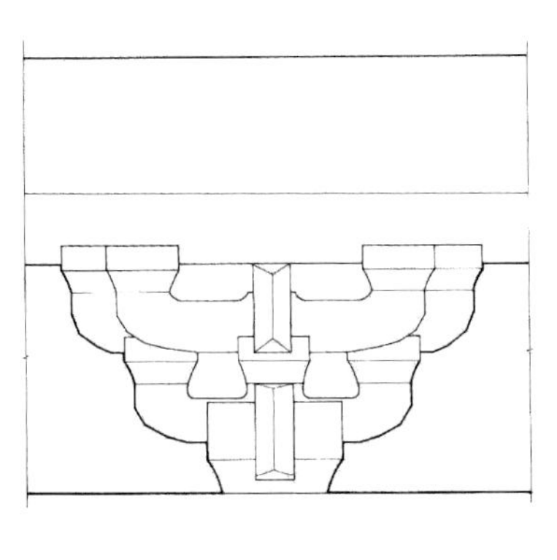

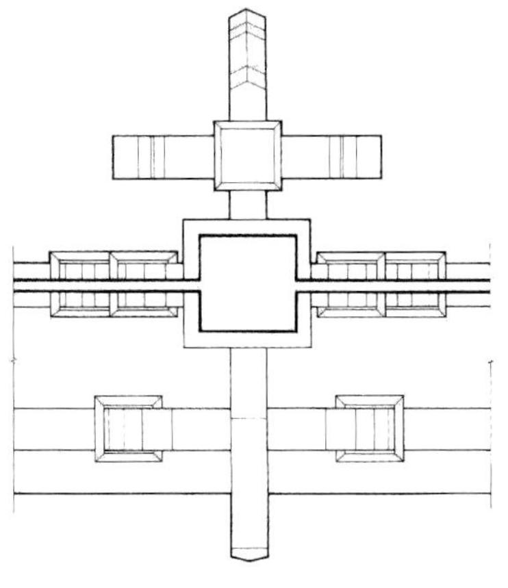

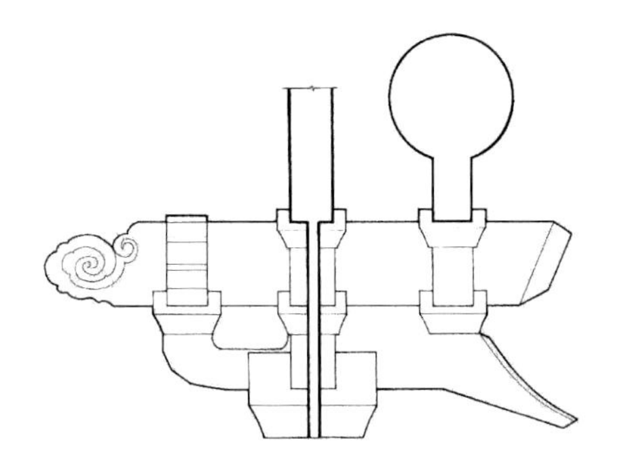

大鼓楼二层平身科斗拱大样图
Pingshenke Dougong of Second Floor of Great Drum Tower

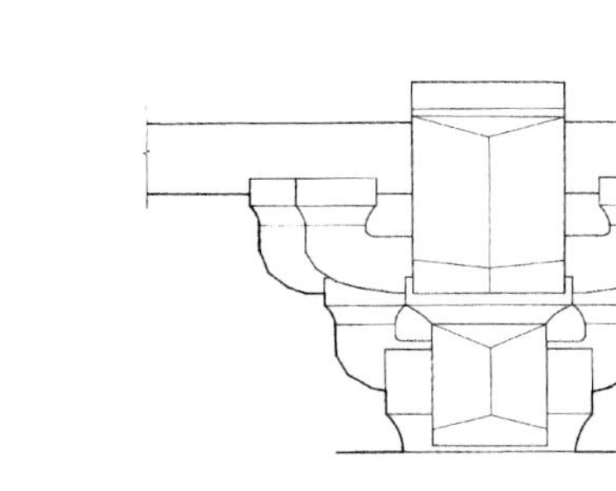

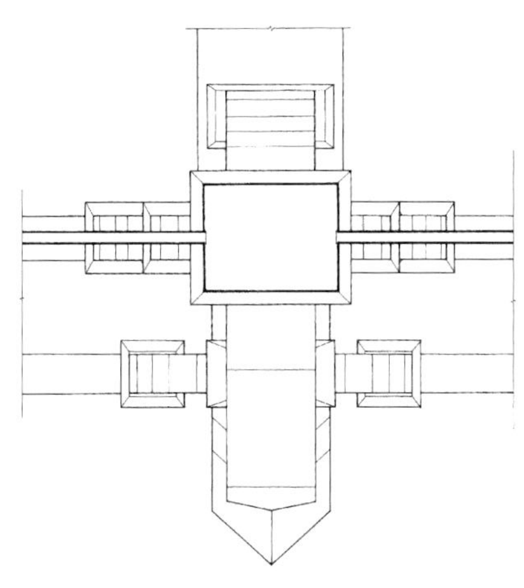

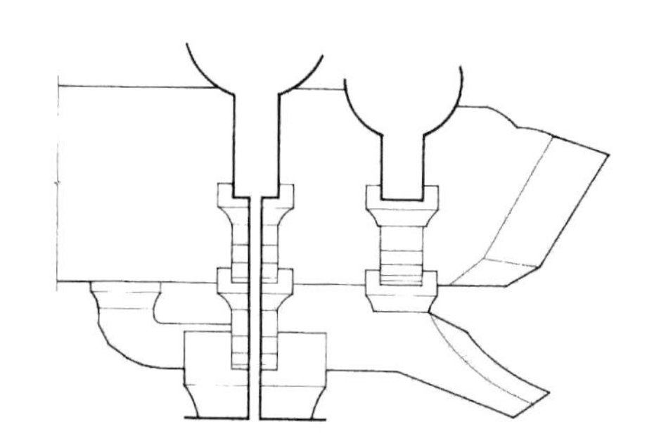

大鼓楼二层柱头科斗拱大样图
Zhutouke Dougong of Second Floor of Great Drum Tower

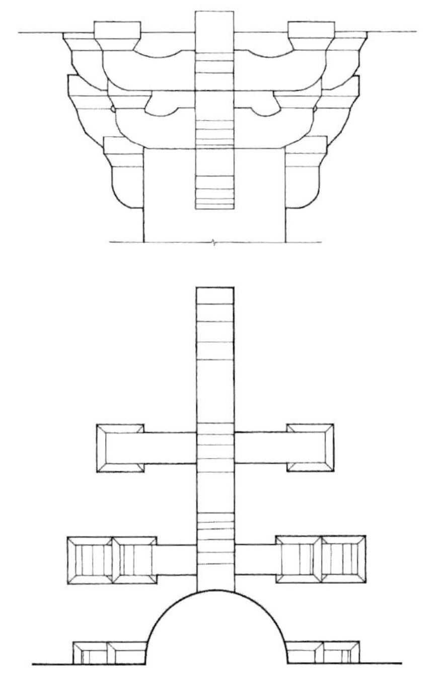

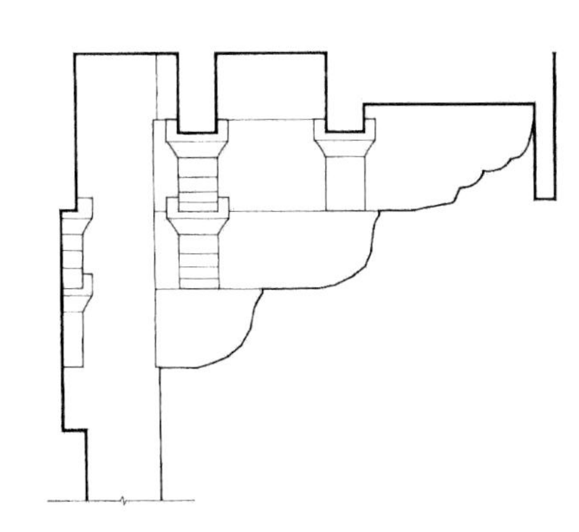

大鼓楼平坐柱头科斗拱大样图
Zhutouke Dougong of Pingzuo Layer of Great Drum Tower

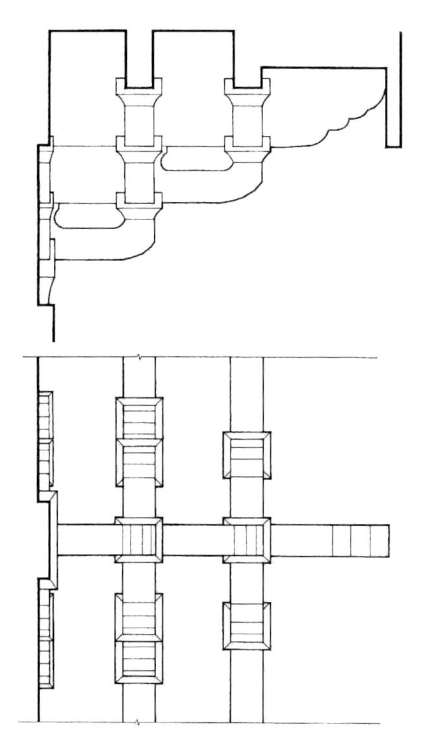

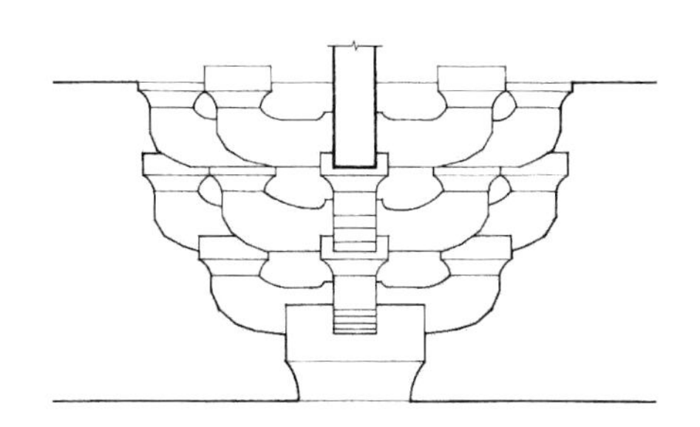

大鼓楼平坐平身科斗拱大样图
Pingshenke Dougong of Pingzuo Layer of Great Drum Tower

大钟楼
Great Bell Tower

大钟楼一层平面图
First Floor Plan of Great Bell Tower

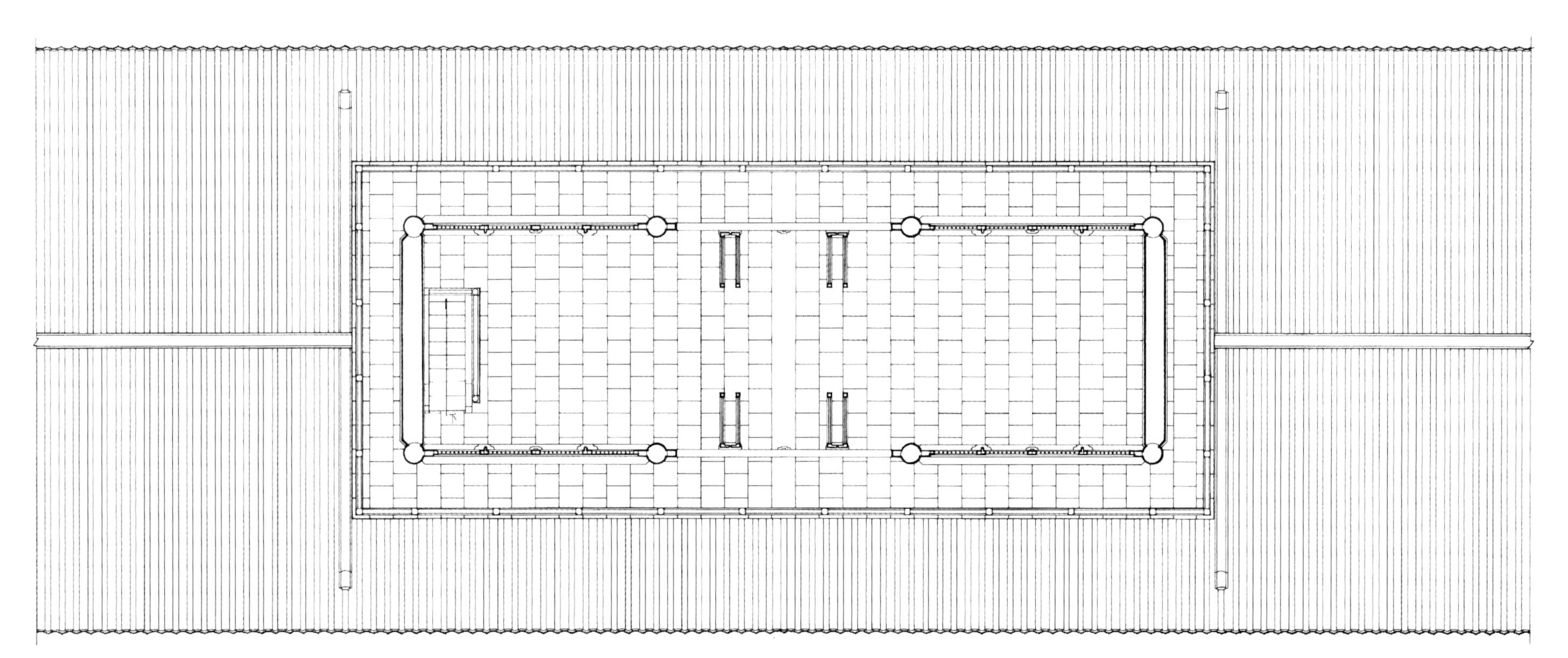

0 1 2 3m

大钟楼二层平面图
Second Floor Plan of Great Bell Tower

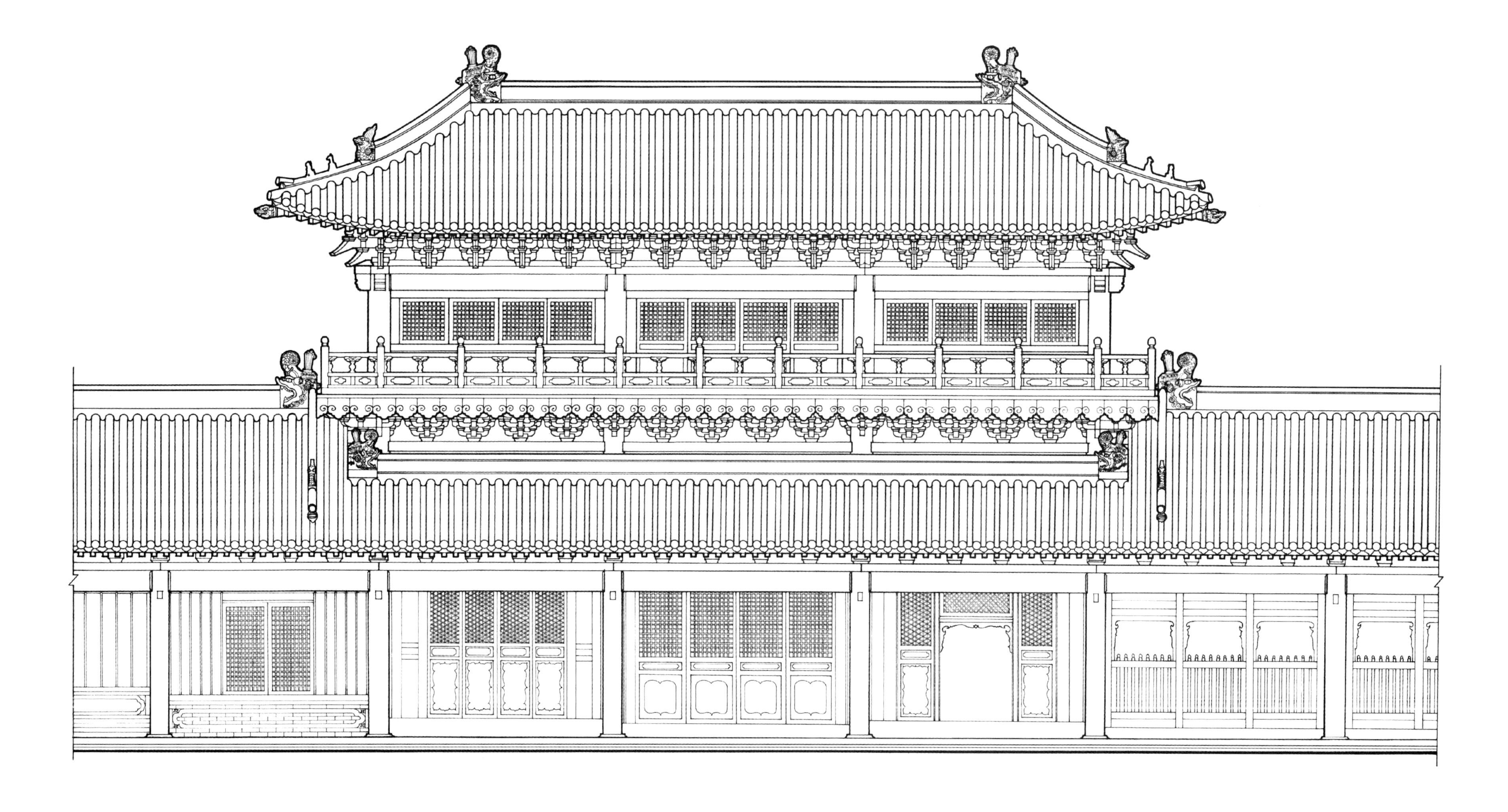

大钟楼东立面图

East Elevation of Great Bell Tower

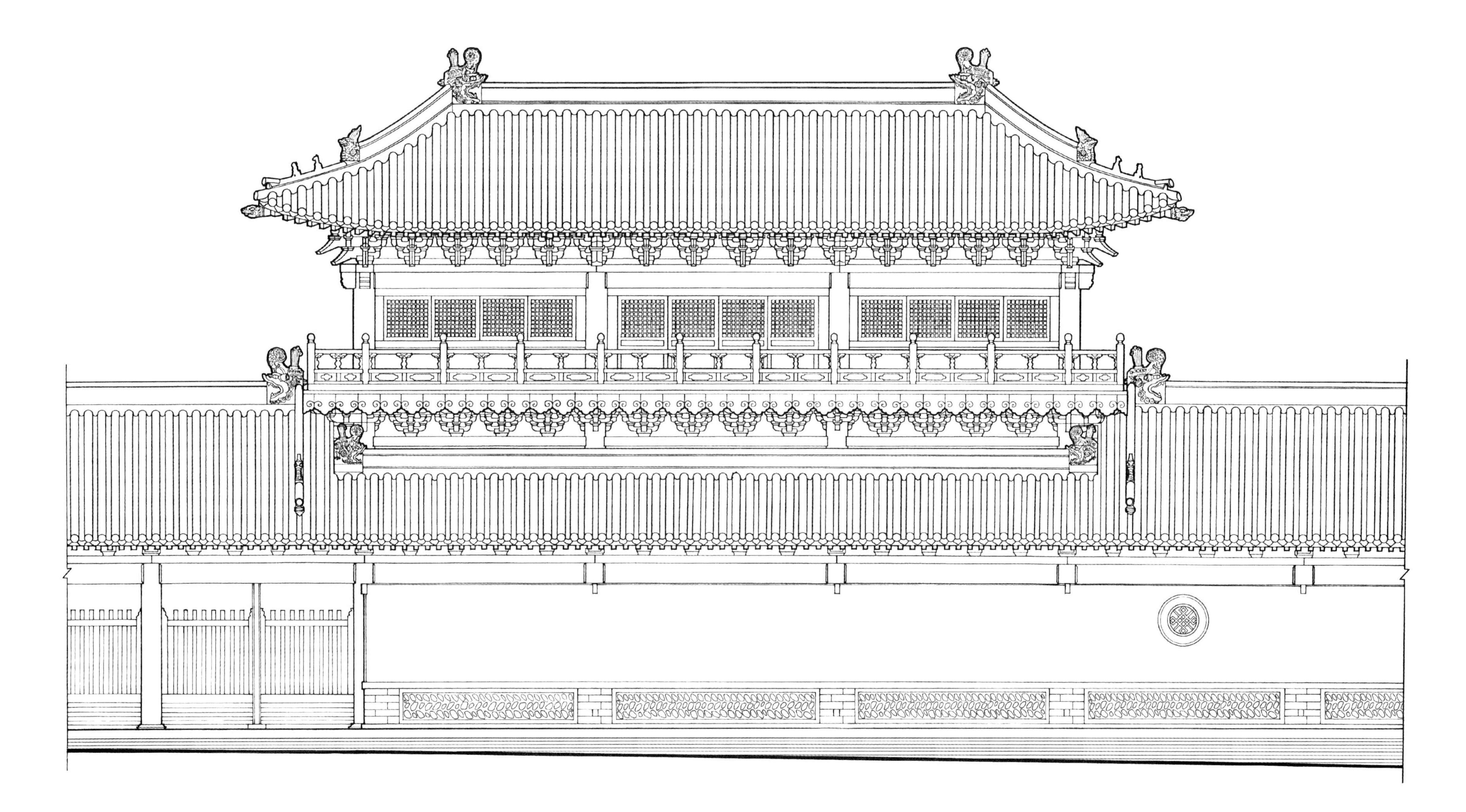

大钟楼西立面图
West Elevation of Great Bell Tower

大钟楼南立面图
South Elevation of Great Bell Tower

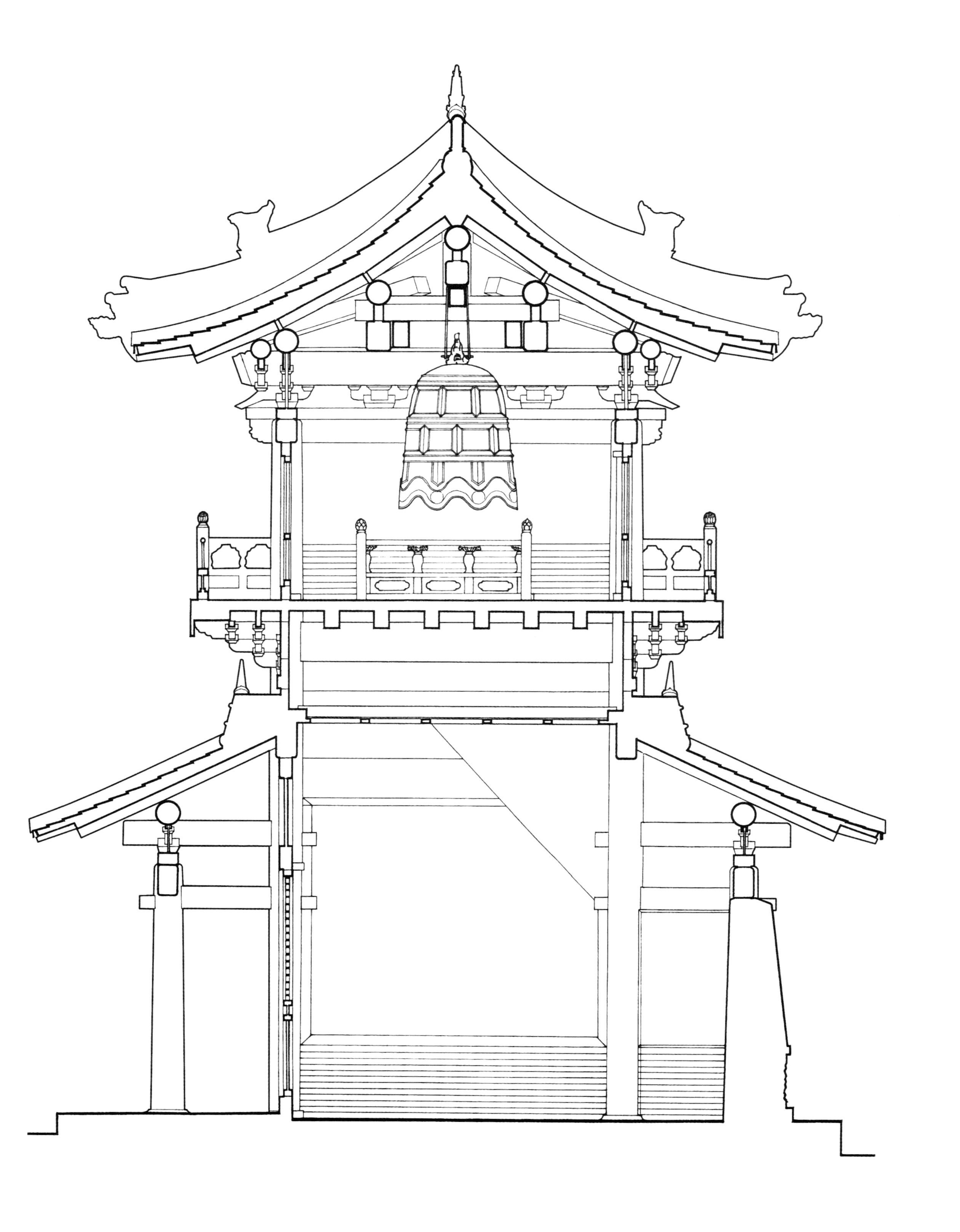

大钟楼明间剖面图
Mingjian Cross-section of Great Bell Tower

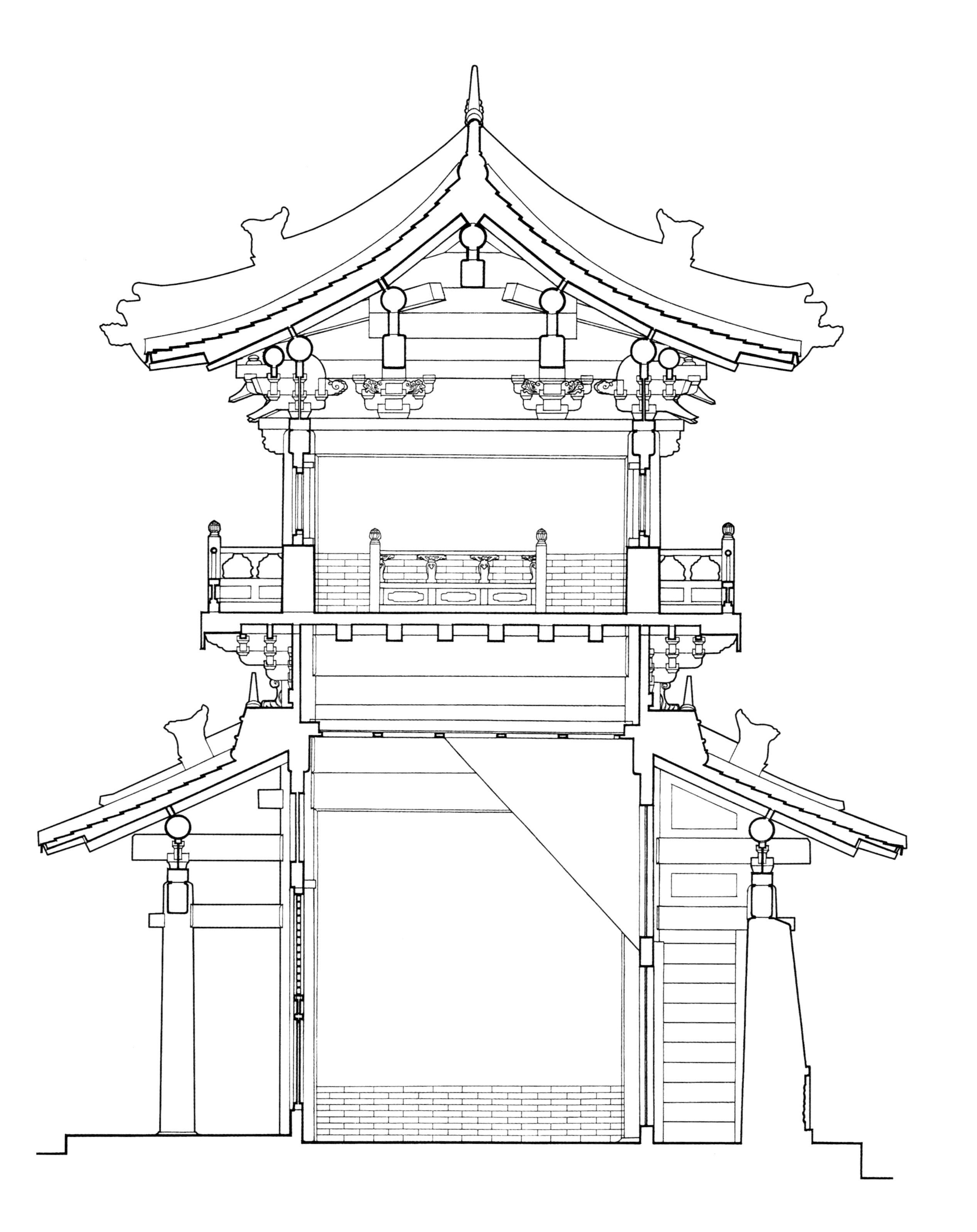

大钟楼次间剖面图
Cijian Cross-section of Great Bell Tower

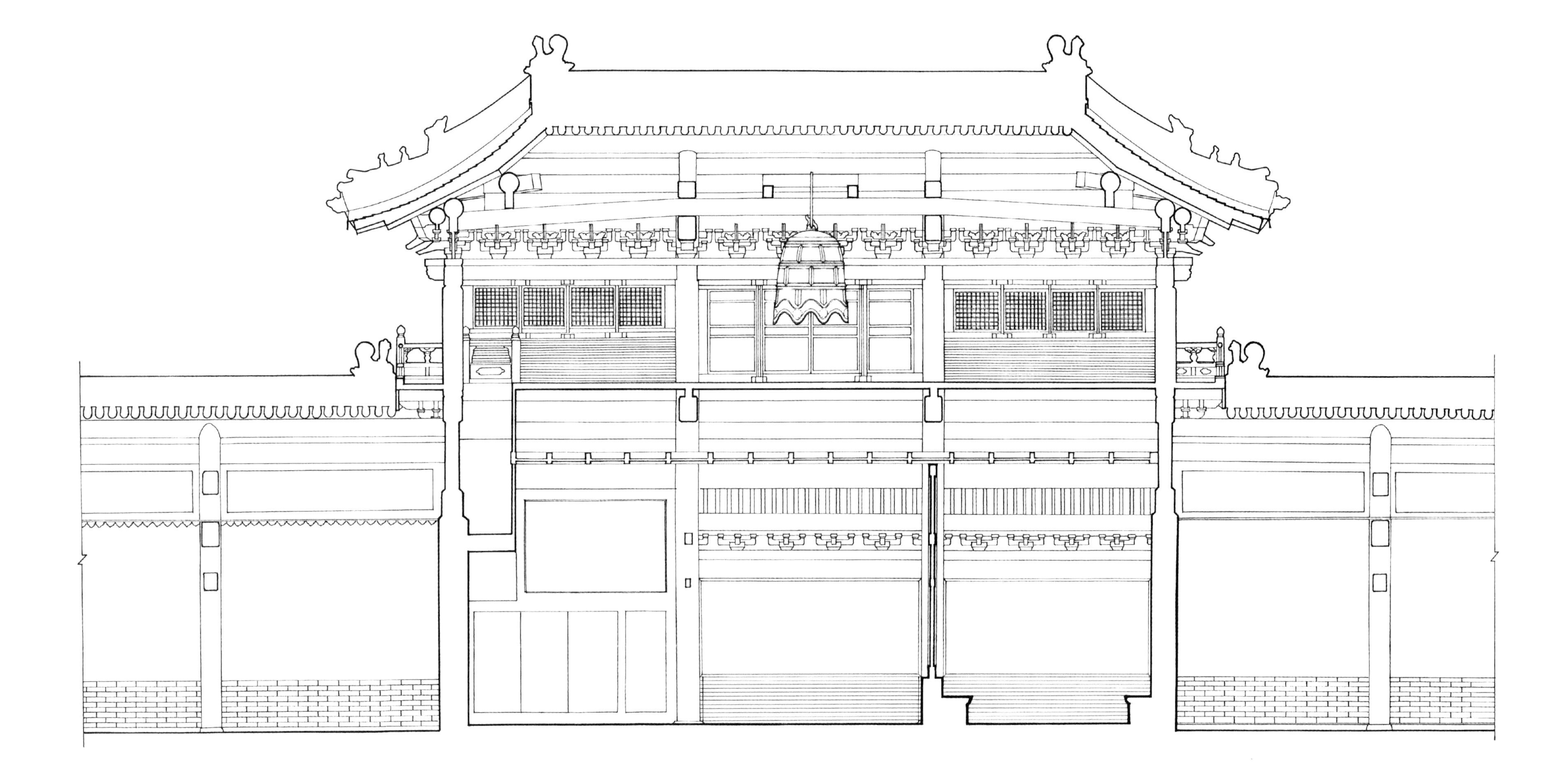

大钟楼纵剖面图
Longitudinal Section of Great Bell Tower

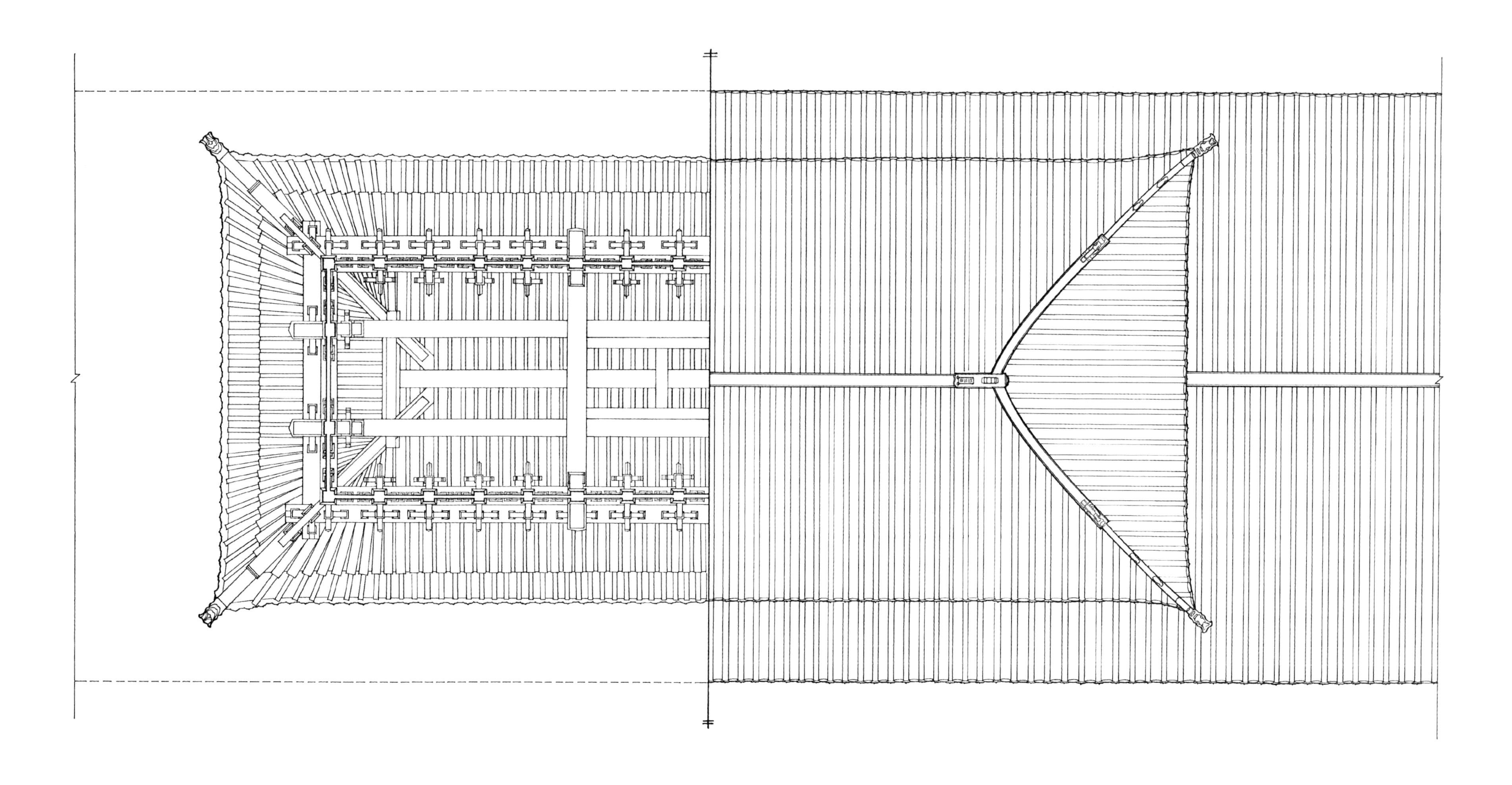

大钟楼二层梁架仰视图及屋顶平面图
Framework Plan and Roof Plan of Second Floor of Great Bell Tower

0 0.5 1 2m

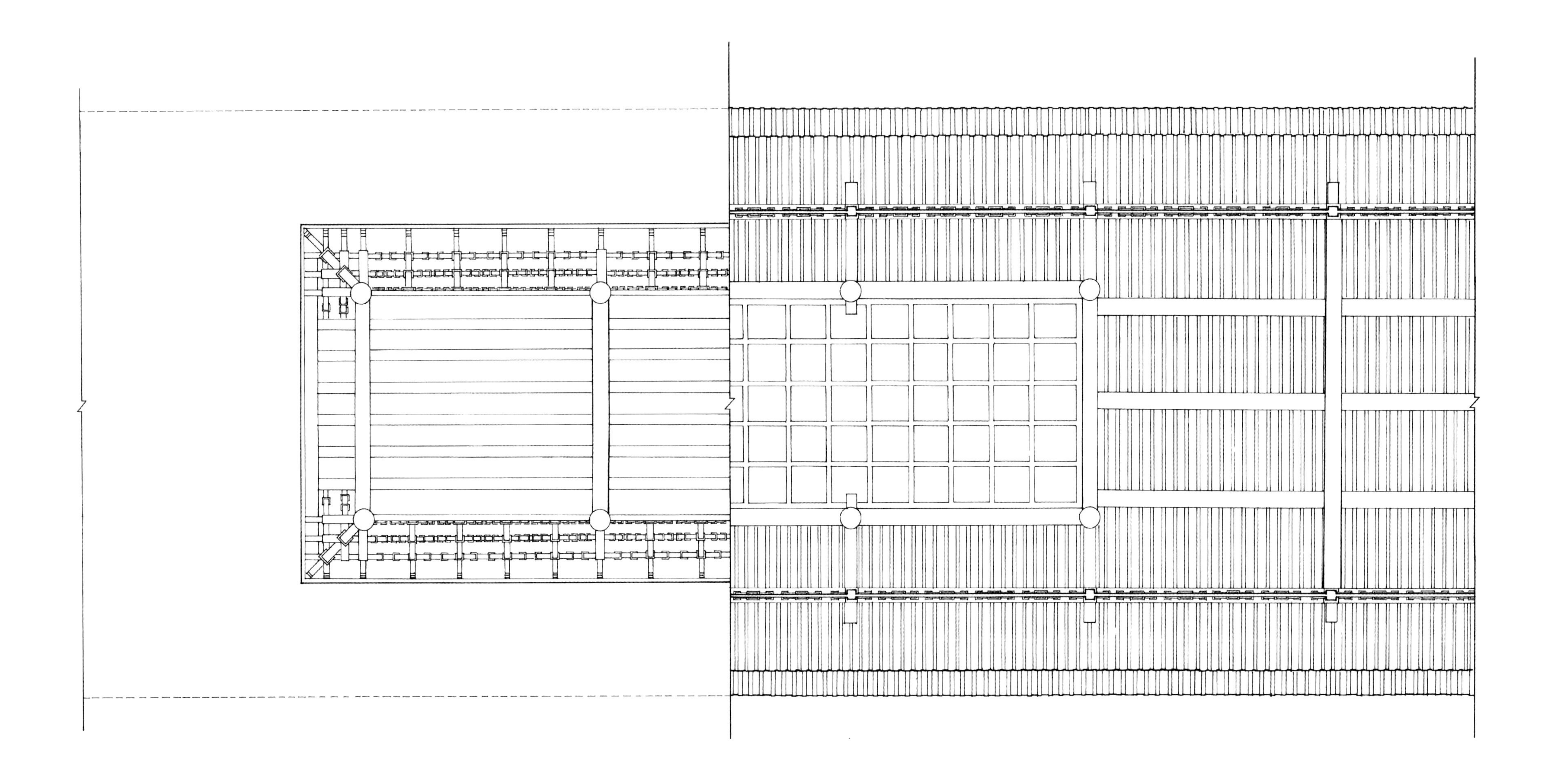

大钟楼一层及平坐梁架仰视图

Framework Plan of First Floor and Pingzuo Layer of Great Bell Tower

0 0.5 1 2m

隆国殿
Longguo Hall

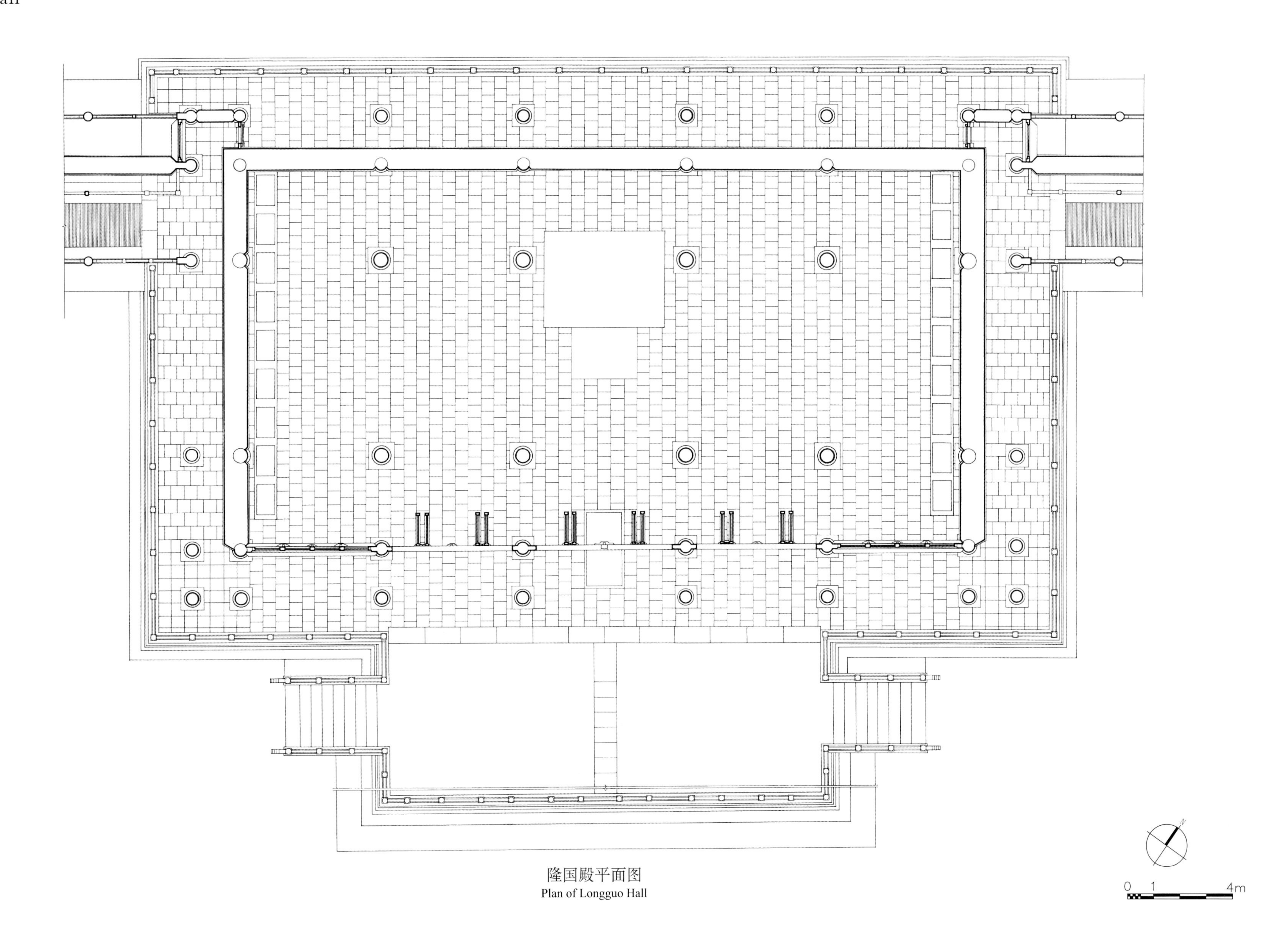

隆国殿平面图
Plan of Longguo Hall

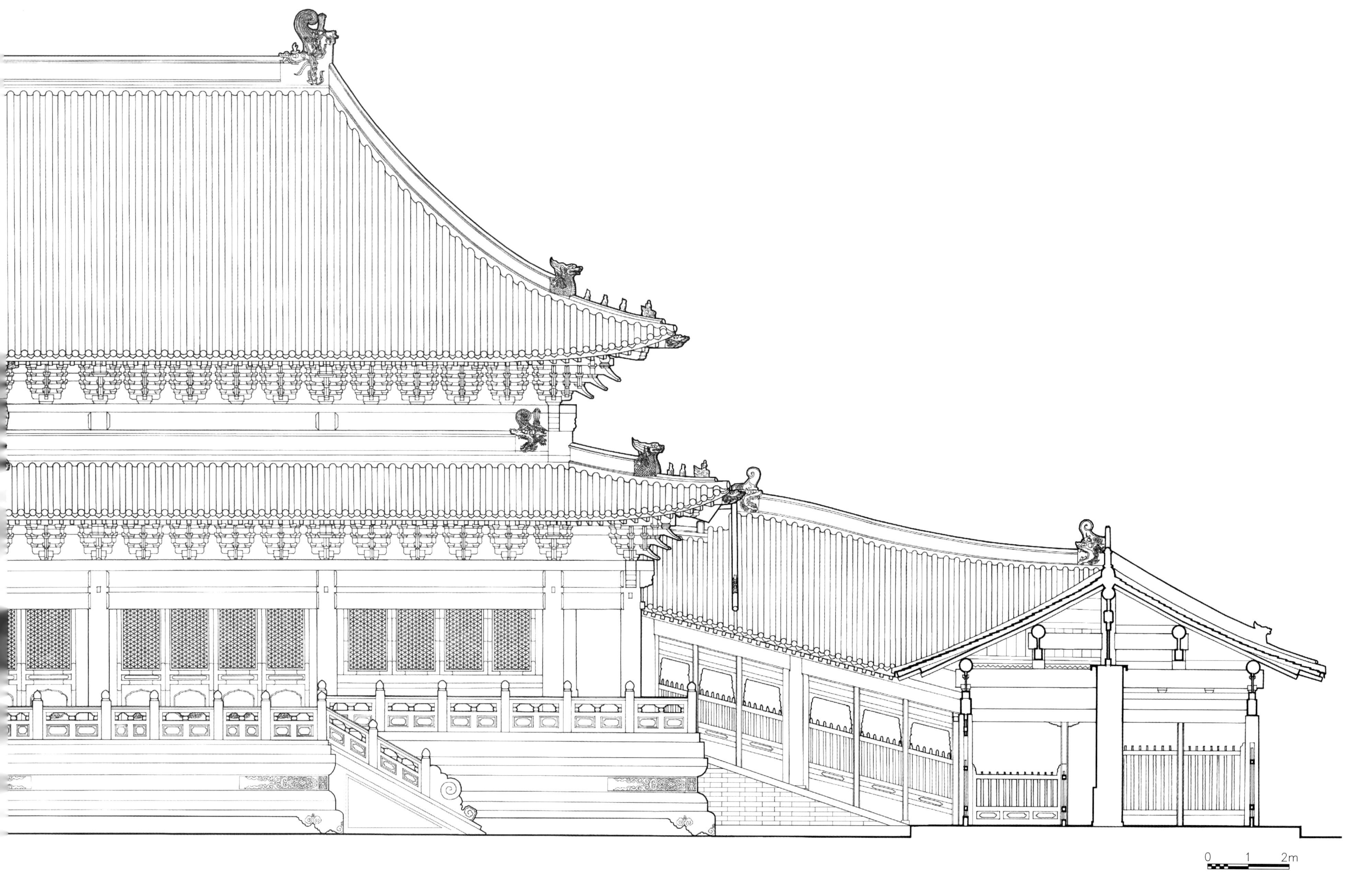
0
1
2m

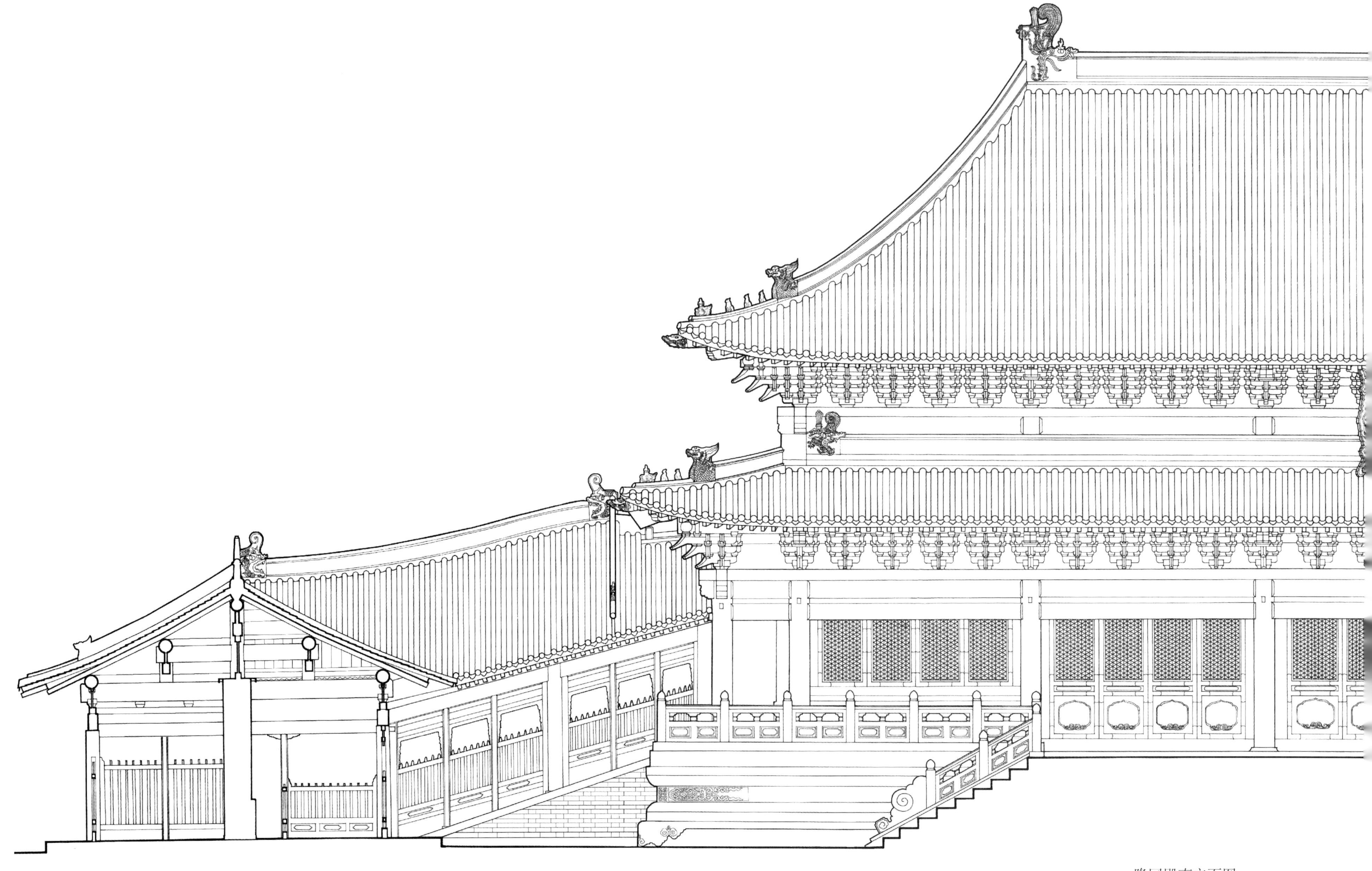

隆国殿南立面图
South Elevation of Longguo Hall

0
1
2m

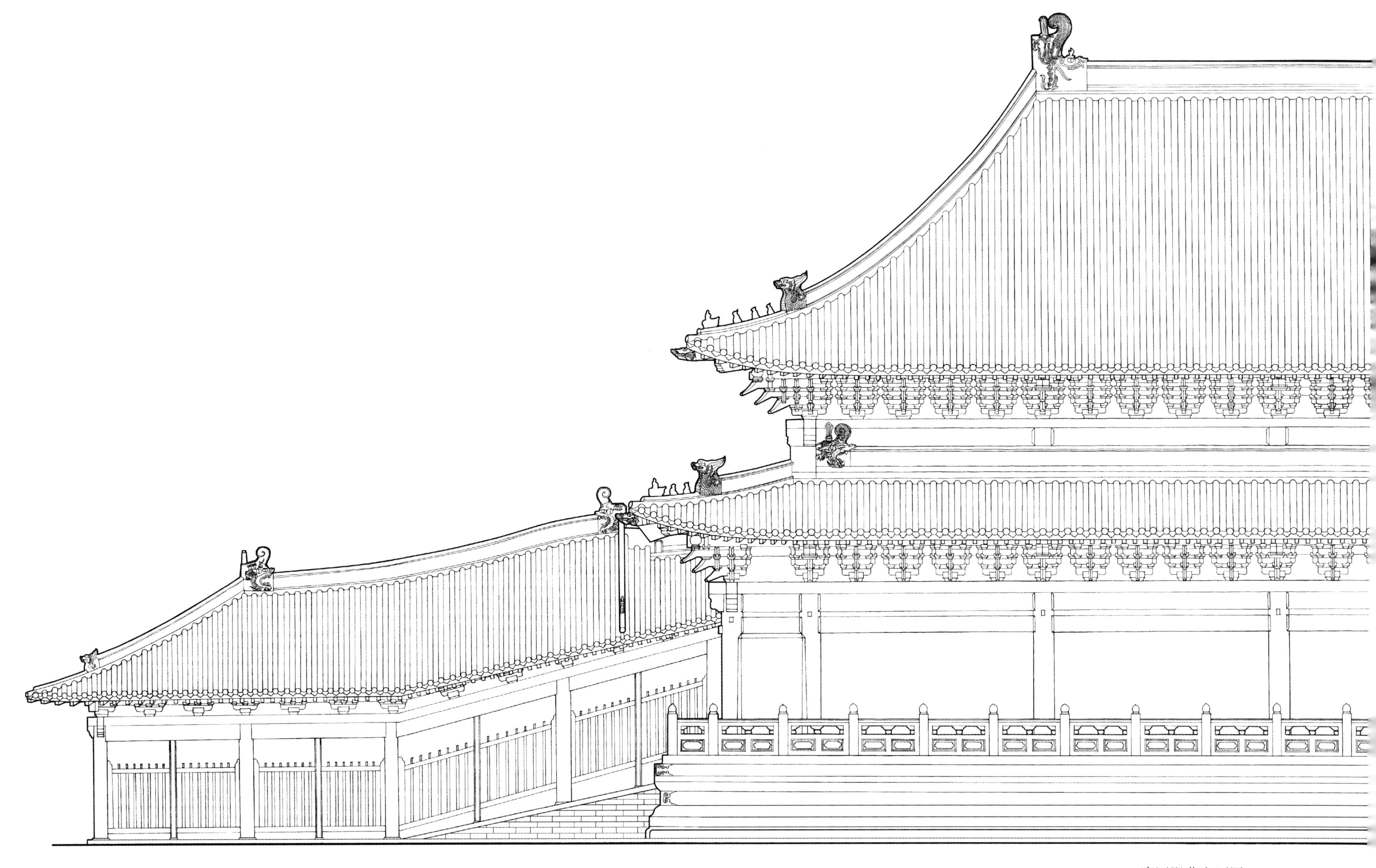

隆国殿北立面图
North Elevation of Longguo Hall

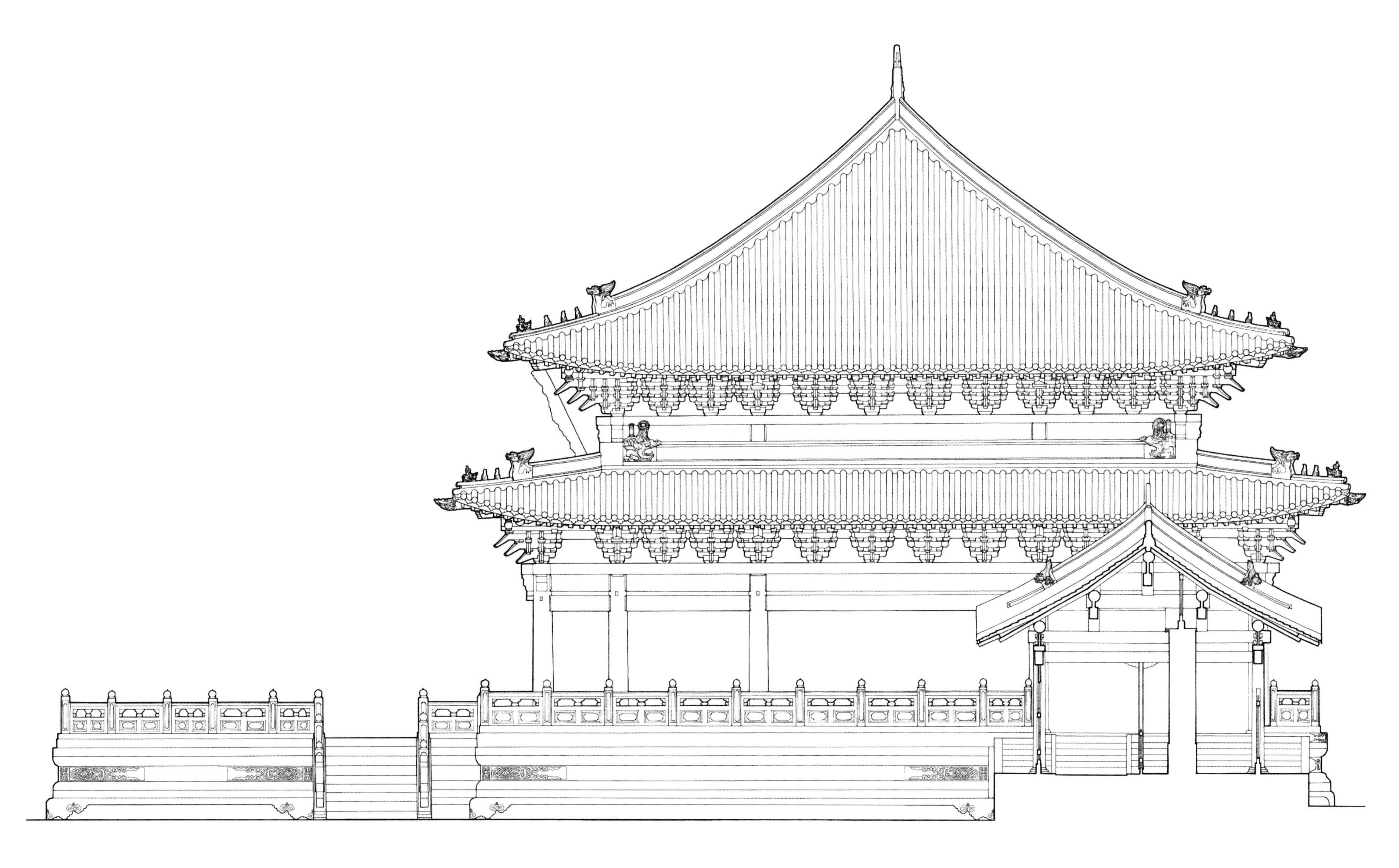

隆国殿东立面图
East Elevation of Longguo Hall

隆国殿明间剖面图

Mingjian Cross-section of Longguo Hall

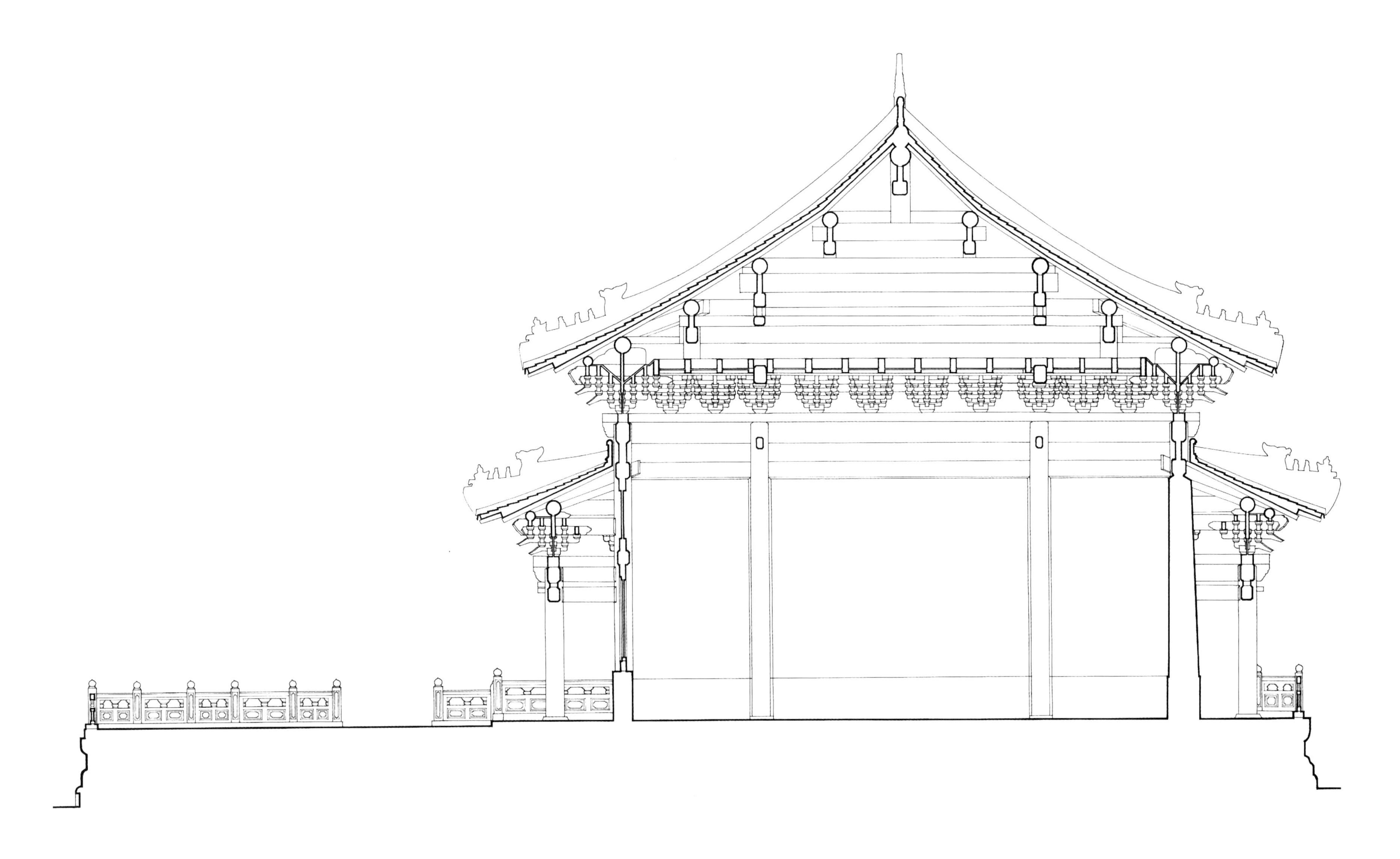

隆国殿次间剖面图

Cijian Cross-section of Longguo Hall

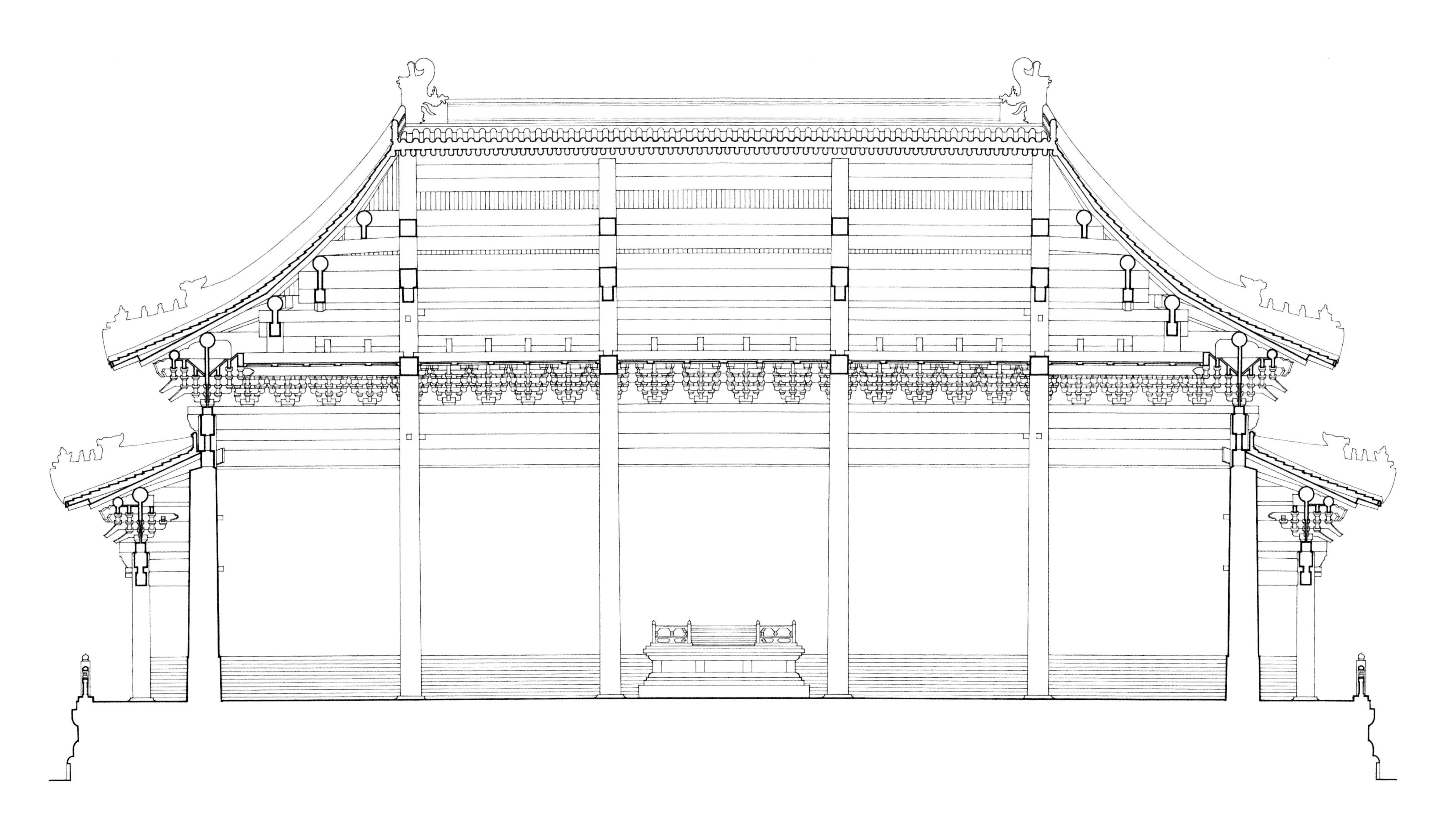

隆国殿纵剖面图

Longitudinal Section of Longguo Hall

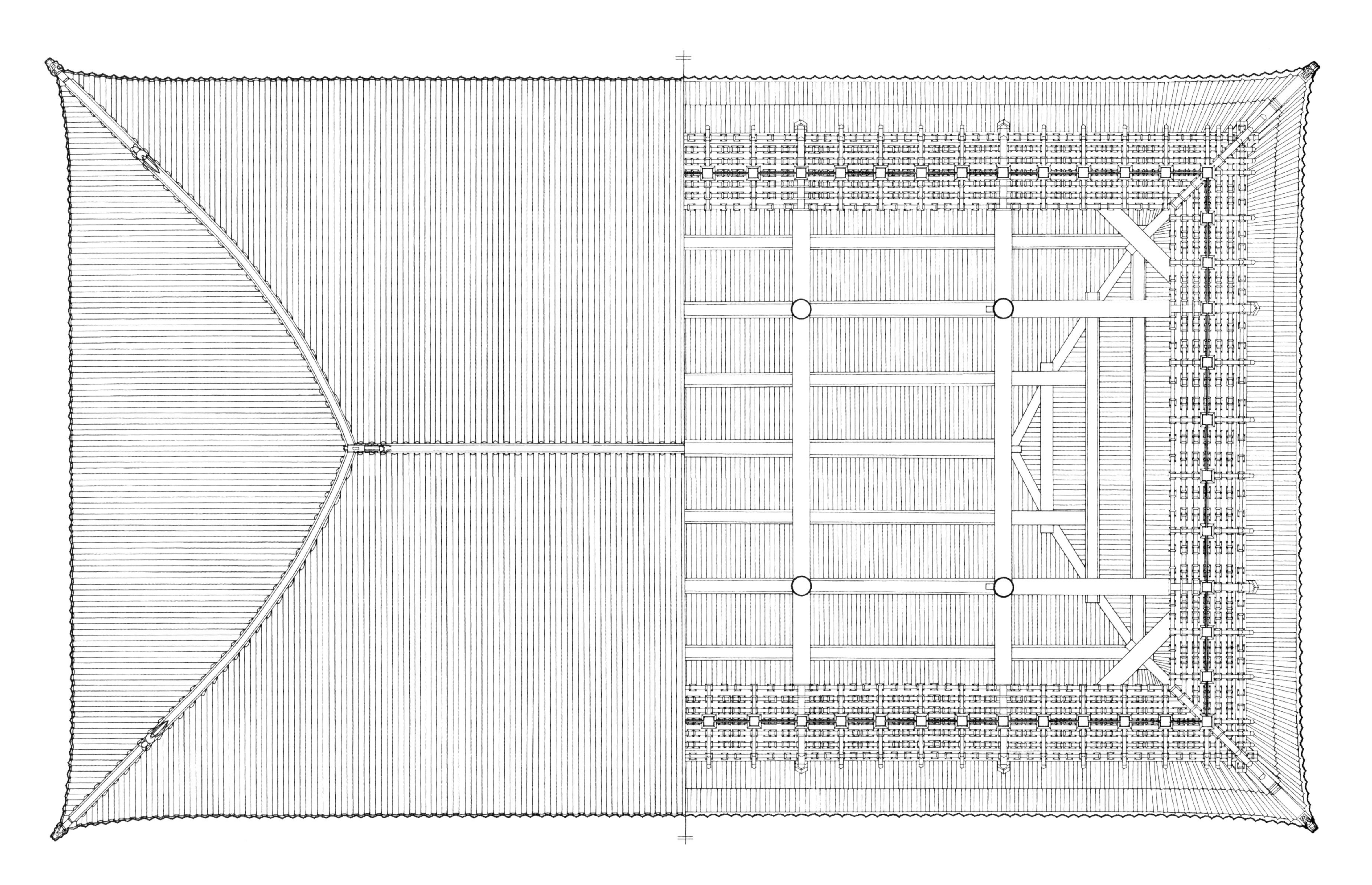

隆国殿上檐屋顶平面及梁架仰视图
Roof Plan and Framework Plan of Upper Eave of Longguo Hall

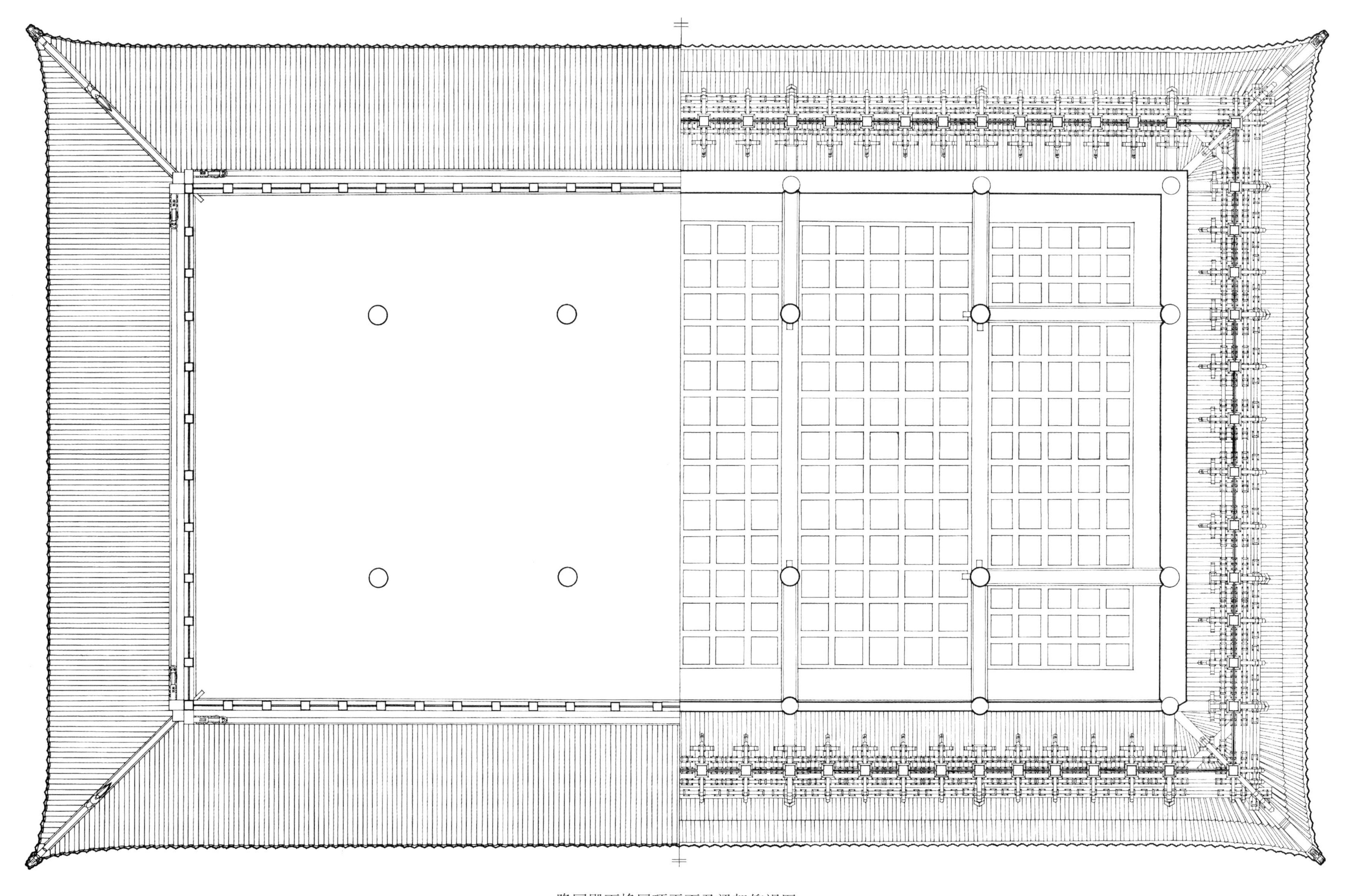

隆国殿下檐屋顶平面及梁架仰视图

Roof Plan and Framework Plan of Lower Eave of Longguo Hall

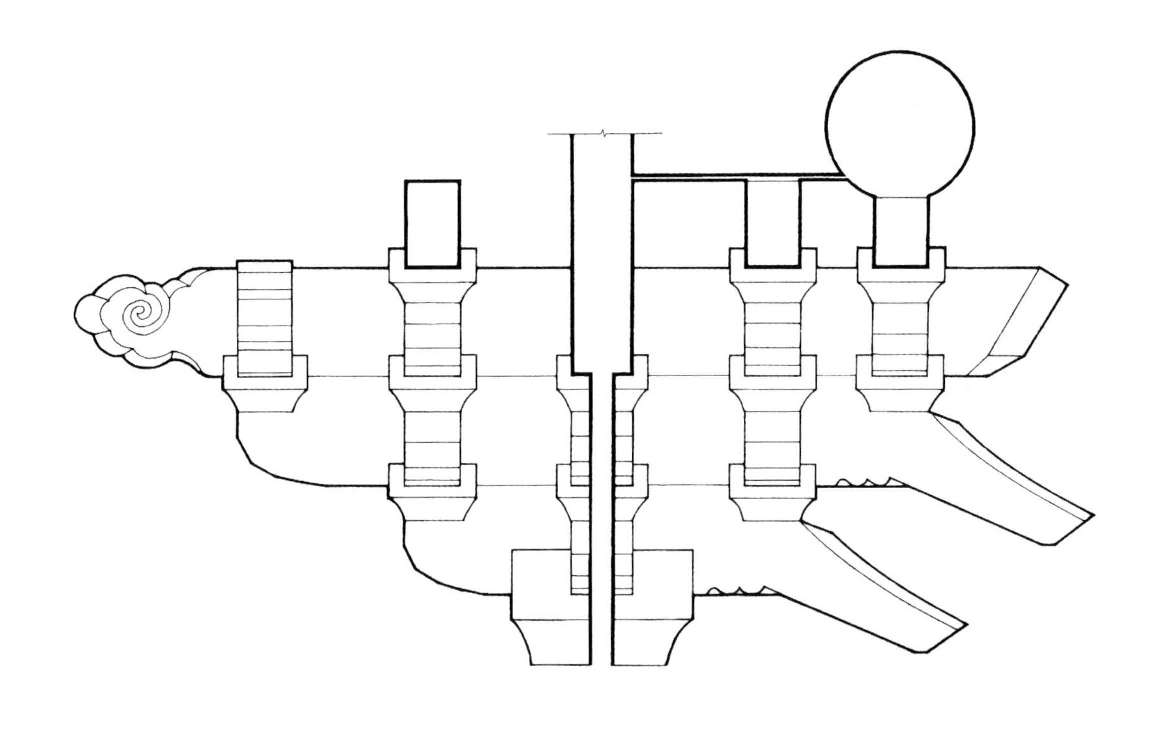

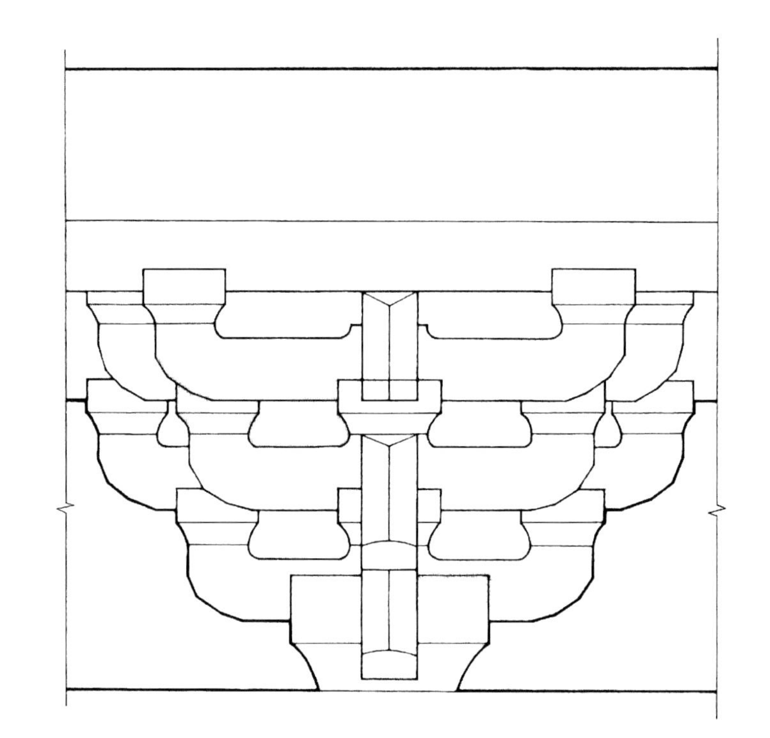

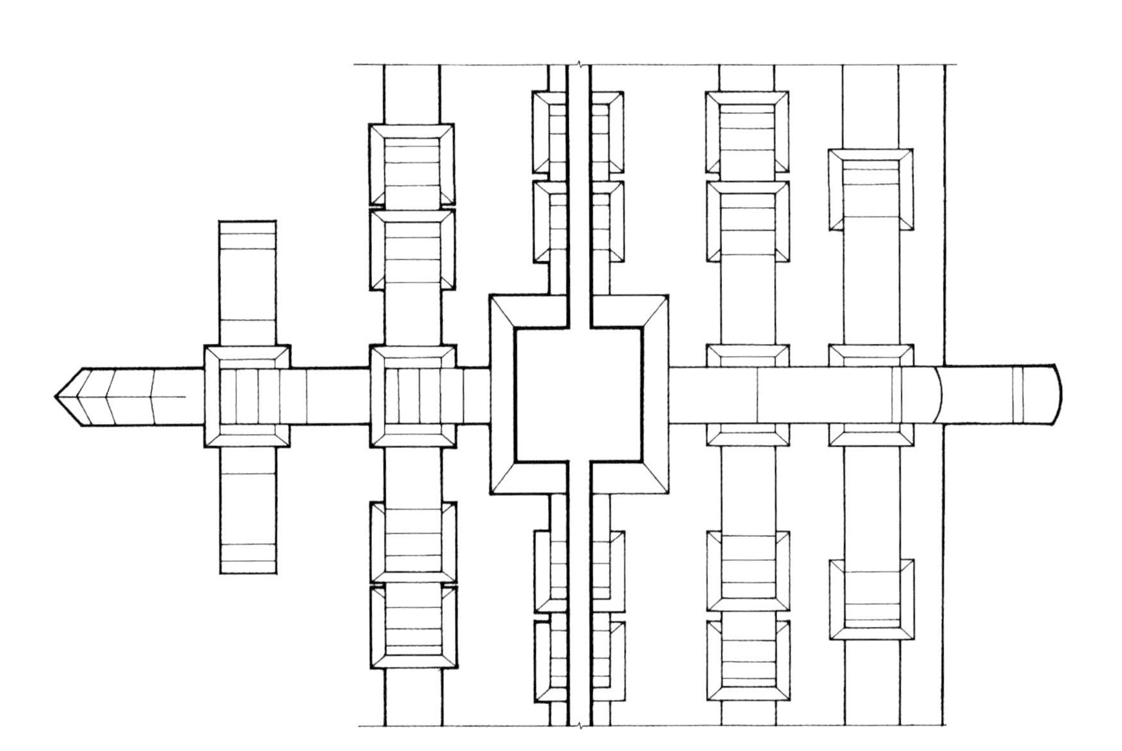

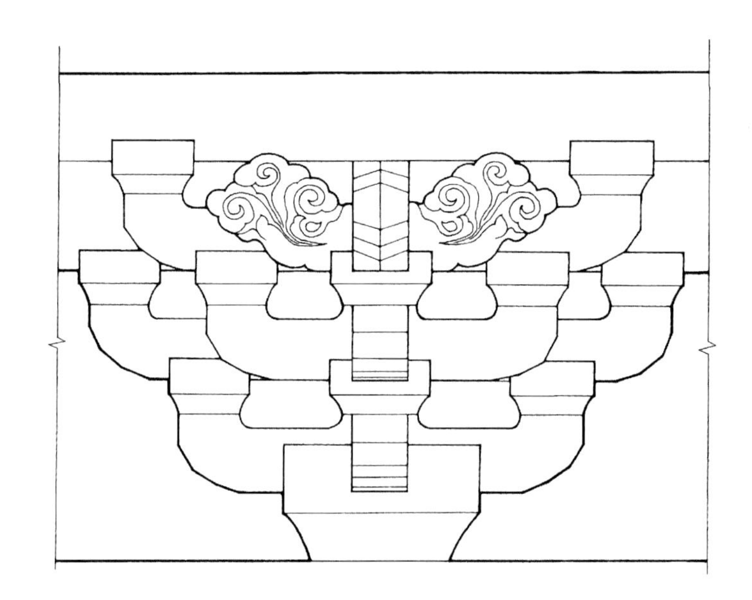

隆国殿下檐平身科斗栱大样图

Pingshenke Dougong Under the Lower Eave of Longguo Hall

0 0.1 0.4m

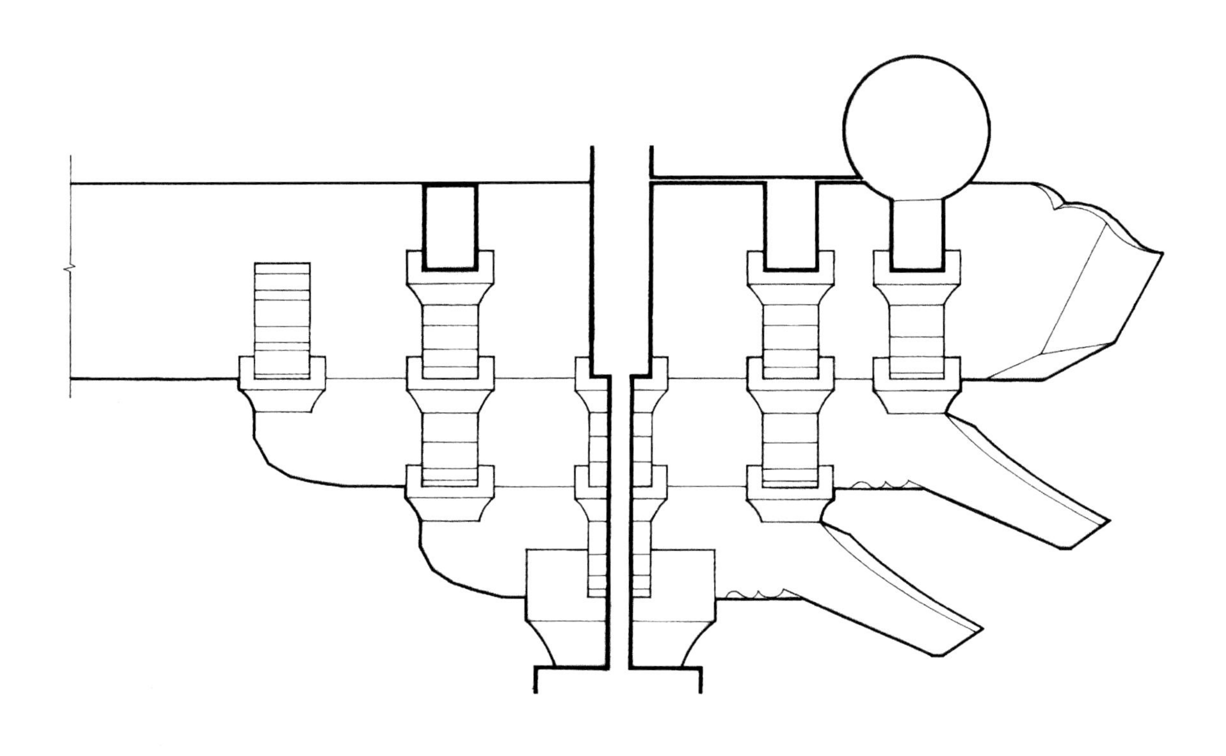

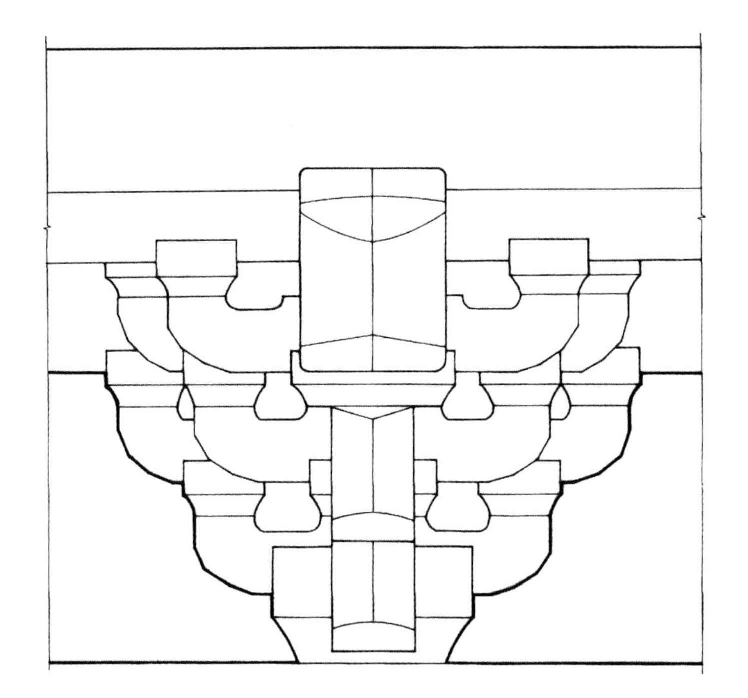

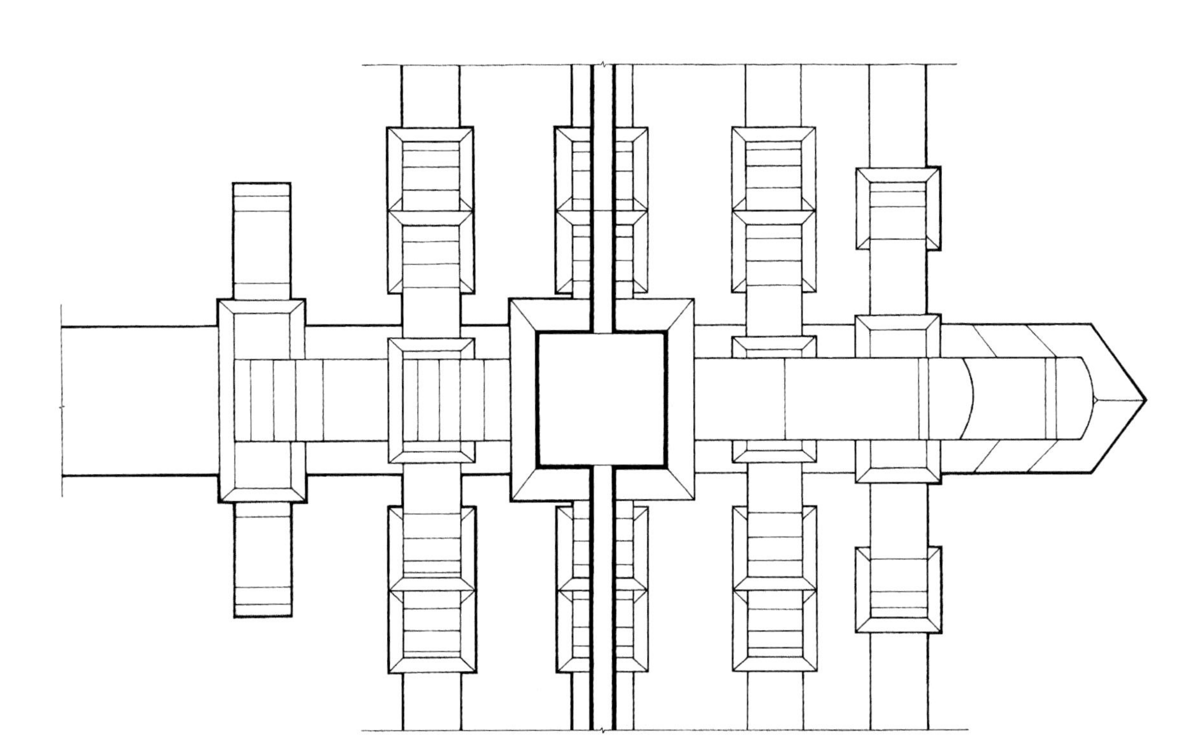

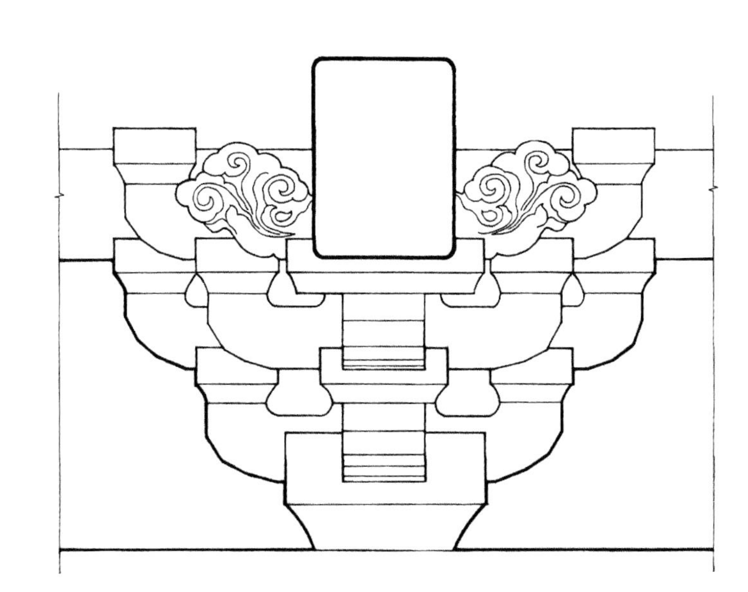

隆国殿下檐柱头科斗栱大样图

Zhutouke Dougong Under the Lower Eave of Longguo Hall

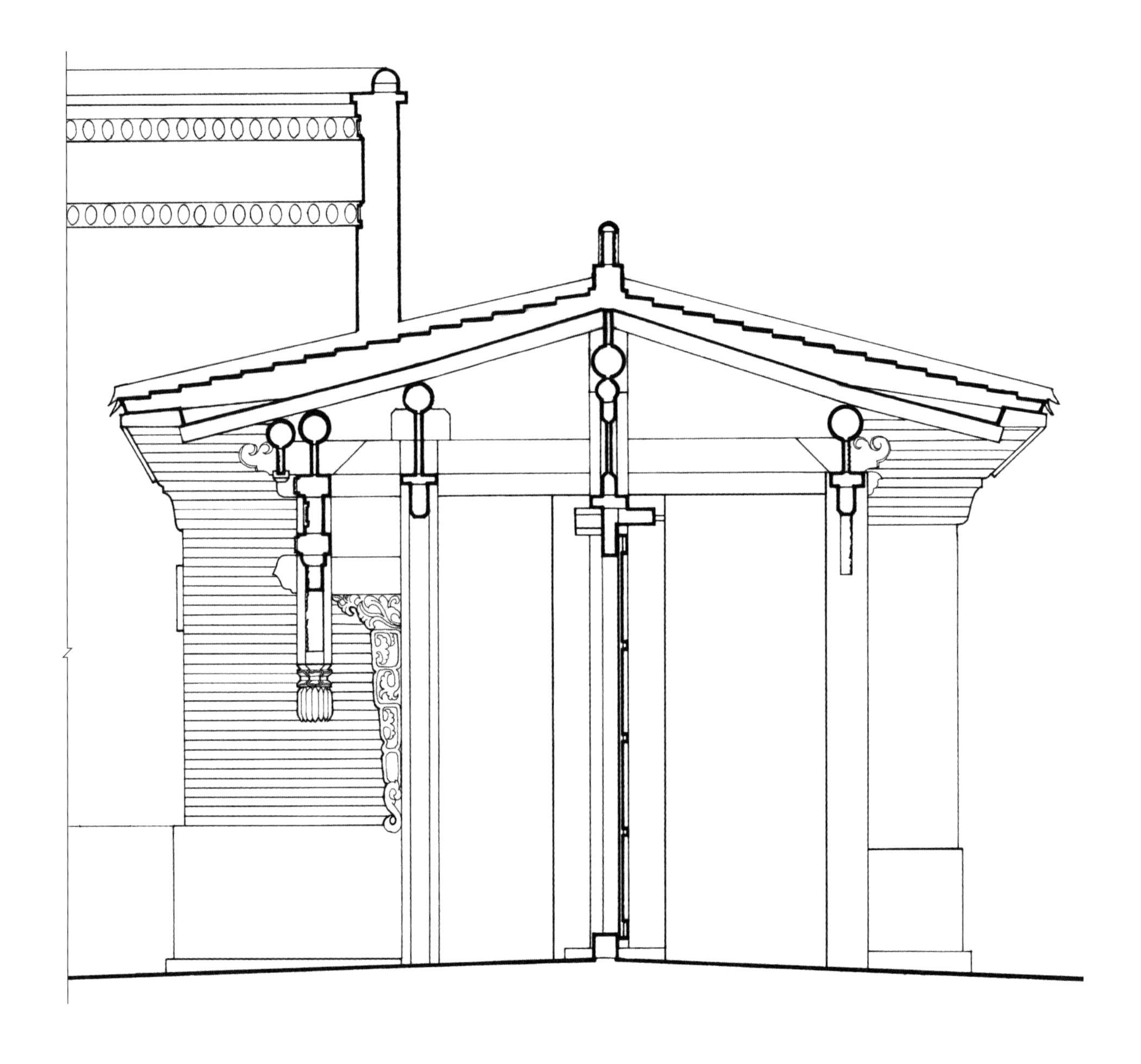

囊谦垂花门横剖面图
Cross-section of Festooned Door of Nangqian

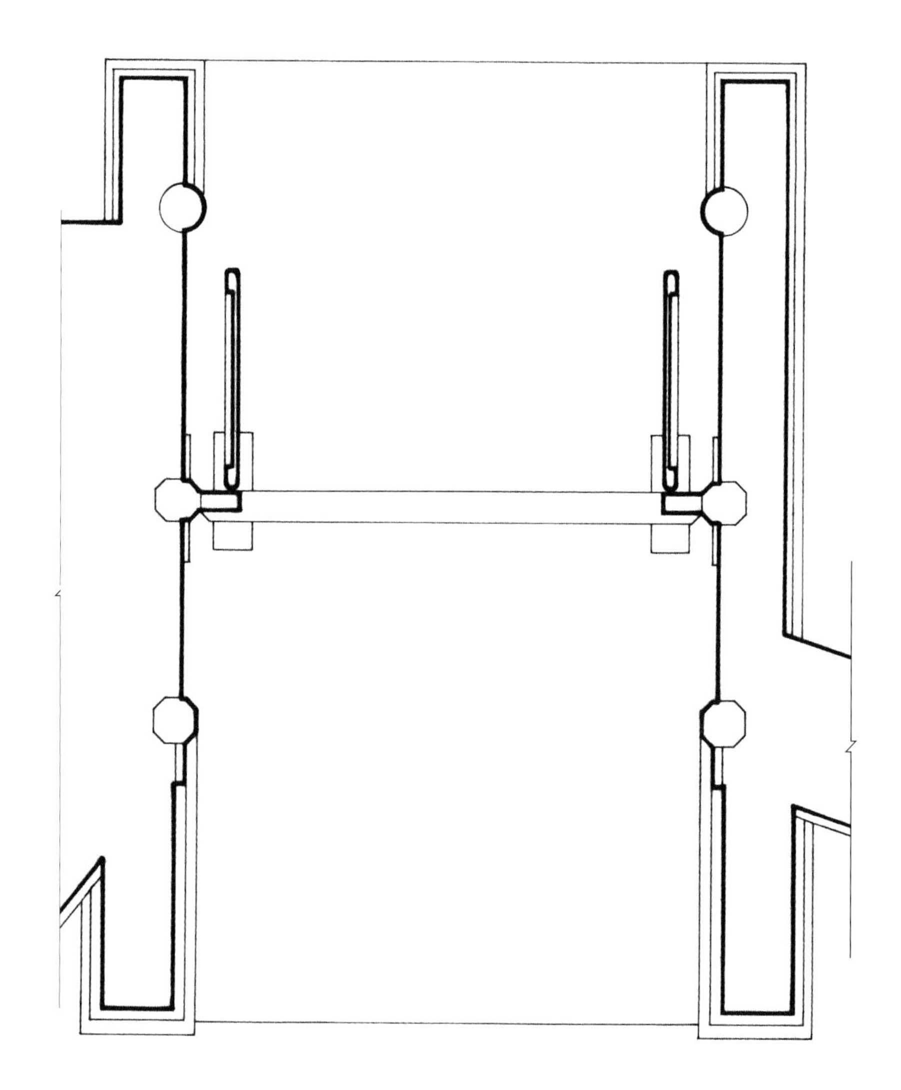

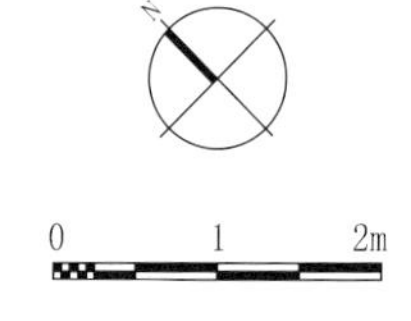

囊谦垂花门平面图
Plan of Festooned Door of Nangqian

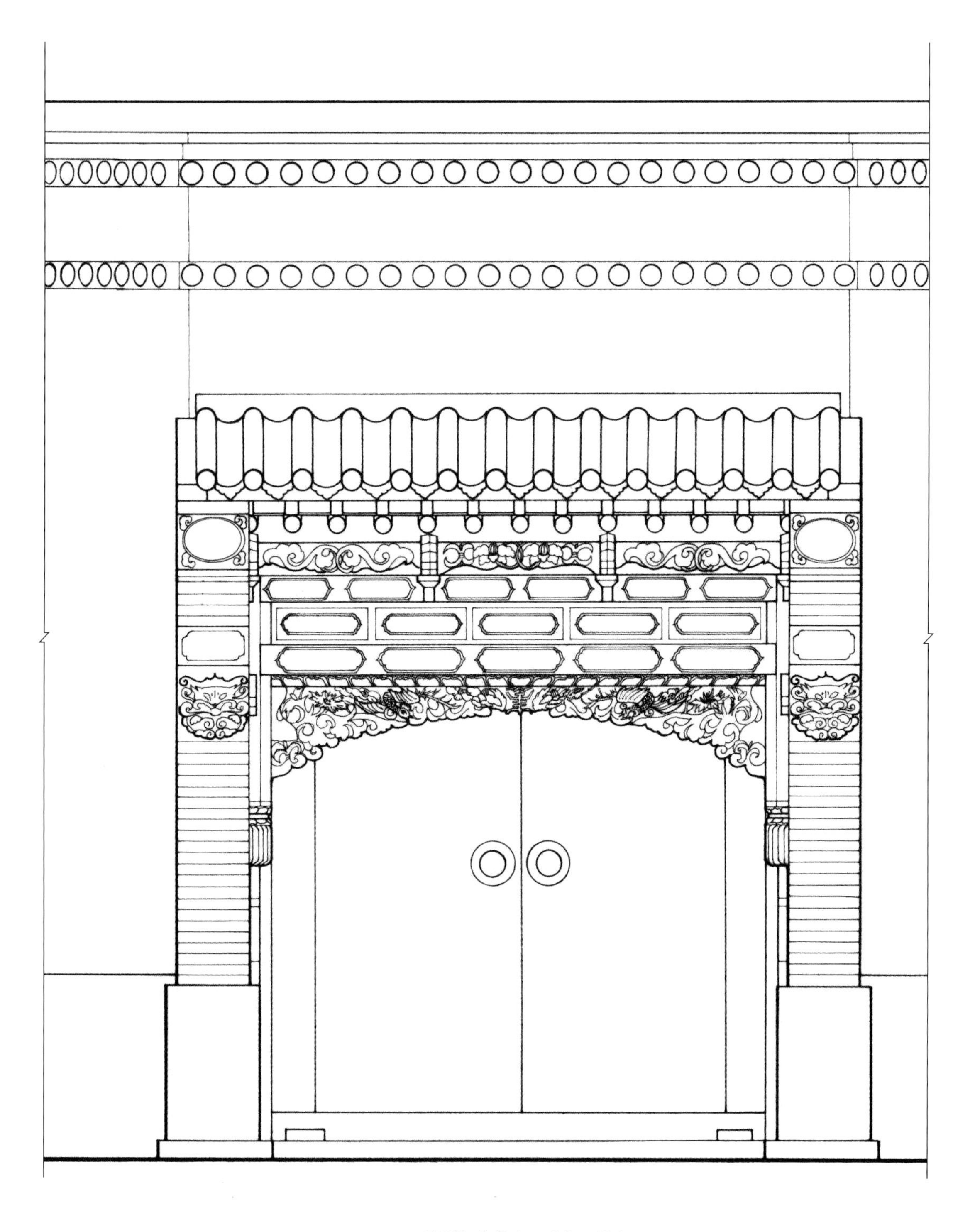

囊谦垂花门西立面图
West Elevation of Festooned Door of Nangqian

囊谦垂花门东立面图
East Elevation of Festooned Door of Nangqian

囊谦倒座横剖面图
Cross-section of Inverted room of Nangqian

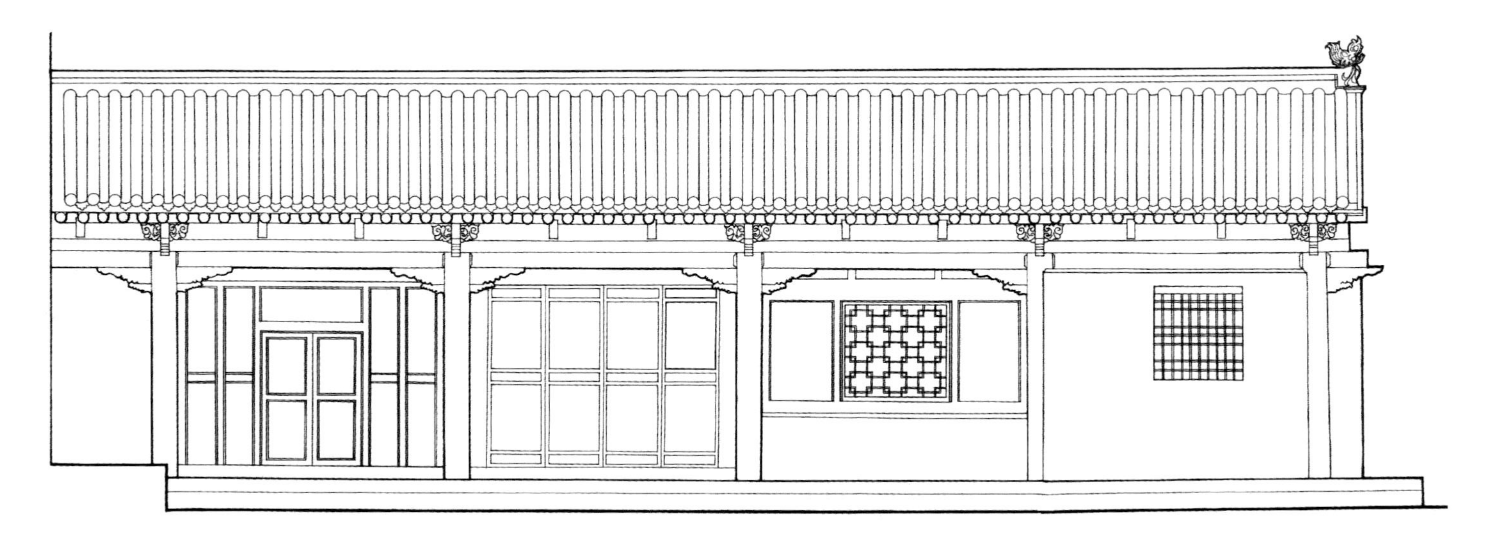

囊谦倒座北立面图
North Elevation of Inverted room of Nangqian

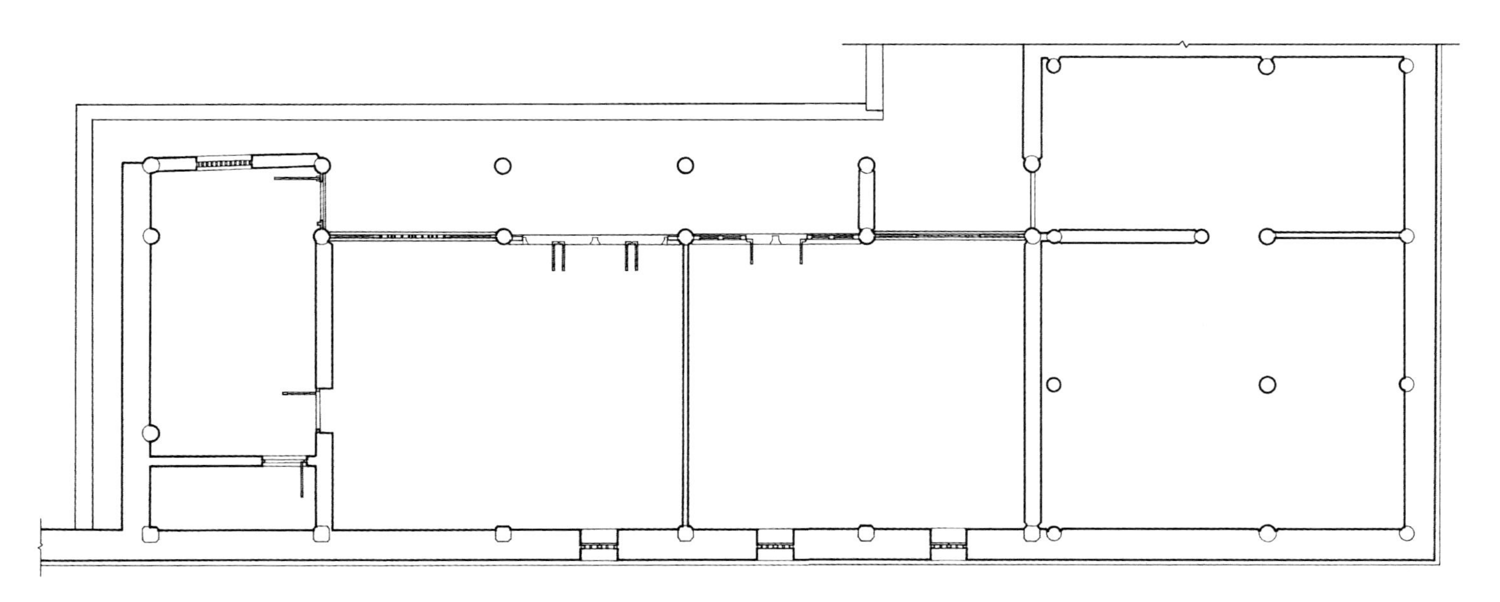

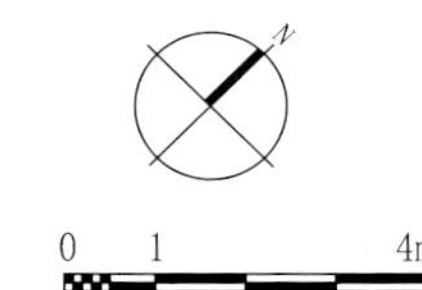

囊谦倒座平面图
Plan of Inverted room of Nangqian

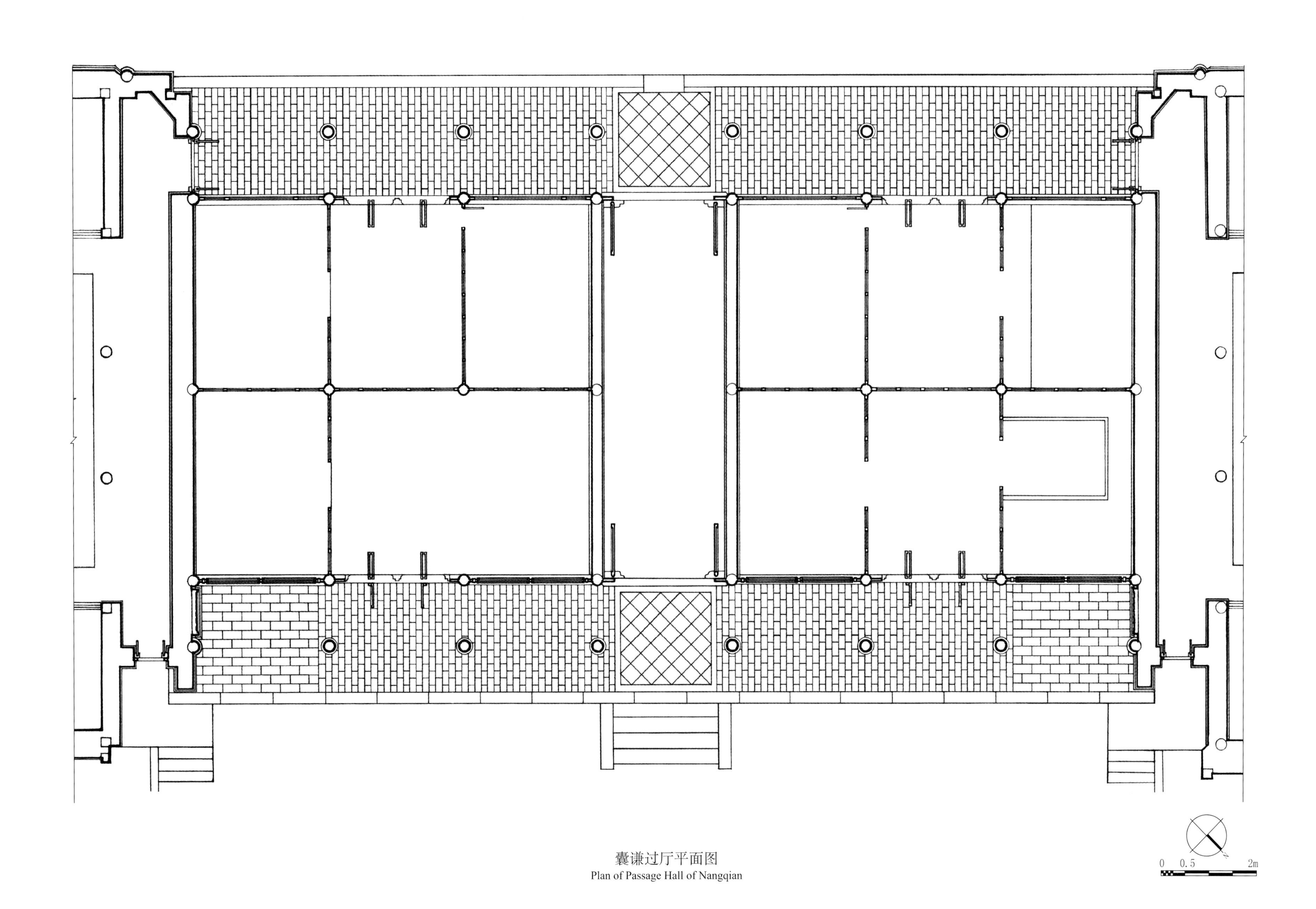

囊谦过厅平面图
Plan of Passage Hall of Nangqian

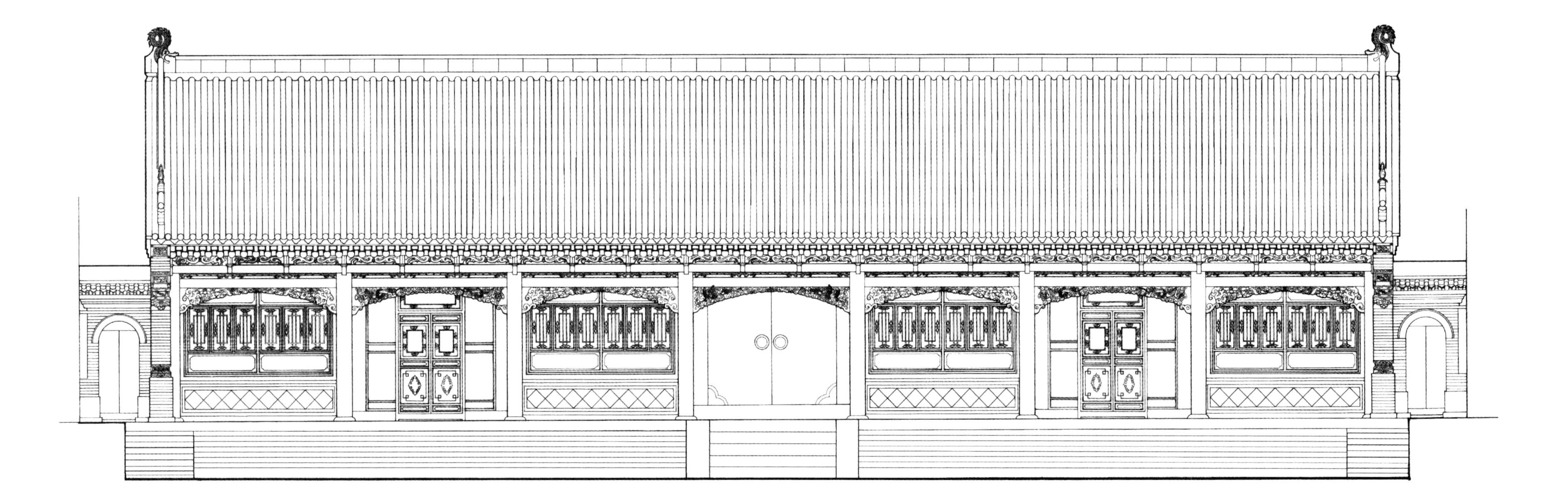

囊谦过厅南立面图

South Elevation of Passage Hall of Nangqian

0 0.5 2m

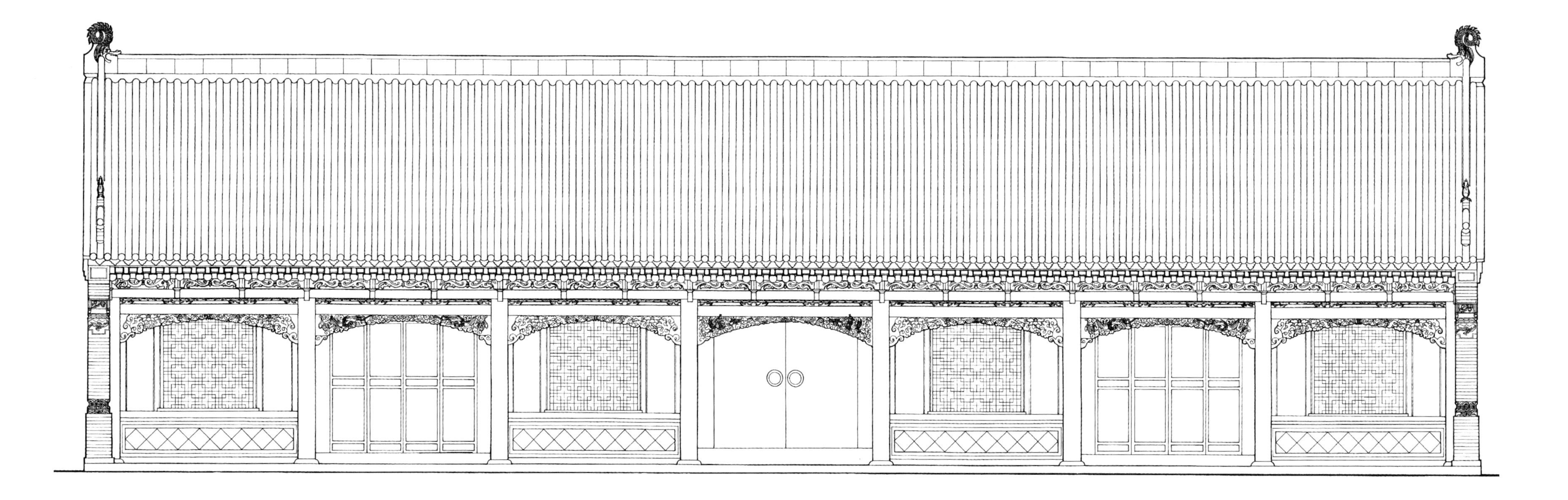

囊谦过厅北立面图

North Elevation of Passage Hall of Nangqian

0 0.5 2m

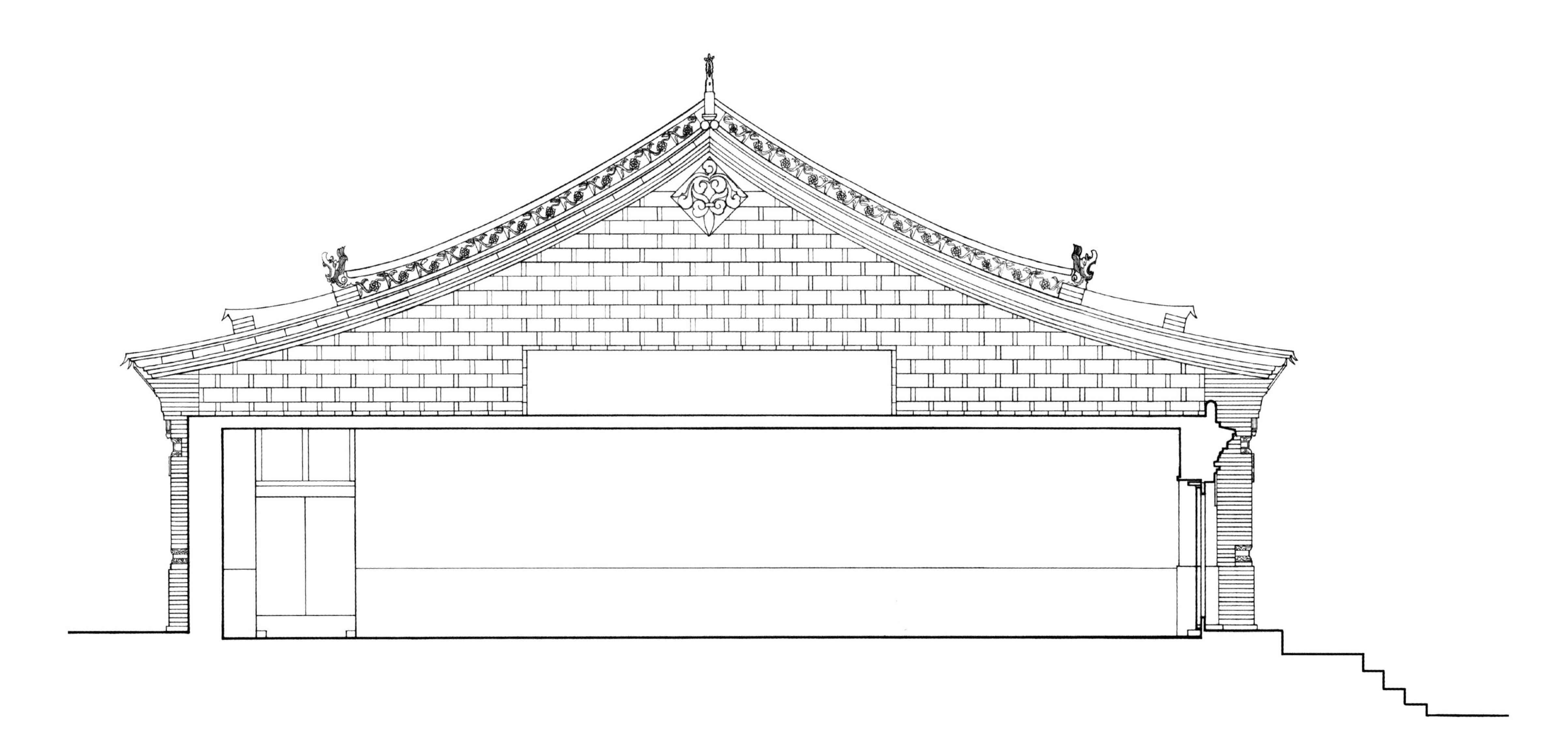

0 0.5 2m

囊谦过厅西立面图
West Elevation of Passage Hall of Nangqian

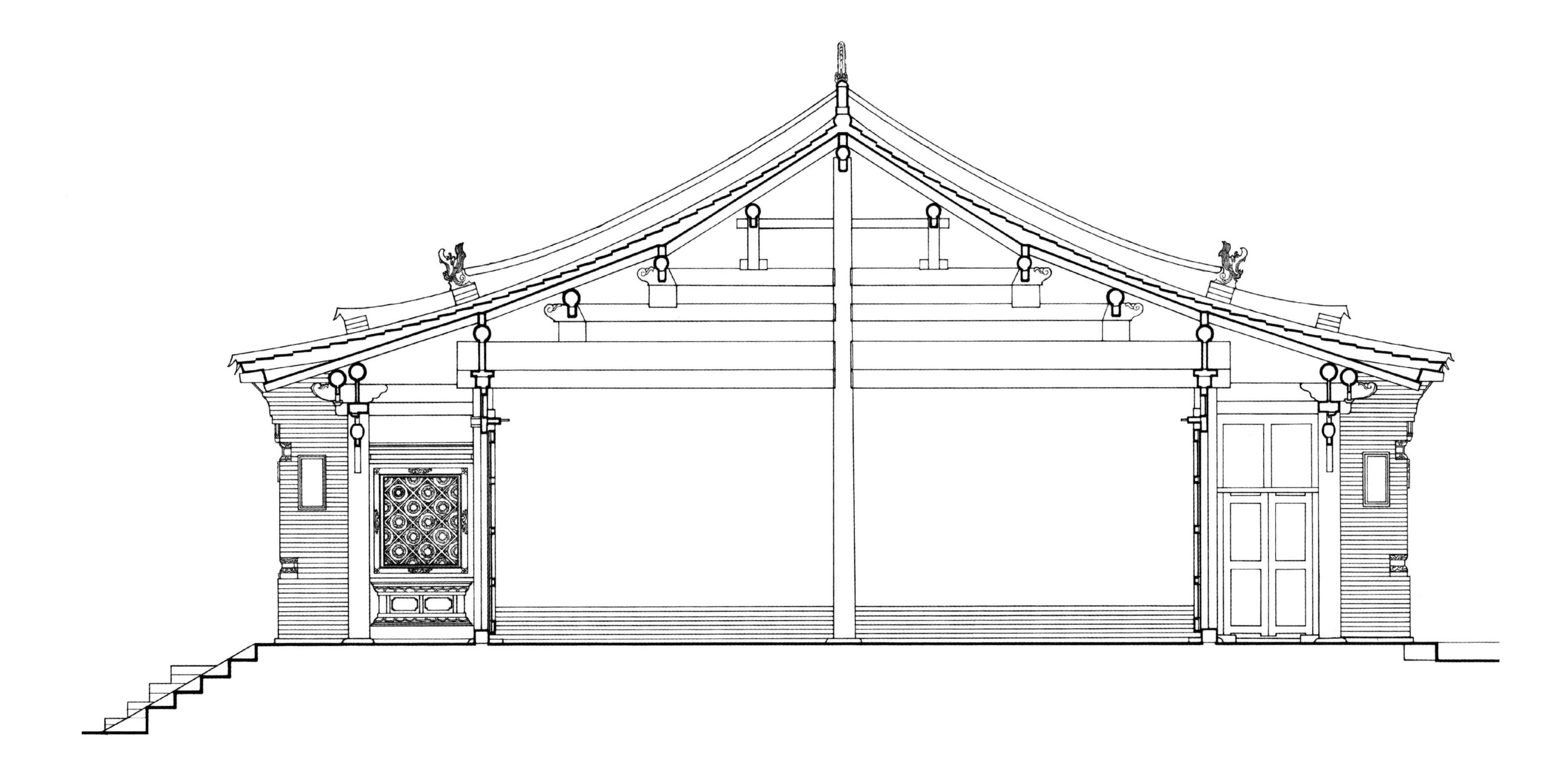

囊谦过厅明间剖面图

Mingjian Cross-section of Passage Hall of Nangqian

0 0.5 2m

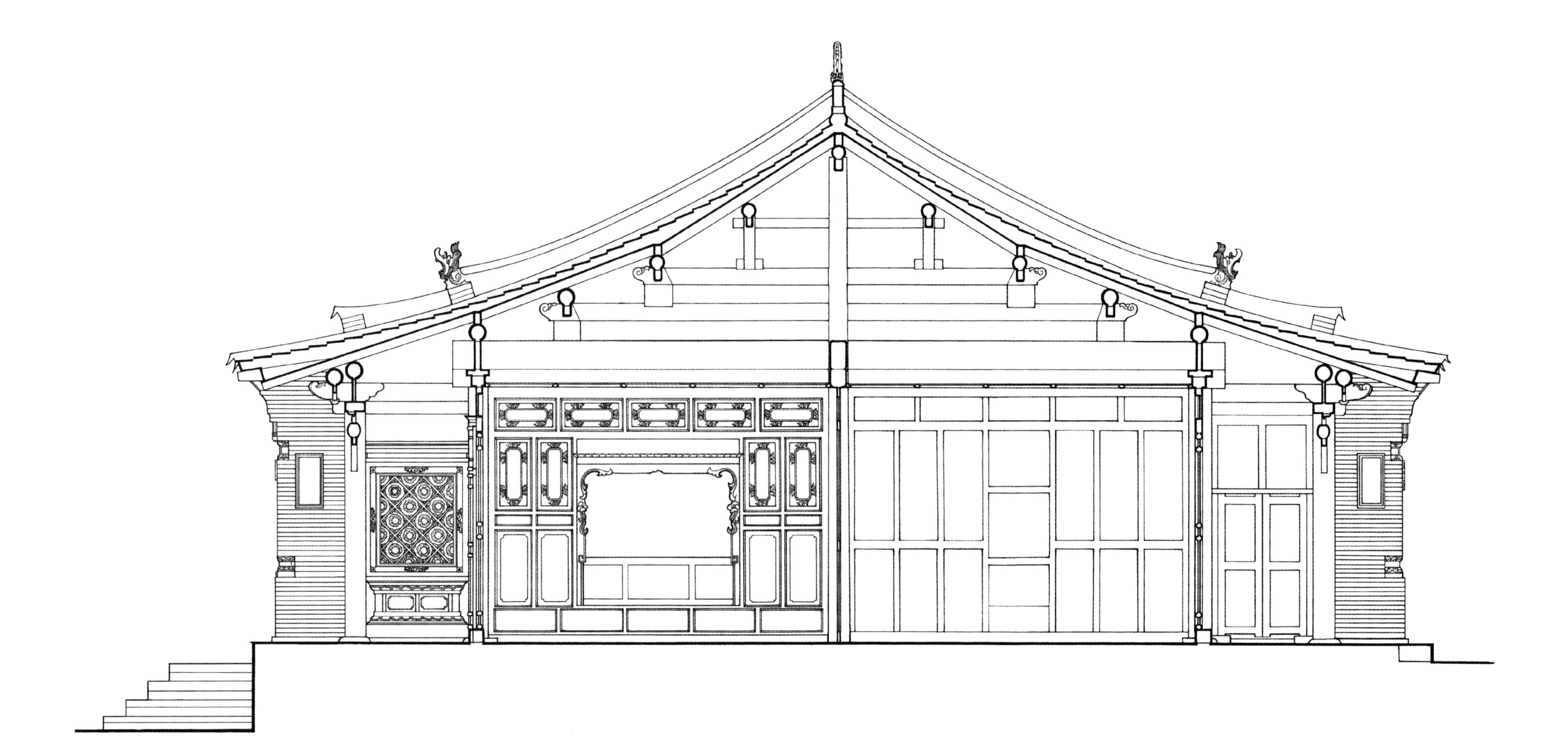

囊谦过厅次间剖面图
Cijian Cross-section of Passage Hall of Nangqian

0 0.5 2m

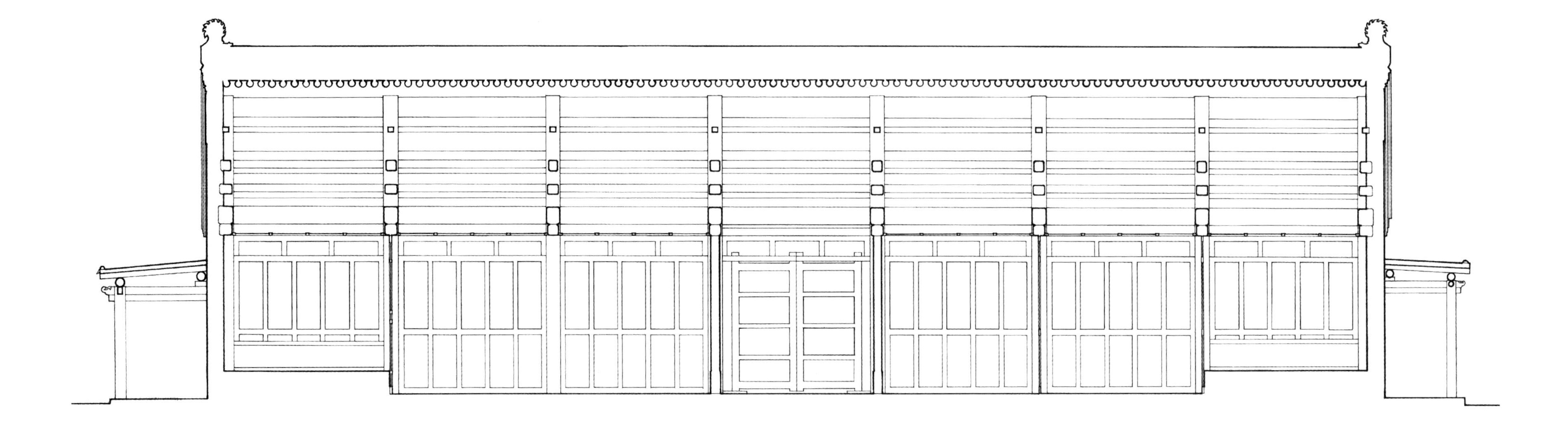

囊谦过厅纵剖面图
Longitudinal Section of Passage Hall of Nangqian

囊谦前东楼二层平面图
Second Floor Plan of Qiandonglou of Nangqian

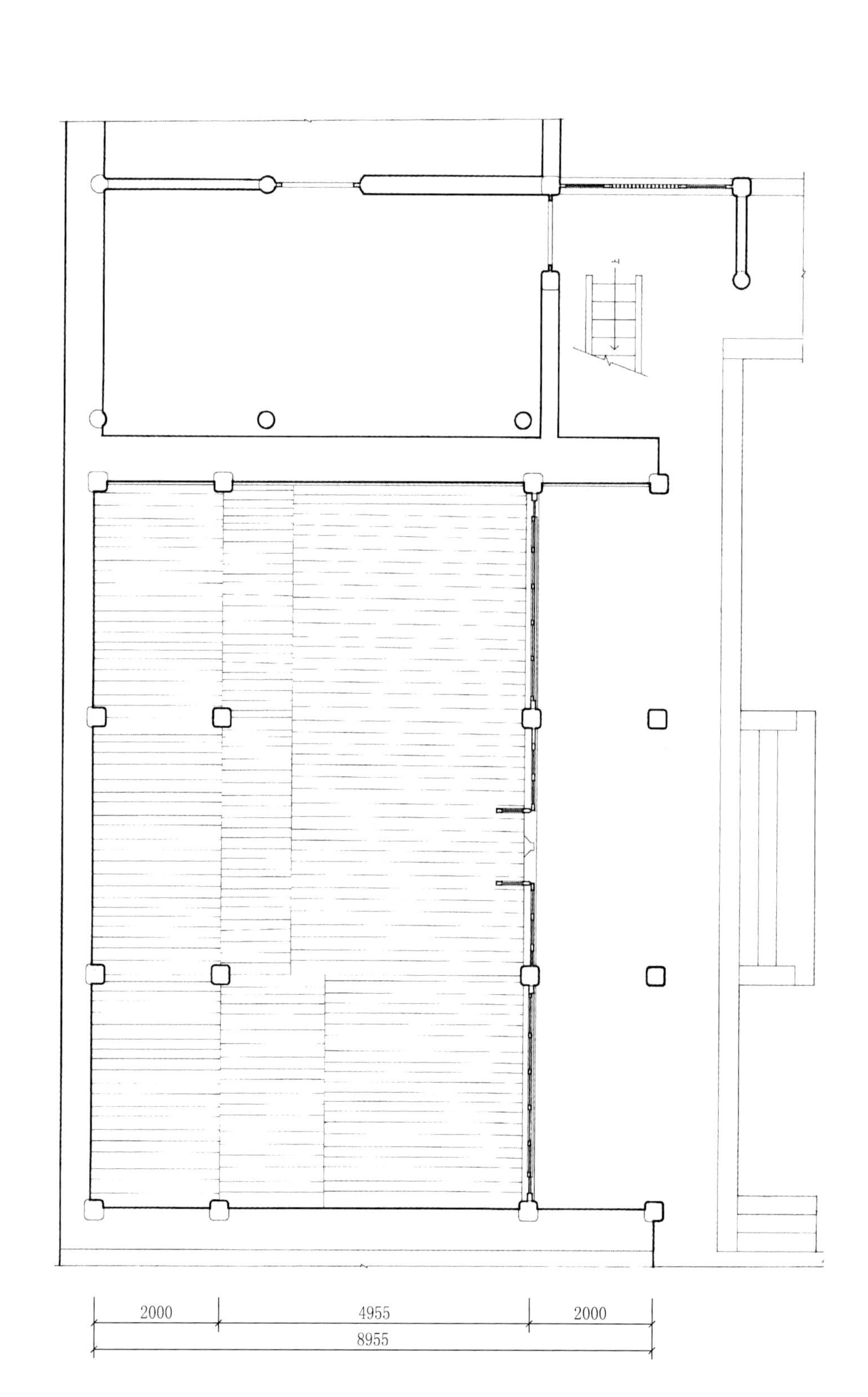

囊谦前东楼一层平面图
First Floor of Qiandonglou of Nangqian

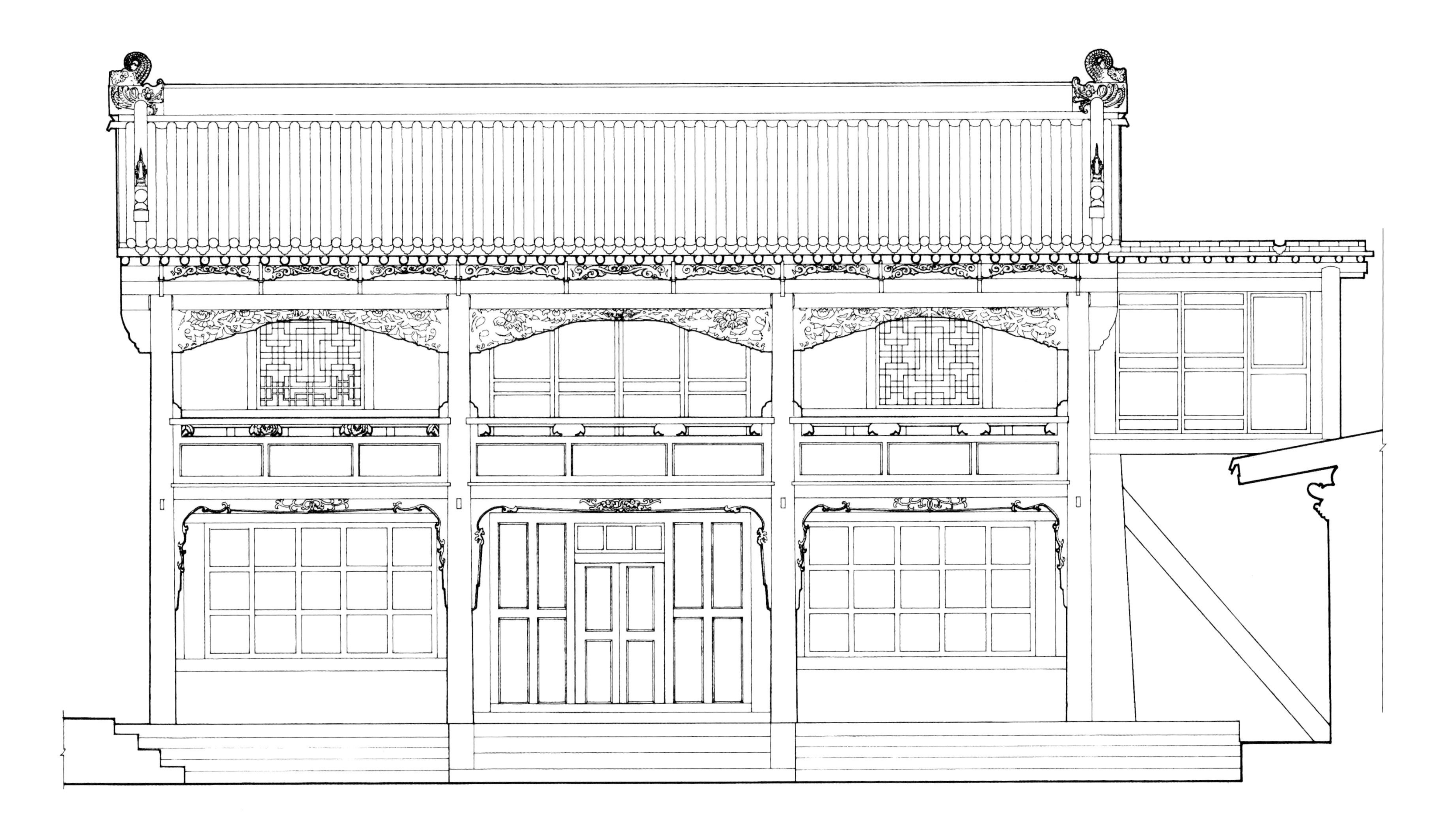

囊谦前东楼西立面图

West Elevation of Qiandonglou of Nangqian

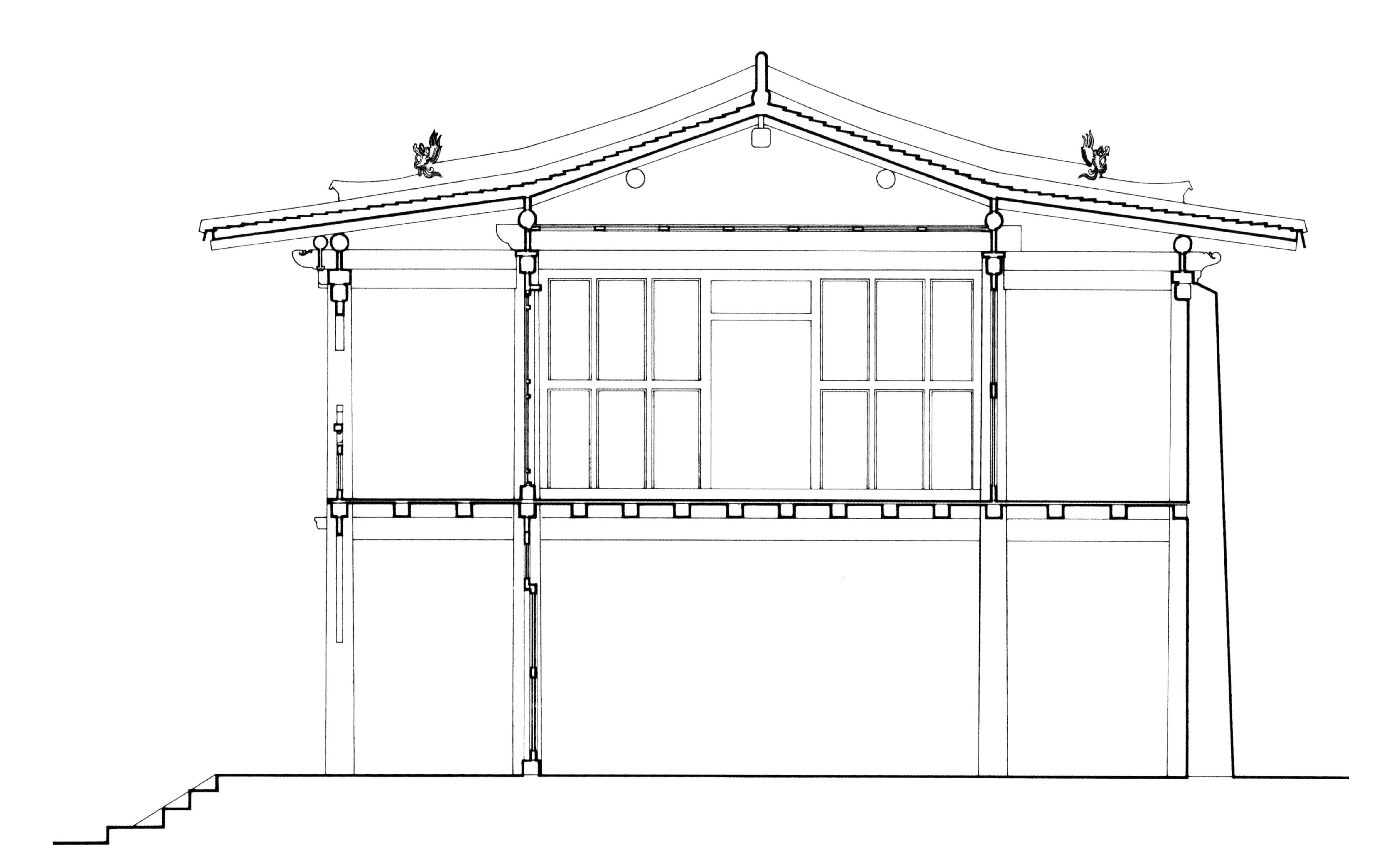

囊谦前东楼横剖面图
Cross-section of Qiandonglou of Nangqian

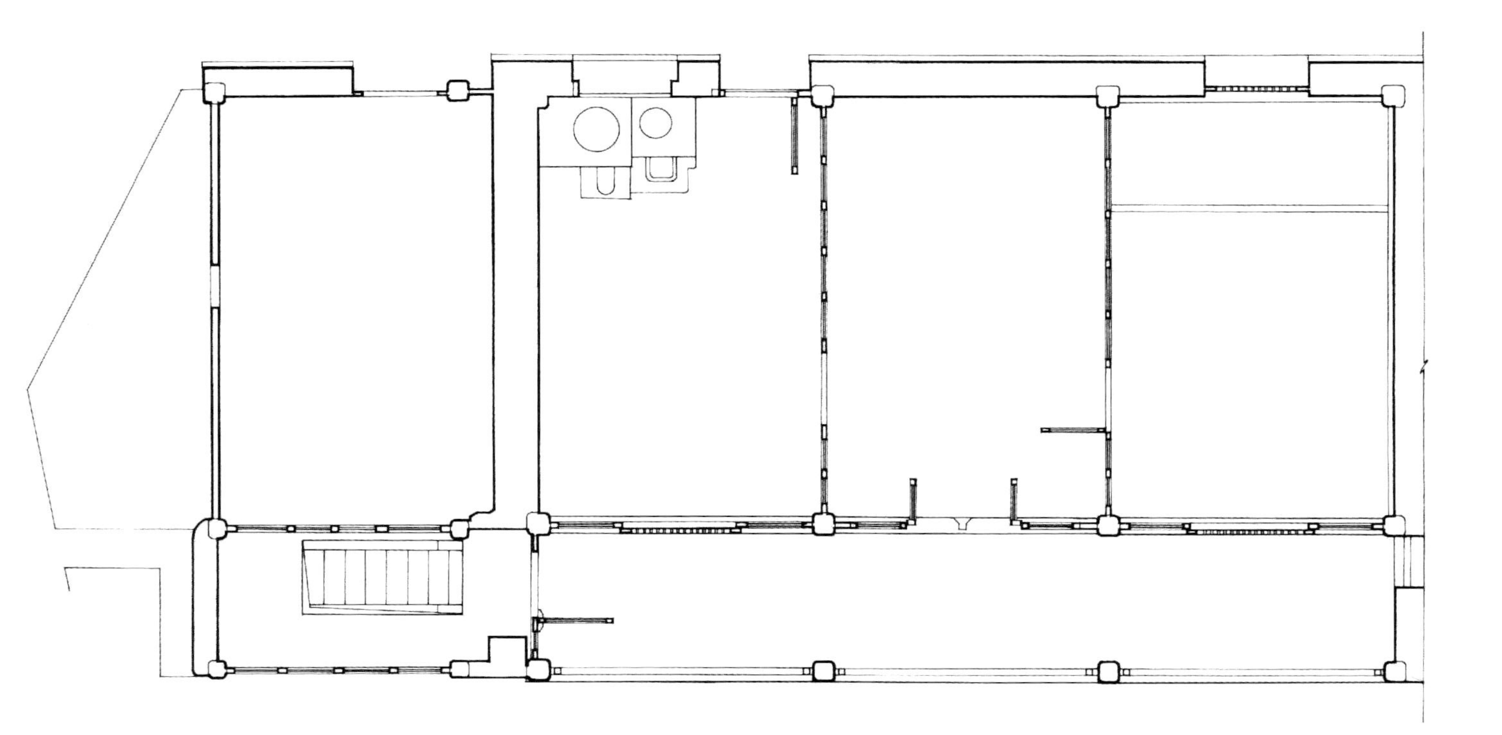

囊谦前西楼二层平面图
Second Floor Plan of Qianxilou of Nangqian

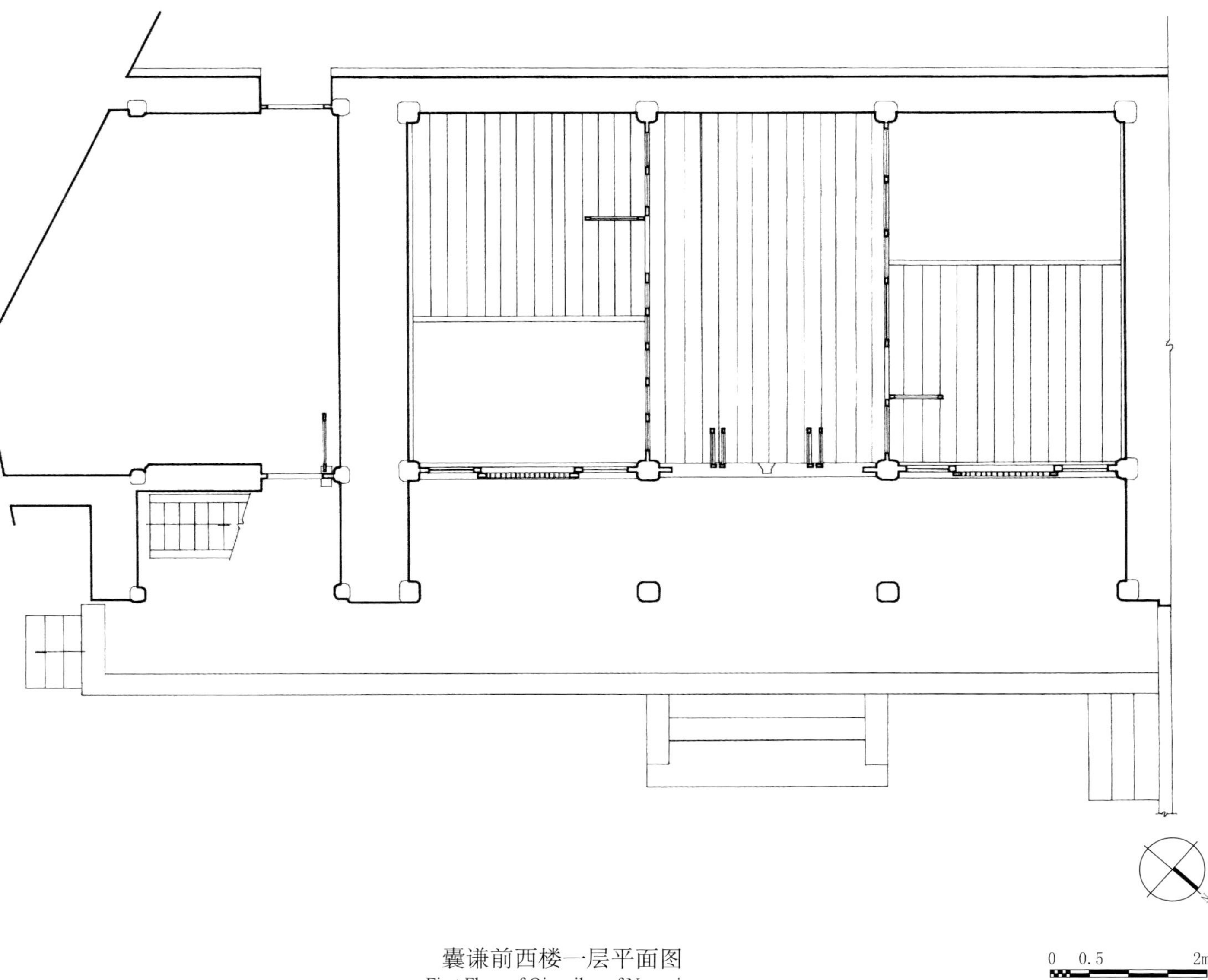

囊谦前西楼一层平面图
First Floor of Qianxilou of Nangqian

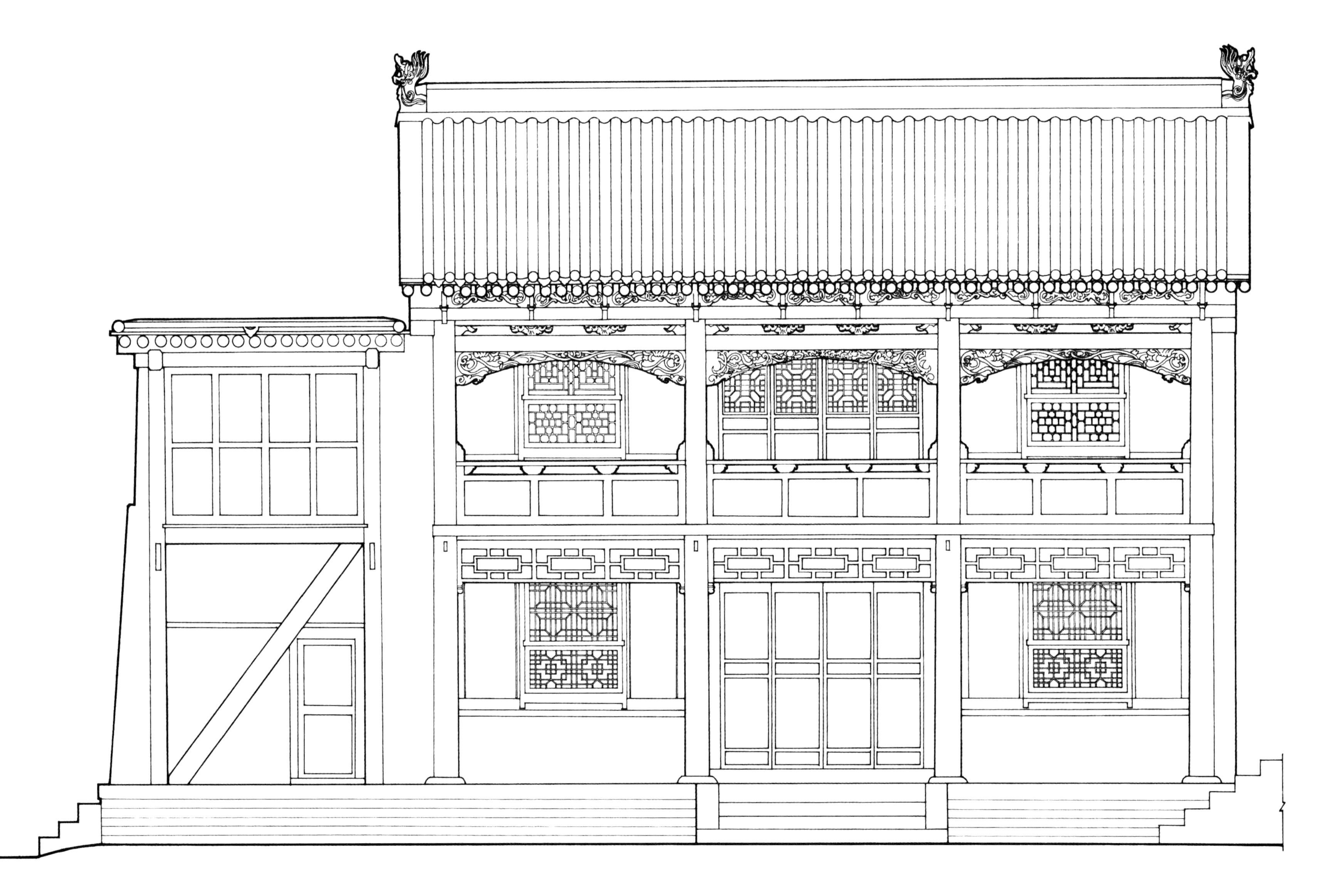

囊谦前西楼东立面图
East Elevation of Qianxilou of Nangqian

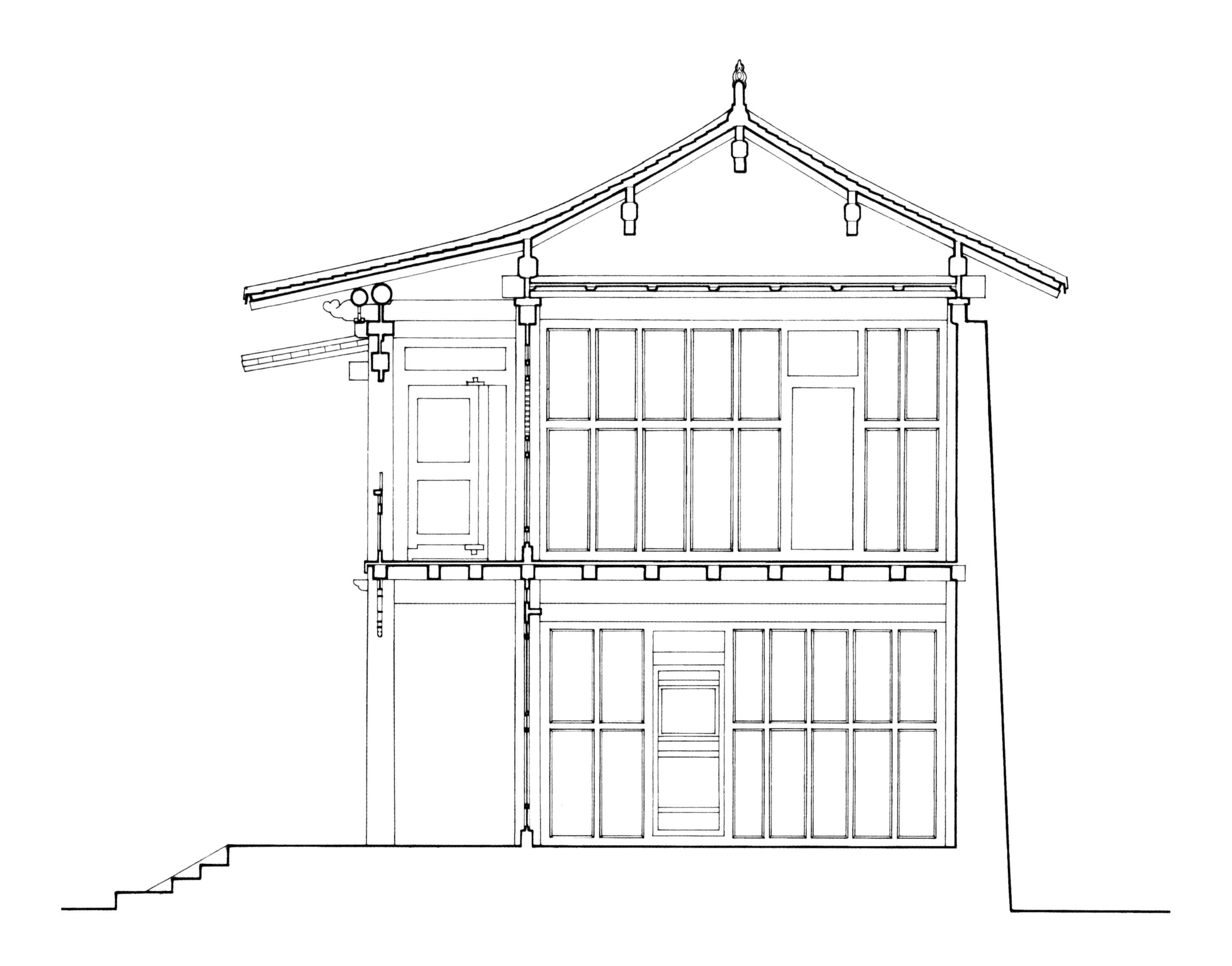

囊谦前西楼横剖面图

Cross-section of Qianxilou of Nangqian

囊谦后东楼二层平面图
Second Floor Plan of Houdonglou of Nangqian

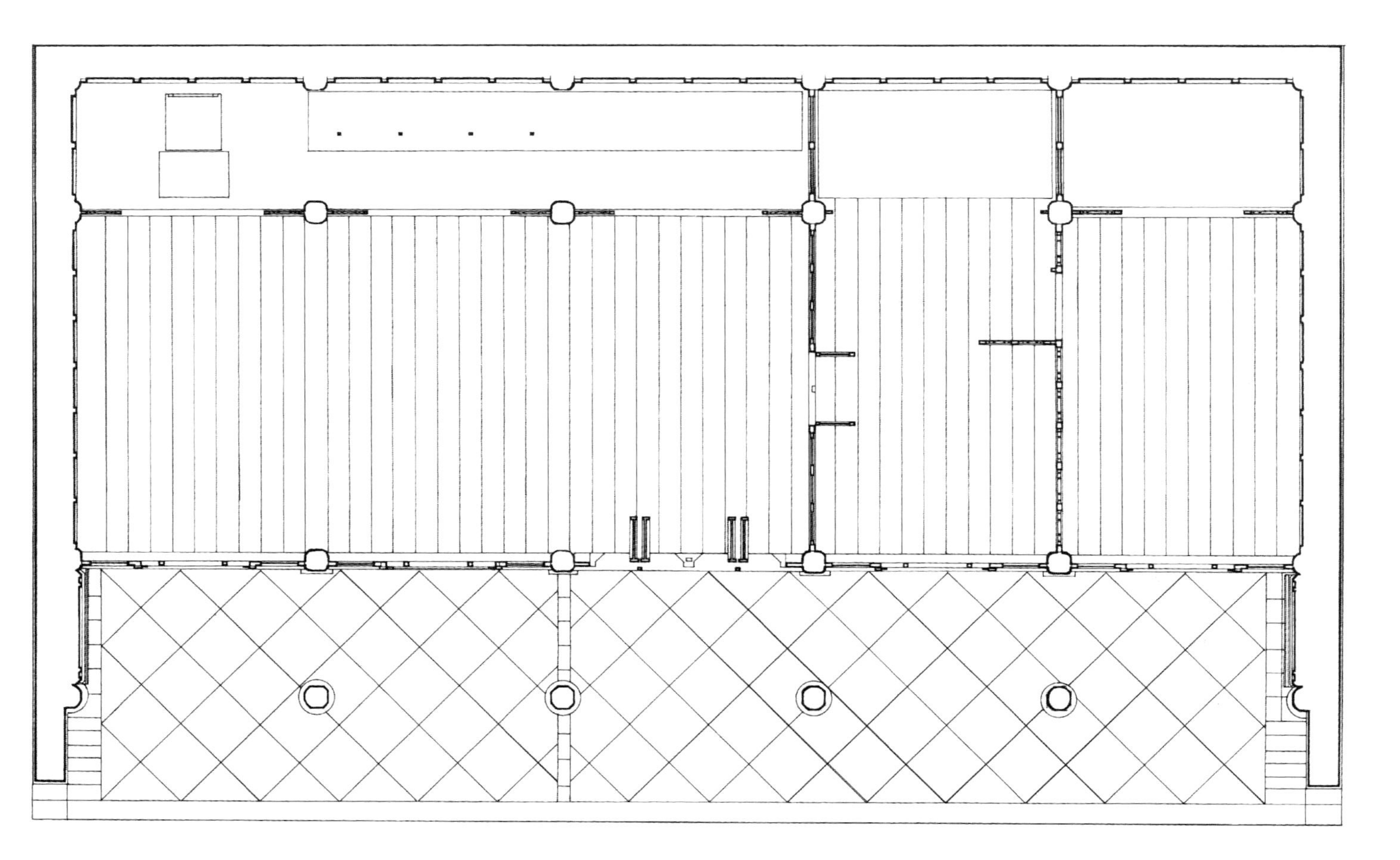

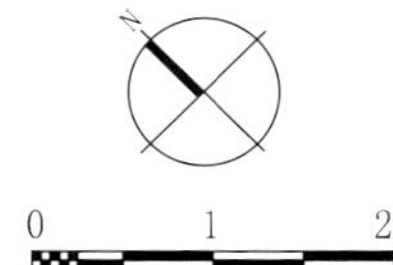

囊谦后东楼一层平面图
First Floor of Houdonglou of Nangqian

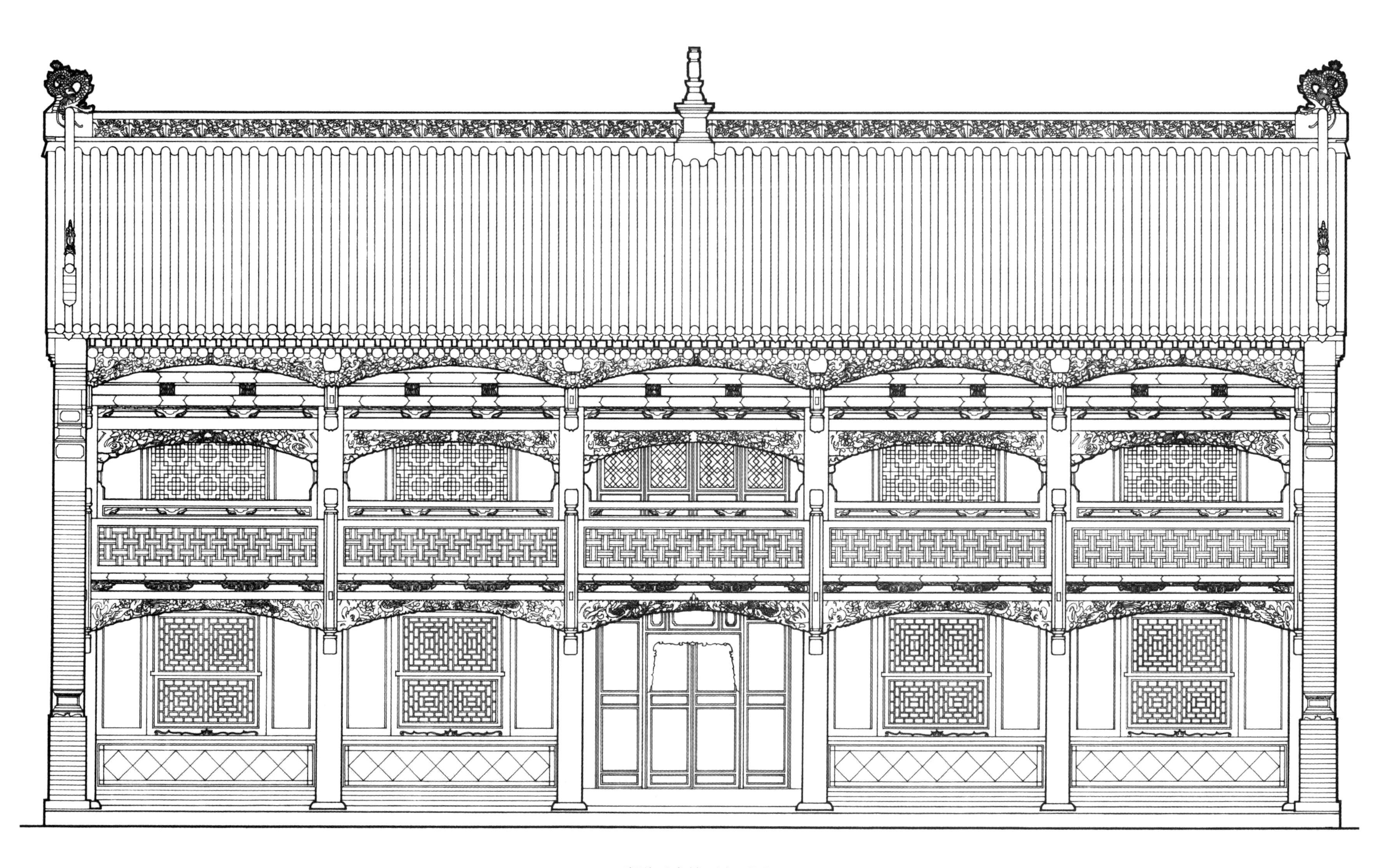

囊谦后东楼西立面图
West Elevation of Houdonglou of Nangqian

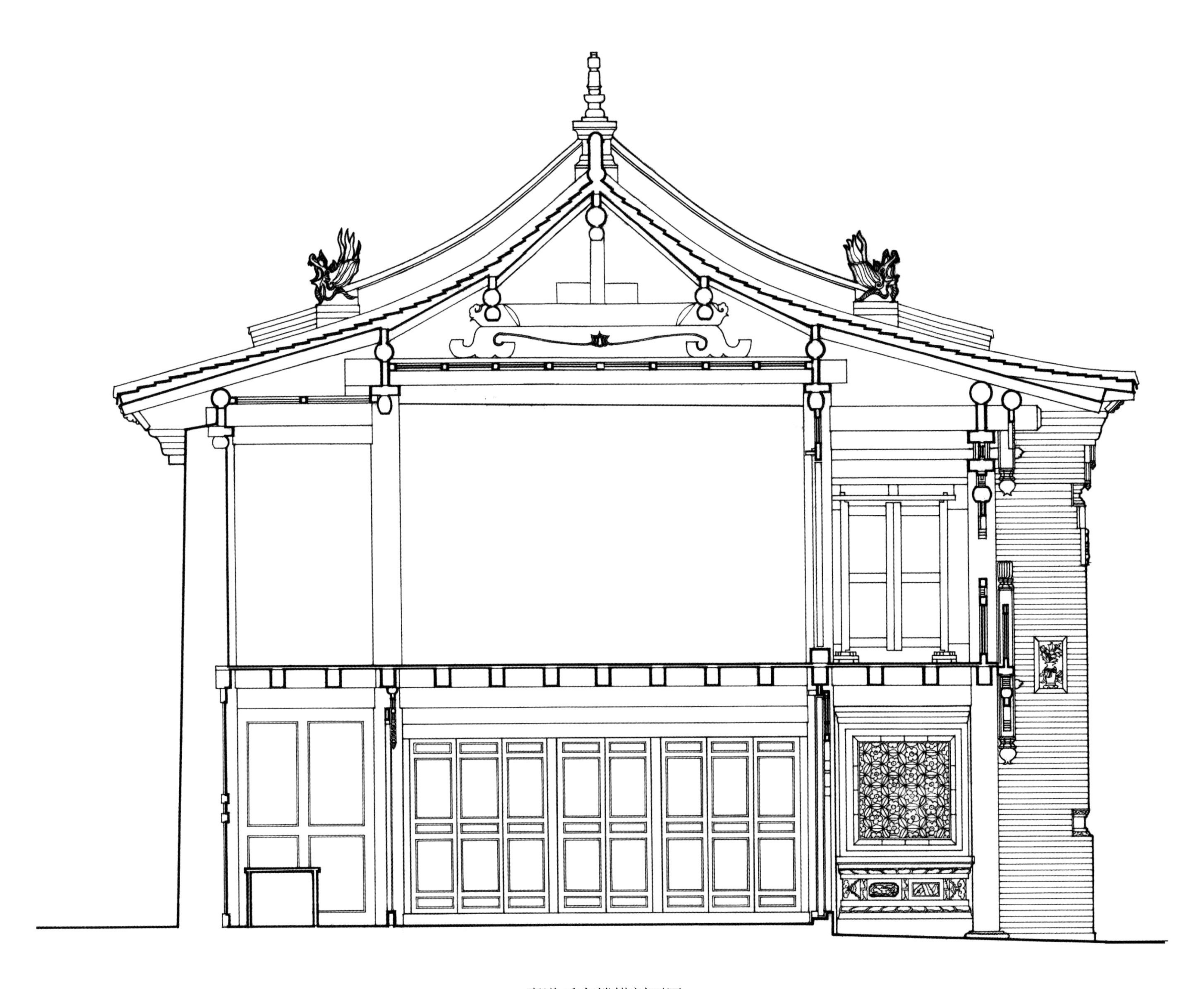

囊谦后东楼横剖面图
Cross-section of Houdonglou of Nangqian

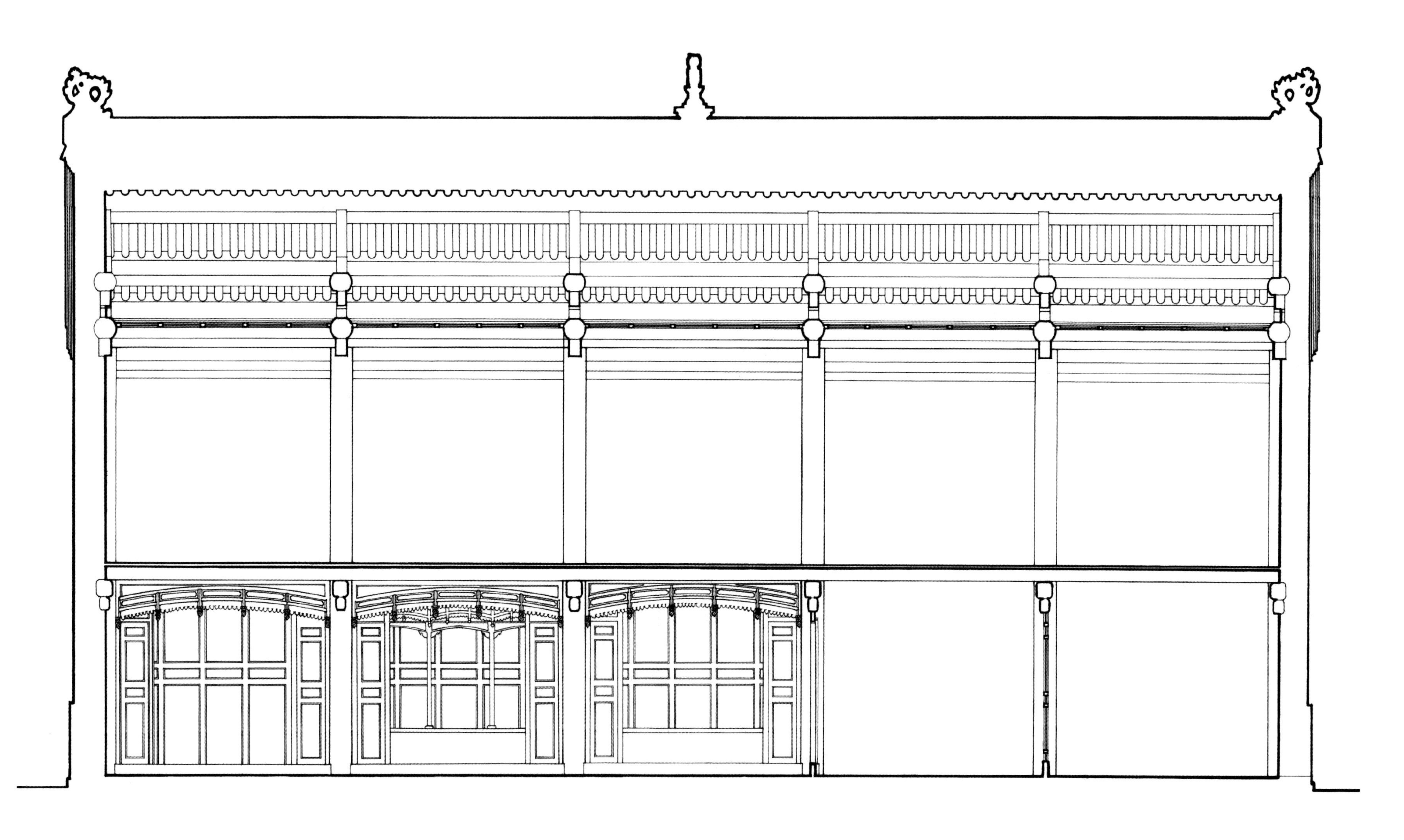

囊谦后东楼纵剖面图
Longitudinal Section of Houdonglou of Nangqian

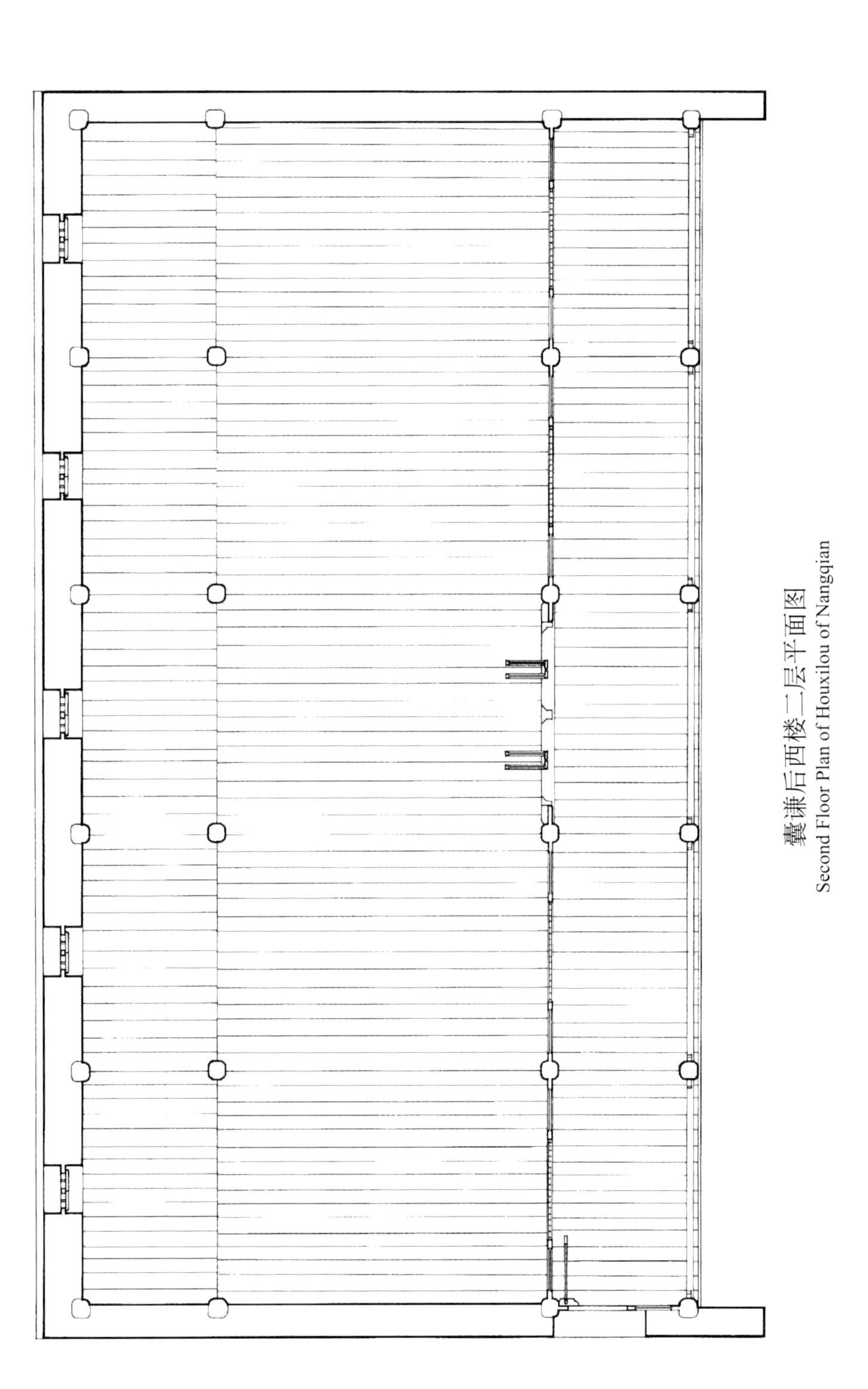

囊谦后西楼二层平面图
Second Floor Plan of Houxilou of Nangqian

囊谦后西楼一层平面图
First Floor of Houxilou of Nangqian

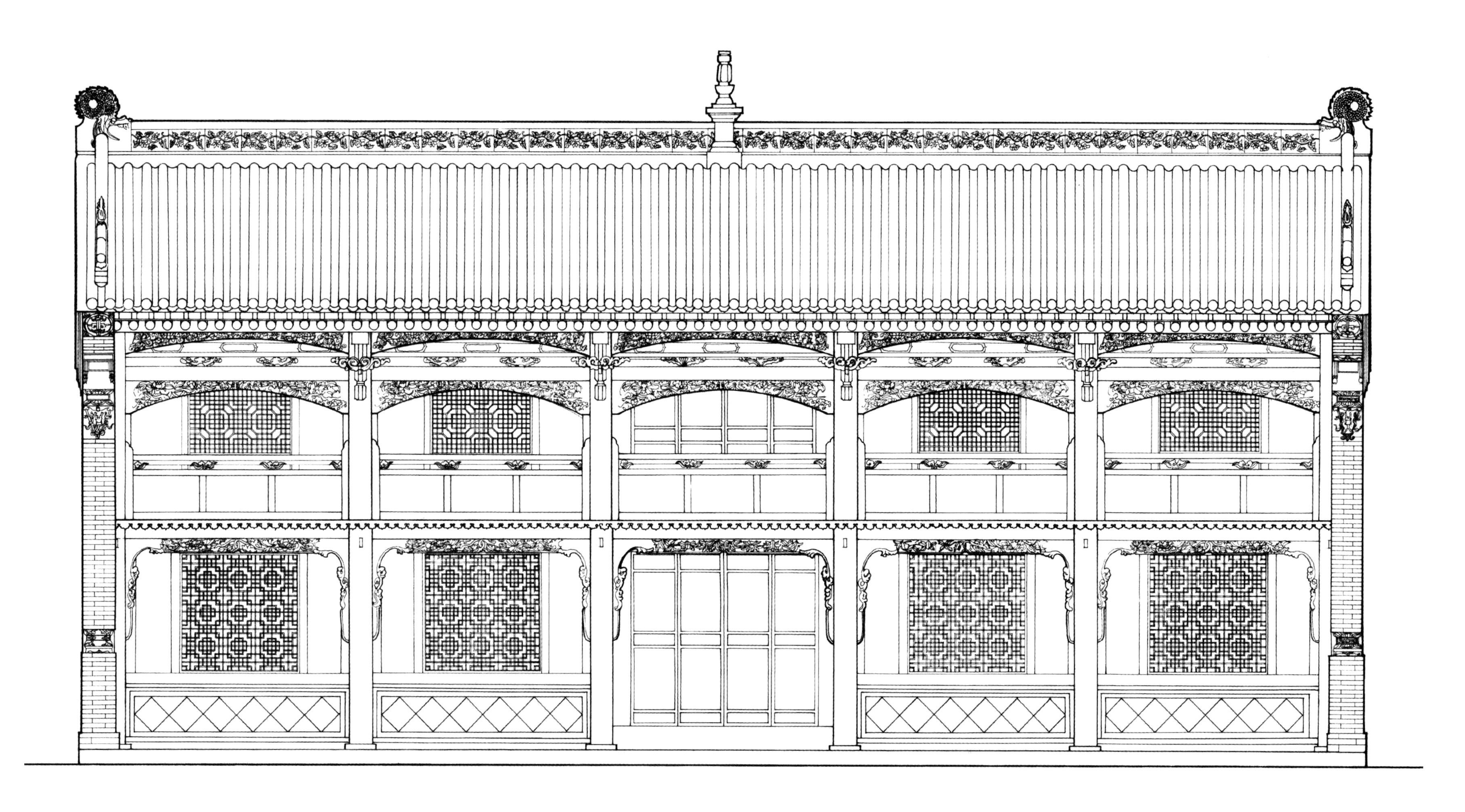

囊谦后西楼东立面图

East Elevation of Houxilou of Nangqian

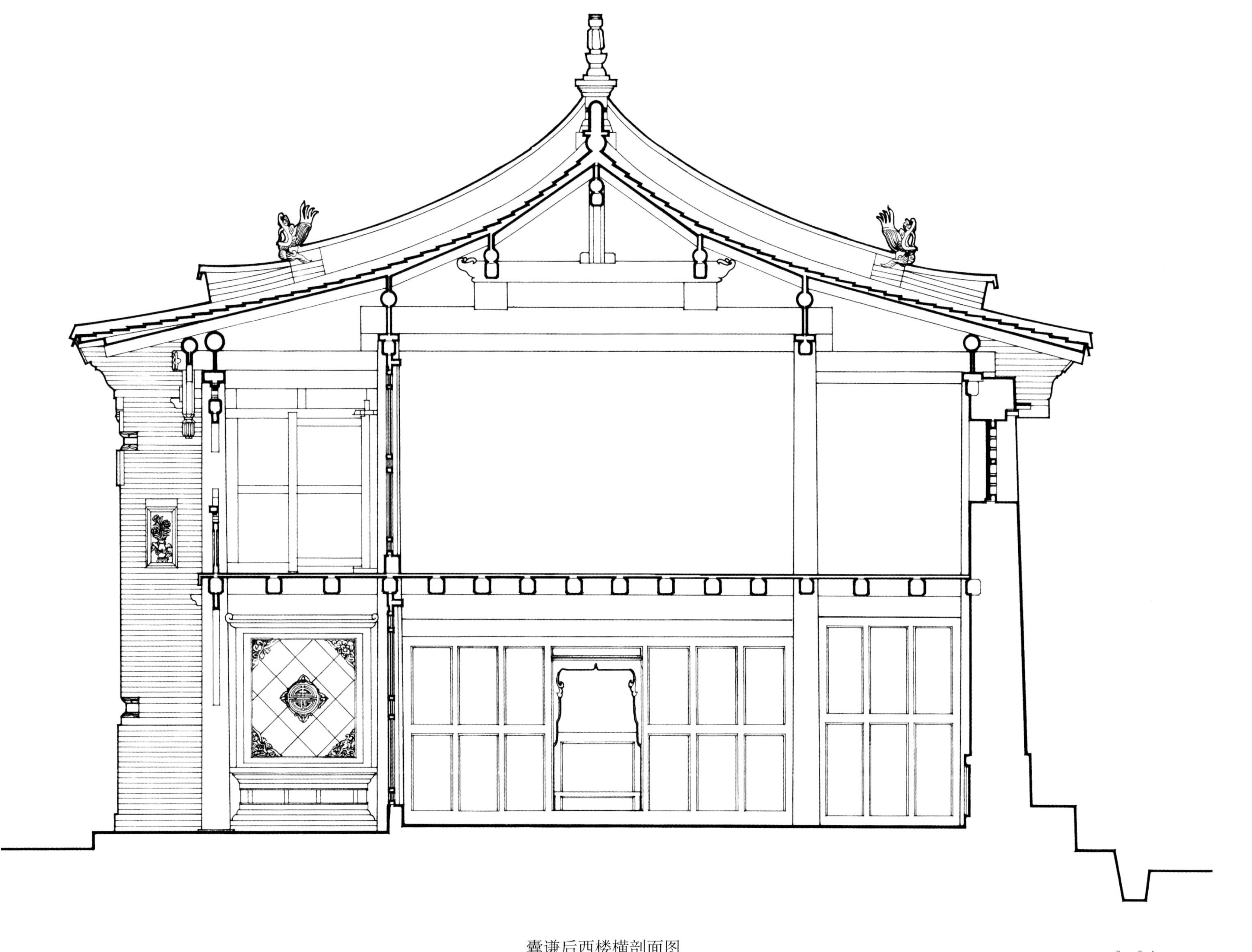

囊谦后西楼横剖面图
Cross-section of Houxilou of Nangqian

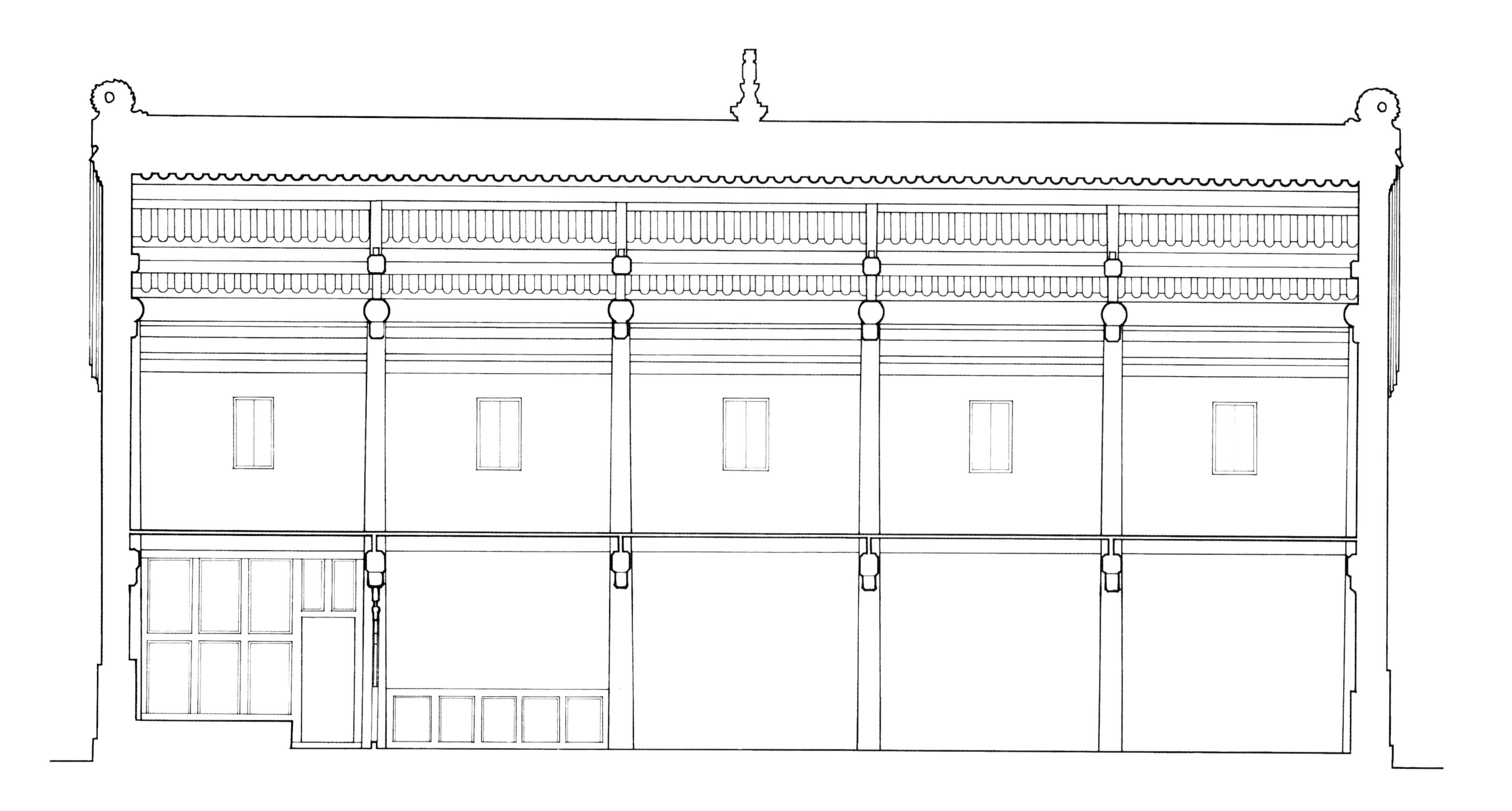

囊谦后西楼纵剖面图
Longitudinal Section of Houxilou of Nangqian

0 0.4 2m

囊谦下转楼二层平面图
Second Floor Plan of Xiazhuanlou of Nangqian

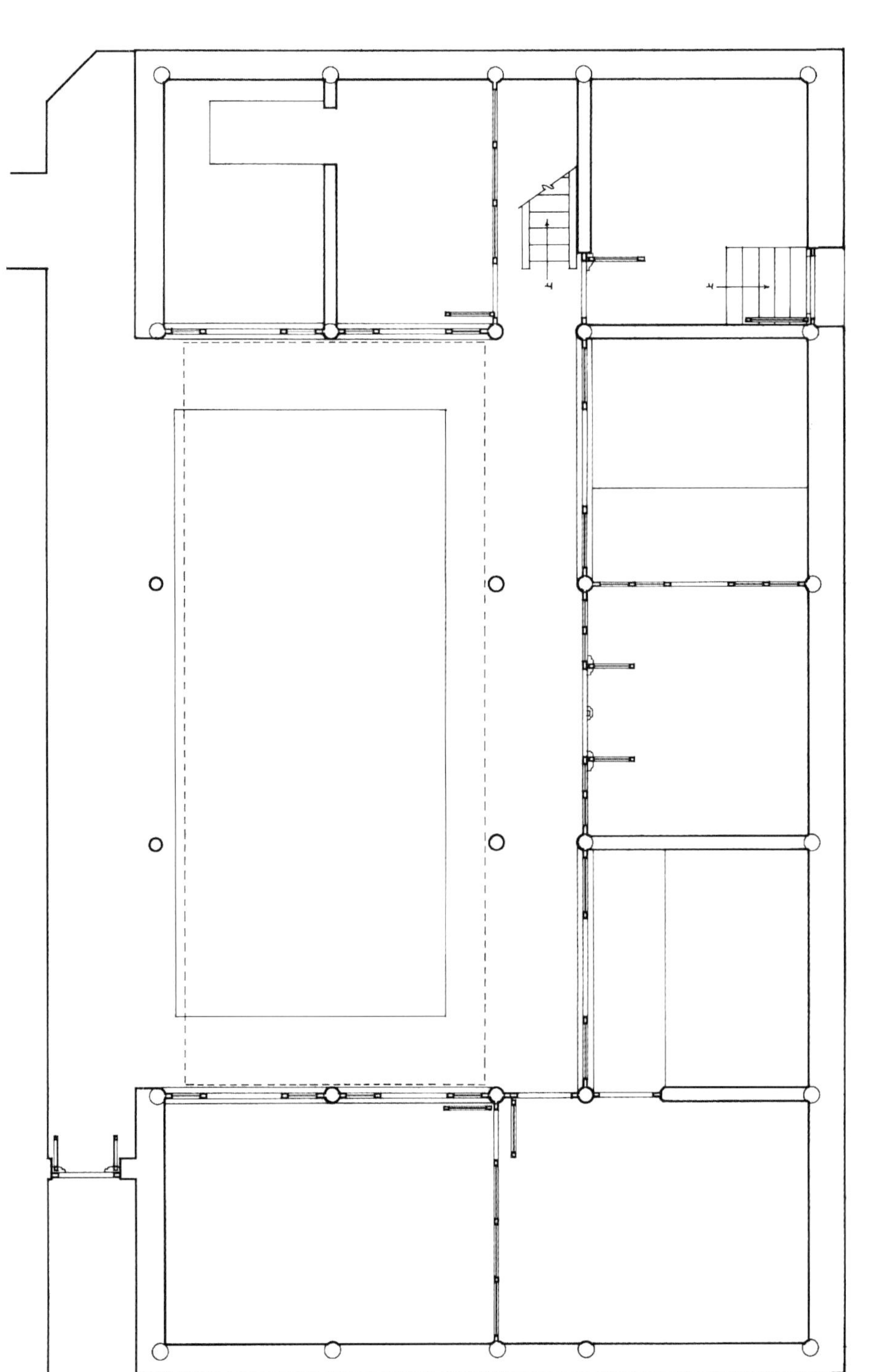

囊谦下转楼一层平面图
First Floor of Xiazhuanlou of Nangqian

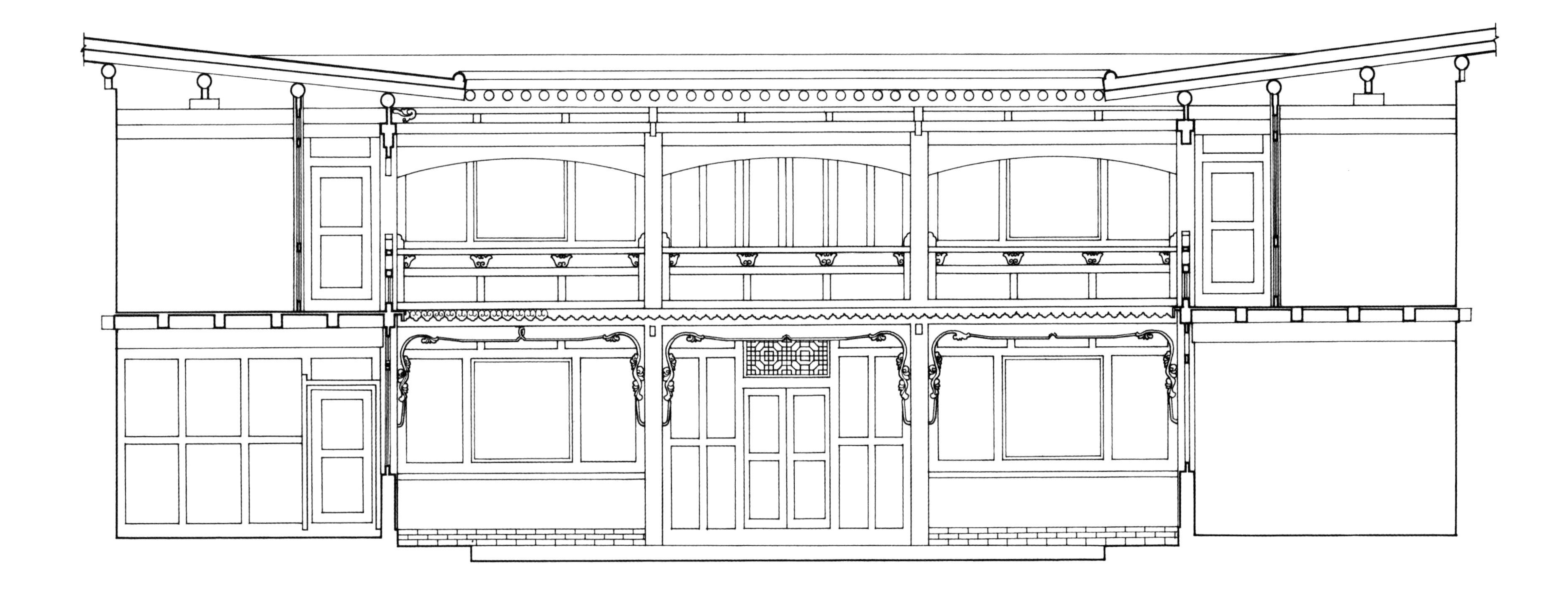

囊谦下转楼西立面图

West Elevation of Xiazhuanlou of Nangqian

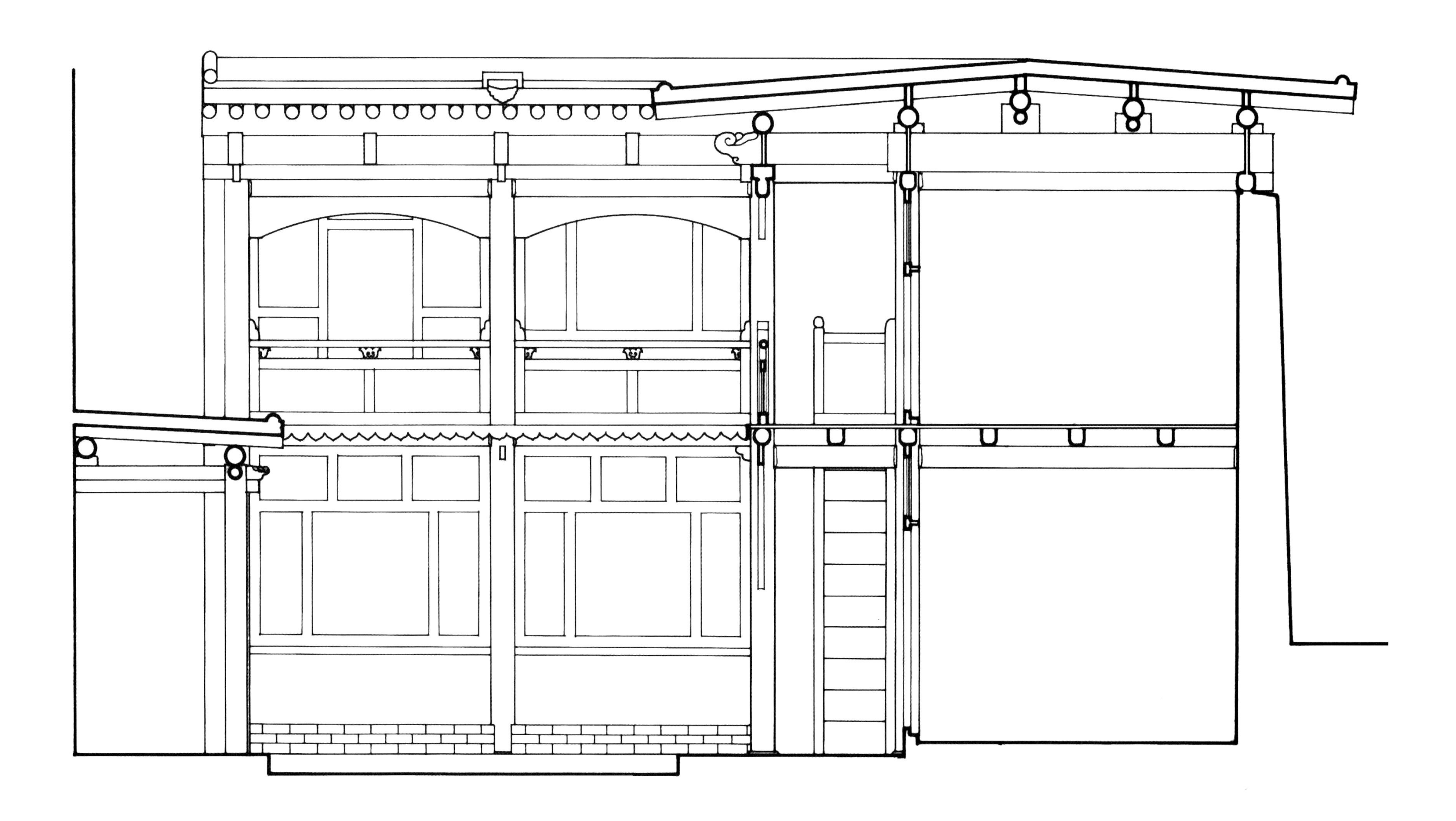

囊谦下转楼横剖面图
Cross-section of Xiazhuanlou of Nangqian

主要参考文献

[一][明]明实录 [M]. 上海：上海书店出版社，2015.

[二][清]张廷玉等. 明史 [M]. 北京：中华书局，1974.

[三][清]世续等. 清实录 [M]. 北京：中华书局，2008.

[四][清]智观巴・贡却乎丹巴饶吉著，吴均，毛继祖，马世林译. 安多政教史 [M] 兰州：甘肃民族出版社，1989.04.

[五]张驭寰，杜仙洲. 青海乐都瞿昙寺调查报告 [J]. 文物，1964（05）.

[六]谢佐. 瞿昙寺补考 [J]. 青海民族学院学报，1981（01）.

[七]谢佐. 瞿昙寺 [M]. 西宁：青海人民出版社，1982（07）.

[八]芈一之. 瞿昙寺及其在明代西宁地区的地位 [J]. 青海考古学会会刊，1983（05）.

[九]赵生琛. 青海乐都瞿昙寺文物调查纪 [J]. 考古与文物，1983（04）.

[十]青海省环境保护局自然保护处，青海省乐都县文化体育局环境保护办公室，瞿昙寺文物管理所. 瞿昙寺 [M]. 1985.

[十一]赵生琛，谢端琚，赵信. 清海古代文化 [M]. 西宁：青海人民出版社，1986.09.

[十二]中国壁画全集编辑委员会. 中国美术分类全集・中国壁画全集. 藏传寺院壁画 1[M]. 天津：天津人民美术出版社，1989.07.

[十三]浦文成. 甘青藏传佛教寺院 [M]. 西宁：青海人民出版社，1990.07.

[十四]吴葱. 青海乐都瞿昙寺建筑研究 [D]. 天津大学，1994.

[十五]钱正坤. 青海乐都瞿昙寺壁画研究 [J]. 美术研究，1995(04).

[十六]宿白. 藏传佛教寺院考古 [M]. 北京：文物出版社，1996.10.

[十七]浦天彪.《耕余琐录》与瞿昙寺史料补遗 [J]. 西藏民族学院学报（哲学社会科学版），2011,32(05).

[十八]谢佐，格桑本，袁复堂. 青海金石录 [M]. 西宁：青海人民出版社，1993.07.

[十九]青海省文化厅. 瞿昙寺 [M]. 成都：四川科学技术出版社；乌鲁木齐：新疆科技卫生出版社，2000.08.

[二十]唐栩. 甘青地区传统建筑工艺特色初探 [D]. 天津大学，2004.

[二十一]陈静微. 甘肃永登连城鲁土司衙门及妙因寺建筑研究：兼论河湟地区明清建筑特征及河州砖雕 [D]. 天津大学，2005.

[二十二]樊非. 青海黄南隆务寺及其附属寺院建筑研究：兼论热贡艺术及藏式建筑装饰 [D]. 天津大学，2005.

[二十三]阴帅可. 青海贵德玉皇阁古建筑群建筑研究 [D]. 天津大学，2006.

[二十四]谢继胜，廖旸. 瞿昙寺回廊佛传壁画内容辨识与风格分析 [J]. 故宫博物院院刊，2006（03）.

[二十五]李江. 明清甘青建筑研究 [D]. 天津大学，2007.

[二十六]陈耀东. 中国藏族建筑 [M]. 北京：中国建筑工业出版社，2007.01.

[二十七]谢继胜主编，谢继胜、熊文彬、罗文华、廖旸等著. 藏传佛教艺术发展史 [M]. 上海：上海书画出版社，2010.12.

[二十八]金萍. 瞿昙寺壁画的艺术考古研究 [D]. 西安美术学院，2012.

[二十九]李江. 明清时期河西走廊建筑研究 [D]. 天津大学，2012.

[三十]格桑本（苏生秀）主编，格桑本、张宝玺、朗钦桑・多杰本、碌逗・旦正加、贾鸿键、刘峥、鄢长青等著，藏族美术集成・绘画艺术・壁画・青海卷 1[M]. 成都：四川民族出版社，2015.08.

[三十一]索端智. 从民间地契印信文献看瞿昙寺建寺及其他 [J]. 中国藏学，2016（04）.

[三十二]苏白. 高原明珠——瞿昙寺 [J]. 收藏，2017（07）.

Main References

(1) (Ming)*Ming Shi Lu (Ming Veritable Records)* [M], Shanghai: Shanghai Bookstore Press, 2015.

(2) (Qing)ZHANG Tingyu eds. *Ming Shi (History of the Ming Dynasty)* [M], Beijing: China Publishing Press, 1974.

(3) (Qing)SHI Xu eds. *Qing Shi Lu (Qing Veritable Records)* [M], Beijing: China Press, 2008.

(4) (Qing)ZHI GUAN BA Gochehudanbajaugi, translated by WU Jun, MAO Jizu, MA Shilin, *Ando Zheng Jiao Shi (The Political and Religious History of Ando)* [M], Lanzhou: Gansu Nationality Press, 1989.

(5) ZHANG Yuhuan, DU Xianzhou, *Qinghai Ledu Qutan Si Diaocha Baogao (Investigation Report on the Qutan Temple in Ledu, Qinghai)* [J], in: *Wenwu* (*Relics*), No.5, 1964.

(6) XIE Zuo, *Qutan Si Bu Kao (Supplementary Research on the Qutan Temple)* [J], in: *Qinghai Minzu Xueyuan Xuebao (Journal of Qinghai Nationalities College)*, No.1, 1981.

(7) XIE Zuo, *Qutan Si (Qutan Temple)* [M], Xining: Qinghai People's Press, July 1982.

(8) MI Yizhi, *Qutan Si jiqi zai Mingdai Xining Diqu de Diwei (Quantan Temple and Its Status in the Xining Area during the Ming Dynasty)*, in: *Qinghai Kaogu Xuehui Huikan (Qinghai Archaeology Society Journal)*, No.5, 1983.

(9) ZHAO Shengchen, *Qinghai Ledu Quantansi Wenwu Diaocha Ji (Investigation Record on the Relics of the Qutan Temple in Ledu, Qinghai)* [J], in: *Kaogu yu Wenwu (Archaeology and Relics)*[J], No.4, 1983.

(10) Qinghai provincial Environmental Protection Bureau, Qinghai Ledu cultural and Sports Bureau Environmental Protection Office, Heritage Management Office of Qutan Temple, *Qutan Si* (*Qutan Temple*)[M], 1985.

(11) ZHAO Shengchen, XIE Duanju, ZHAO Xin, *Qinghai Gudai Wenhua (Ancient Culture in Qinghai)* [M], Xining: Qinghai People's Publishing House, September 1986.

(12) Editorial Committee of Chinese murals, *Zhongguo meishu fenlei quanji - zhongguo bihua quanji - zangchuan siyuan bihua 1 (Complete Works of Classified Chinese Fine Arts - Completed Works of Chinese Frescoes - Frescoes in the Tibetan Buddhist Temples 1)*[M], Tianjin: Tianjin People's Press, July 1989.

(13) PU Wencheng (ed.), *Ganqing Zangchuanfojiao Siyuan (Tibetan Buddhism Temples in Gansu and Qinghai)* [M], Xining: Qinghai People's Press, July 1990.

(14) WU Cong, *Qinghai Ledu Qutansi Jianzhu Yanjiu (Architectural Research on the Qutan Temple in Ledu, Qinghai)* [D],Tianjin University, 1994.

(15) QIAN Zhengkun, *Qinghai ledu qutansi bihua yanjiu (Research on the Frescoes of the Qutan Temple in Ledu, Qinghai Province)*[J], in: *Meishu yanjiu (Fine Art Research)*, No. 4, 1995.

(16) SU Bai, *Zangchuanfojiao Siyuan Kaogu (Archaeology on the Tibetan Buddhist Temples)* [M], Beijing: Relics Press, October 1996.

(17) PU Tianbiao, *Gengyu Suo Lu yu Qutan Si Shiliao Buyi (Trivial Records during Non-farming Hours and the Addendum to the Historical Documents on the Qutan Temple)* [J], in: *Tibet Minzu Xueyuan Xuebao [Journal of Tibet Nationalities College (Philosophy and Social Sciences Edition)]*, Vol.32, No.5, 2011.

(18) XIE Zuo, Kelsang Ben, YUAN Futang, (eds.) *Qinghai Jin Shi Lu (Records on the Relics in Qinghai)* [M], Xining: Qinghai People's Press, July 1993.

(19) Qinghai Province Department of Culture, *Qutan Si (Qutan Temple)* [M], Chengdu: Sichuan Science and Technology Press; Urumqi: Xinjiang Science Technology and Hygiene Press, August 2000.

(20) TANG Xu, *Ganqing Diqu Chuantong Jianzhu Gongyi Tese Chutan (Preliminary Investigation on the Craftsmanship Features of Traditional Buildings in Gansu and Qinghai Area)* [D], Tianjin University, 2004.

(21) CHEN Jingwei, *Gansu Yongdeng Liancheng LU Tusi Yamen ji Miaoyin Si Jianzhu Yanjiu (Research on the architecture of the Miaoyin Temple and the Administration of Lu Chieftain in Liancheng, Yongdeng, Gansu)* [D], Tianjin University, 2005.

(22) FAN Fei, *Qinghai Huangnan Longwu Si jiqi Fushu Siyuan Jianzhu Yanjiu: Jianlun Regong Yishu ji Zangshi Jianzhu Zhuangshi (Research on the architecture of the Longwu Temple and Its Affiliated Temples in Huangna, Qinghai: Concurrent Discussion on the Regong Art and the Decoration on Tibetan Architecture)* [D], Tianjin University, 2005.

(23) YIN Shuaike, *Qinghai Guide Yuhuang Ge Gu Jianzhu Qun Jianzhu Yanjiu (Research on the Architecture of the Ancient Building Complex of Yuhuang Ge in Guide, Qinghai)* [D], Tianjin University, 2006.

(24) XIE Jisheng, LIAO Yang, *Quantansi huilang fochuan bihua neirong bianshi yu fengfe fenxi* (Content Identification and Style Analysis on the Buddhist Frescoes of the Side Corridors in the Qutan Temple) [J], in: *Gugong bowuyuan yuankan (Palace Museum Journal)* , No.3, 2006.

(25) LI Jiang, *Mingqing Ganqing Jianzhu Yanjiu (Research on the Ming and Qing Architecture in Gansu and Qinghai Area)* [D], Tianjin University, 2007.

(26) CHEN Yaodong, *Zhongguo zangzu jianzhu (Tibetan Architecture in China)*[M],Beijing：*Zhongguo jianzhu gongye chubanshe,* China Architecture & Building Press, January, 2007.

(27) XIE Jisheng, XIONG Wenbin, LUO Wenhua, LIAO Yang, eds. *Zangchuan fojiao yishu fazhanshi (History of Tibetan Buddhist Art Development)*, [M],Shanghai：*Shanghai shuhua chubanshe* (Shanghai Calligraphy and Paintings Press) , December, 2010.

(28) JIN Ping, *Qutansi bihua de yishu kaogu yanjiu (Archaeological Research on the Fresco Art of the Qutan Temple)*[D], Xi'an Academy of Fine Arts, 2012.

(29) LI Jiang, *Mingqing Shiqi Hexi Zoulang Jianzhu Yanjiu (Research on the Architecture in the Hexi Corridor Area during the Ming and Qing Dynasties)* [D], Tianjin University, 2012

(30) Kelsang Ben (SU Shengxiu) editor in chief, Kelsang Ben, ZHANG Baoxi, Langqinsang- Duojieben, Ludou-danzhengjia, JIA Hongjian, LIU Zheng, YAN Changqing, eds., *Zangzu meishu jicheng - huihuayishu - bihua -qinghaijuan 1 (Collection of Tibetan Fine Arts, Art of Paintings, Frescoes, Qinghai Volume 1)* [M], Chengdu: Sichuan Minorities Press, August 2015.

(31) SUO Duanzhi, *Cong minjian diqi yinxin wenxian kan qutansi ji qita* (Research on the Qutan Temple and Others based on non-governmental title deed and sealed contract literature)[J], in: *Zhongguo zangxue (China Tibetology)*, No. 4, 2016.

(32) SU Bai, *Gaoyuan mingzhu: Qutansi* (Pearl on the Plateau: Qutan Temple)[J], in: Shoucang (Collections), No.7, 2017.

参与瞿昙寺古建筑测绘、修缮设计的人员名单

1993年测绘

指导教师：王其亨 杨昌鸣

硕士研究生（1991级）：黄波 李倩枚

硕士研究生（1992级）：吴葱

本科生（1989级、1993届）：高晓峰 王重 张雪峰

本科生（1989级、1994届）：段猛 刘端阳 刘方磊 刘英杰 王海康

万千 谢国杰 徐磊 张江涛

土木系教授：吴家珣

建筑系教工：苏其盛

1994年修缮设计

指导教师：王其亨

硕士研究生（1991级）：黄波 李倩枚

硕士研究生（1992级）：吴葱

硕士研究生（1993级）：官嵬 舒平

硕士研究生（1994级）：戴建新 盛梅 谢国杰 徐磊

本科生（1992级）：卜雪旸 费曦强 高伦

1995年测绘

指导教师：王其亨

硕士研究生（1992级）：吴葱

硕士研究生（1993级）：何捷

本科生（1993级）：苏怡 詹晟

1996年测绘

王其亨

1997年测绘

指　导　教　师：王其亨

硕士研究生（1992级）：吴　葱

硕士研究生（1993级）：何　捷　官　嵬

本科生（1992级）：卜雪旸　费曦强　高　伦

天津大学设计院工程师：任泽军

修缮工程顾问：单士元　罗哲文　冯建逵　于倬云　杜仙洲　傅连兴　孙儒僩

郭　旃　晋宏逵　傅清远

图纸整理

审阅

王其亨　张凤梧　张　龙

整理修改

王笑石　常振宁　李　倩　刘生雨　刘婉琳　刘雄伟　石　铮　王奥怡

王方捷　谢家良　严　谨

英文翻译

赵春兰

Participating Staff in Survey and Repairing Design of Qutan Temple

Measured & Drawn by (1993)

Supervisors: WANG Qiheng, YANG Changming

Master Students (Class 1991): HUANG Bo, LI Qianmei

Master Student (Class 1992): WU Cong

Undergraduate Students (Class 1989, graduated in 1993): GAO Xiaofeng, WANG Chong, ZHANG Xuefeng

Undergraduate Students (Class 1989, graduated in 1994): DUAN Meng, LIU Duanyang, LIU Fanglei, LIU Yingjie, WANG Haikang, WAN Qian, XIE Guojie, XU Lei, ZHANG Jiangtao

Professor from Dept. of Civil Engineering: WU Jiaxun

Logistic staff from Dept. of Architecture: SU Qisheng

Repairing Design

Supervisor: WANG Qiheng

Master Students (Class 1991): HUANG Bo, LI Qianmei

Master Student (Class 1992): WU Cong

Master Students (Class 1993): GUAN Wei, SHU Ping

Master Students (Class 1994): DAI Jianxin, SHENG Mei, XIE Guojie, XU Lei

Undergraduate Students (Class 1992): BU Xueyang, FEI Xiqiang, GAO Lun

Measured & Drawn by (1995)

Supervisor: WANG Qiheng

Master Student (Class 1992): WU Cong

Master Student (Class 1993): HE Jie

Undergraduate Students (Class 1993): SU Yi, ZHAN Sheng

Measured & Drawn by (1996)

WANG Qiheng

Measured & Drawn by (1997)

Supervisor: WANG Qiheng

Master Student (Class 1992): WU Cong

Master Students (Class 1993): HE Jie, GUAN Wei

Undergraduate Students (Class 1992): BU Xueyang, FEI Xiqiang, GAO Lun

Tianjin University Design Institute Engineer: REN Zejun

Consultants to the Repair Project

SHAN Shiyuan, LUO Zhewen, FENG Jiankui, YU Zhuoyun, DU Xianzhou, FU Lianxing, SUN Ruxian, GUO Zhan, JIN Hongkui, FU Qingyuan

Drawings Arrangement

Check and Proof

WANG Qiheng, ZHANG Fengwu, ZHANG Long

Collation and Revision

WANG Xiaoshi, CHANG Zhenning, LI Qian, LIU Shengyu, LIU Wanlin, LIU Xiongwei, SHI Zheng, WANG Aoyi, WANG Fangjie, XIE Jialiang, YAN Jin

English Translation

ZHAO Chunlan

图书在版编目(CIP)数据

瞿昙寺/王其亨，吴葱主编；天津大学建筑学院，青海省文物考古研究所合作编写.—北京：中国建筑工业出版社，2019.9
（中国古建筑测绘大系·宗教建筑）
ISBN 978-7-112-23934-4

Ⅰ.①瞿… Ⅱ.①王… ②吴… ③天… ④青… Ⅲ.①佛教－寺庙－建筑艺术－乐都县－图集 Ⅳ.①TU-885

中国版本图书馆CIP数据核字（2019）第131438号

丛书策划：王莉慧
责任编辑／李 鸽 李 婧
书籍设计／付金红
责任校对／王 烨

中国古建筑测绘大系·宗教建筑
瞿昙寺
天津大学建筑学院 青海省文物考古研究所 合作编写
王其亨 吴 葱 主编

＊

中国建筑工业出版社出版、发行（北京海淀三里河路9号）
各地新华书店、建筑书店经销
北京方舟正佳图文设计有限公司制版
北京雅昌艺术印刷有限公司印刷

＊

开本：787×1092毫米 横1／8 印张：29 字数：792千字
2020年10月第一版 2020年10月第一次印刷
定价：278.00元
ISBN 978-7-112-23934-4
（34219）